LONDON MATHEMATICAL SOCIETY LECTURE NOTE SERIES

Managing Editor: Professor M. Reid, Mathematics Institute, University of Warwick, Coventry, CV4 7AL, United Kingdom

The titles below are available from booksellers, or from Cambridge University Press at www.cambridge.org/mathematics.

190 Polynomial invariants of finite groups, D. J. BENSON
191 Finite geometry and combinatorics, F. DE CLERCK et al
192 Symplectic geometry, D. SALAMON (ed.)
194 Independent random variables and rearrangement invariant spaces, M. BRAVERMAN
195 Arithmetic of blowup algebras, W. VASCONCELOS
196 Microlocal analysis for differential operators, A. GRIGIS & J. SJÖSTRAND
197 Two-dimensional homotopy and combinatorial group theory, C. HOG-ANGELONI et al
198 The algebraic characterization of geometric 4-manifolds, J. A. HILLMAN
199 Invariant potential theory in the unit ball of C^n, M. STOLL
200 The Grothendieck theory of dessins d'enfant, L. SCHNEPS (ed.)
201 Singularities, J.-P. BRASSELET (ed.)
202 The technique of pseudodifferential operators, H. O. CORDES
203 Hochschild cohomology of von Neumann algebras, A. SINCLAIR & R. SMITH
204 Combinatorial and geometric group theory, A. J. DUNCAN, N. D. GILBERT & J. HOWIE (eds)
205 Ergodic theory and its connections with harmonic analysis, K. PETERSEN & I. SALAMA (eds)
207 Groups of Lie type and their geometries, W. M. KANTOR & L. DI MARTINO (eds)
208 Vector bundles in algebraic geometry, N. J. HITCHIN, P. NEWSTEAD & W. M. OXBURY (eds)
209 Arithmetic of diagonal hypersurfaces over infite fields, F. Q. GOUVÉA & N. YUI
210 Hilbert C*-modules, E. C. LANCE
211 Groups 93 Galway / St Andrews I, C. M. CAMPBELL et al (eds)
212 Groups 93 Galway / St Andrews II, C. M. CAMPBELL et al (eds)
214 Generalised Euler–Jacobi inversion formula and asymptotics beyond all orders, V. KOWALENKO et al
215 Number theory 1992–93, S. DAVID (ed.)
216 Stochastic partial differential equations, A. ETHERIDGE (ed.)
217 Quadratic forms with applications to algebraic geometry and topology, A. PFISTER
218 Surveys in combinatorics, 1995, P. ROWLINSON (ed.)
220 Algebraic set theory, A. JOYAL & I. MOERDIJK
221 Harmonic approximation, S. J. GARDINER
222 Advances in linear logic, J.-Y. GIRARD, Y. LAFONT & L. REGNIER (eds)
223 Analytic semigroups and semilinear initial boundary value problems, KAZUAKI TAIRA
224 Computability, enumerability, unsolvability, S. B. COOPER, T. A. SLAMAN & S. S. WAINER (eds)
225 A mathematical introduction to string theory, S. ALBEVERIO et al
226 Novikov conjectures, index theorems and rigidity I, S. FERRY, A. RANICKI & J. ROSENBERG (eds)
227 Novikov conjectures, index theorems and rigidity II, S. FERRY, A. RANICKI & J. ROSENBERG (eds)
228 Ergodic theory of Z^d actions, M. POLLICOTT & K. SCHMIDT (eds)
229 Ergodicity for infinite dimensional systems, G. DA PRATO & J. ZABCZYK
230 Prolegomena to a middlebrow arithmetic of curves of genus 2, J. W. S. CASSELS & E. V. FLYNN
231 Semigroup theory and its applications, K. H. HOFMANN & M. W. MISLOVE (eds)
232 The descriptive set theory of Polish group actions, H. BECKER & A. S. KECHRIS
233 Finite fields and applications, S. COHEN & H. NIEDERREITER (eds)
234 Introduction to subfactors, V. JONES & V. S. SUNDER
235 Number theory 1993–94, S. DAVID (ed.)
236 The James forest, H. FETTER & B. G. DE BUEN
237 Sieve methods, exponential sums, and their applications in number theory, G. R. H. GREAVES et al
238 Representation theory and algebraic geometry, A. MARTSINKOVSKY & G. TODOROV (eds)
240 Stable groups, F. O. WAGNER
241 Surveys in combinatorics, 1997, R. A. BAILEY (ed.)
242 Geometric Galois actions I, L. SCHNEPS & P. LOCHAK (eds)
243 Geometric Galois actions II, L. SCHNEPS & P. LOCHAK (eds)
244 Model theory of groups and automorphism groups, D. EVANS (ed.)
245 Geometry, combinatorial designs and related structures, J. W. P. HIRSCHFELD et al
246 p-Automorphisms of finite p-groups, E. I. KHUKHRO
247 Analytic number theory, Y. MOTOHASHI (ed.)
248 Tame topology and o-minimal structures, L. VAN DEN DRIES
249 The atlas of finite groups: ten years on, R. CURTIS & R. WILSON (eds)
250 Characters and blocks of finite groups, G. NAVARRO
251 Gröbner bases and applications, B. BUCHBERGER & F. WINKLER (eds)
252 Geometry and cohomology in group theory, P. KROPHOLLER, G. NIBLO, R. STÖHR (eds)
253 The q-Schur algebra, S. DONKIN
254 Galois representations in arithmetic algebraic geometry, A. J. SCHOLL & R. L. TAYLOR (eds)
255 Symmetries and integrability of difference equations, P. A. CLARKSON & F. W. NIJHOFF (eds)
256 Aspects of Galois theory, H. VÖLKLEIN et al
257 An introduction to noncommutative differential geometry and its physical applications 2ed, J. MADORE
258 Sets and proofs, S. B. COOPER & J. TRUSS (eds)
259 Models and computability, S. B. COOPER & J. TRUSS (eds)
260 Groups St Andrews 1997 in Bath, I, C. M. CAMPBELL et al
261 Groups St Andrews 1997 in Bath, II, C. M. CAMPBELL et al
262 Analysis and logic, C. W. HENSON, J. IOVINO, A. S. KECHRIS & E. ODELL
263 Singularity theory, B. BRUCE & D. MOND (eds)
264 New trends in algebraic geometry, K. HULEK, F. CATANESE, C. PETERS & M. REID (eds)
265 Elliptic curves in cryptography, I. BLAKE, G. SEROUSSI & N. SMART
267 Surveys in combinatorics, 1999, J. D. LAMB & D. A. PREECE (eds)
268 Spectral asymptotics in the semi-classical limit, M. DIMASSI & J. SJÖSTRAND
269 Ergodic theory and topological dynamics, M. B. BEKKA & M. MAYER
270 Analysis on Lie Groups, N. T. VAROPOULOS & S. MUSTAPHA
271 Singular perturbations of differential operators, S. ALBEVERIO & P. KURASOV
272 Character theory for the odd order theorem, T. PETERFALVI

273 Spectral theory and geometry, E. B. DAVIES & Y. SAFAROV (eds)
274 The Mandelbrot set, theme and variations, TAN LEI (ed.)
275 Descriptive set theory and dynamical systems, M. FOREMAN et al
276 Singularities of plane curves, E. CASAS-ALVERO
277 Computational and geometric aspects of modern algebra, M. D. ATKINSON et al
278 Global attractors in abstract parabolic problems, J. W. CHOLEWA & T. DLOTKO
279 Topics in symbolic dynamics and applications, F. BLANCHARD, A. MAASS & A. NOGUEIRA (eds)
280 Characters and automorphism groups of compact Riemann surfaces, T. BREUER
281 Explicit birational geometry of 3-folds, A. CORTI & M. REID (eds)
282 Auslander–Buchweitz approximations of equivariant modules, M. HASHIMOTO
283 Nonlinear elasticity, Y. FU & R. OGDEN (eds)
284 Foundations of computational mathematics, R. DEVORE, A. ISERLES & E. SÜLI (eds)
285 Rational points on curves over finite fields, H. NIEDERREITER & C. XING
286 Clifford algebras and spinors 2ed, P. LOUNESTO
287 Topics on Riemann surfaces and Fuchsian groups, E. BUJALANCE et al
288 Surveys in combinatorics, 2001, J. HIRSCHFELD (ed.)
289 Aspects of Sobolev-type inequalities, L. SALOFF-COSTE
290 Quantum groups and Lie theory, A. PRESSLEY (ed.)
291 Tits buildings and the model theory of groups, K. TENT (ed.)
292 A quantum groups primer, S. MAJID
293 Second order partial differential equations in Hilbert spaces, G. DA PRATO & J. ZABCZYK
294 Introduction to operator space theory, G. PISIER
295 Geometry and Integrability, L. MASON & YAVUZ NUTKU (eds)
296 Lectures on invariant theory, I. DOLGACHEV
297 The homotopy category of simply connected 4-manifolds, H.-J. BAUES
298 Higher operands, higher categories, T. LEINSTER
299 Kleinian Groups and Hyperbolic 3-Manifolds, Y. KOMORI, V. MARKOVIC & C. SERIES (eds)
300 Introduction to Möbius Differential Geometry, U. HERTRICH-JEROMIN
301 Stable Modules and the D(2)-Problem, F. E. A. JOHNSON
302 Discrete and Continuous Nonlinear Schrödinger Systems, M. J. ABLOWITZ, B. PRINARI & A. D. TRUBATCH
303 Number Theory and Algebraic Geometry, M. REID & A. SKOROBOGATOV (eds)
304 Groups St Andrews 2001 in Oxford Vol. 1, C. M. CAMPBELL, E. F. ROBERTSON & G. C. SMITH (eds)
305 Groups St Andrews 2001 in Oxford Vol. 2, C. M. CAMPBELL, E. F. ROBERTSON & G. C. SMITH (eds)
306 Peyresq lectures on geometric mechanics and symmetry, J. MONTALDI & T. RATIU (eds)
307 Surveys in Combinatorics 2003, C. D. WENSLEY (ed.)
308 Topology, geometry and quantum field theory, U. L. TILLMANN (ed.)
309 Corings and Comdules, T. BRZEZINSKI & R. WISBAUER
310 Topics in Dynamics and Ergodic Theory, S. BEZUGLYI & S. KOLYADA (eds)
311 Groups: topological, combinatorial and arithmetic aspects, T. W. MÜLLER (ed.)
312 Foundations of Computational Mathematics, Minneapolis 2002, FELIPE CUCKER et al (eds)
313 Transcendantal aspects of algebraic cycles, S. MÜLLER-STACH & C. PETERS (eds)
314 Spectral generalizations of line graphs, D. CVETKOVIC, P. ROWLINSON & S. SIMIC
315 Structured ring spectra, A. BAKER & B. RICHTER (eds)
316 Linear Logic in Computer Science, T. EHRHARD et al (eds)
317 Advances in elliptic curve cryptography, I. F. BLAKE, G. SEROUSSI & N. SMART
318 Perturbation of the boundary in boundary-value problems of Partial Differential Equations, DAN HENRY
319 Double Affine Hecke Algebras, I. CHEREDNIK
320 L-Functions and Galois Representations, D. BURNS, K. BUZZARD & J. NEKOVÁŘ (eds)
321 Surveys in Modern Mathematics, V. PRASOLOV & Y. ILYASHENKO (eds)
322 Recent perspectives in random matrix theory and number theory, F. MEZZADRI, N. C. SNAITH (eds)
323 Poisson geometry, deformation quantisation and group representations, S. GUTT et al (eds)
324 Singularities and Computer Algebra, C. LOSSEN & G. PFISTER (eds)
325 Lectures on the Ricci Flow, P. TOPPING
326 Modular Representations of Finite Groups of Lie Type, J. E. HUMPHREYS
328 Fundamentals of Hyperbolic Manifolds, R. D. CANARY, A. MARDEN & D. B. A. EPSTEIN (eds)
329 Spaces of Kleinian Groups, Y. MINSKY, M. SAKUMA & C. SERIES (eds)
330 Noncommutative Localization in Algebra and Topology, A. RANICKI (ed.)
331 Foundations of Computational Mathematics, Santander 2005, L. PARDO, A. PINKUS, E. SULI & M. TODD (eds)
332 Handbook of Tilting Theory, L. ANGELERI HÜGEL, D. HAPPEL & H. KRAUSE (eds)
333 Synthetic Differential Geometry 2ed, A. KOCK
334 The Navier–Stokes Equations, P. G. DRAZIN & N. RILEY
335 Lectures on the Combinatorics of Free Probability, A. NICA & R. SPEICHER
336 Integral Closure of Ideals, Rings, and Modules, I. SWANSON & C. HUNEKE
337 Methods in Banach Space Theory, J. M. F. CASTILLO & W. B. JOHNSON (eds)
338 Surveys in Geometry and Number Theory, N. YOUNG (ed.)
339 Groups St Andrews 2005 Vol. 1, C. M. CAMPBELL, M. R. QUICK, E. F. ROBERTSON & G. C. SMITH (eds)
340 Groups St Andrews 2005 Vol. 2, C. M. CAMPBELL, M. R. QUICK, E. F. ROBERTSON & G. C. SMITH (eds)
341 Ranks of Elliptic Curves and Random Matrix Theory, J. B. CONREY, D. W. FARMER, F. MEZZADRI & N. C. SNAITH (eds)
342 Elliptic Cohomology, H. R. MILLER & D. C. RAVENEL (eds)
343 Algebraic Cycles and Motives Vol. 1, J. NAGEL & C. PETERS (eds)
344 Algebraic Cycles and Motives Vol. 2, J. NAGEL & C. PETERS (eds)
345 Algebraic and Analytic Geometry, A. NEEMAN
346 Surveys in Combinatorics, 2007, A. HILTON & J. TALBOT (eds)
347 Surveys in Contemporary Mathematics, N. YOUNG & Y. CHOI (eds)
348 Transcendental Dynamics and Complex Analysis, P. RIPPON & G. STALLARD (eds)
349 Model Theory with Applications to Algebra and Analysis Vol 1, Z. CHATZIDAKIS, D. MACPHERSON, A. PILLAY & A. WILKIE (eds)
350 Model Theory with Applications to Algebra and Analysis Vol 2, Z. CHATZIDAKIS, D. MACPHERSON, A. PILLAY & A. WILKIE (eds)
351 Finite von Neumann Algebras and Masas, A. SINCLAIR & R. SMITH
352 Number Theory and Polynomials, J. MCKEE & C. SMYTH (eds)
353 Trends in Stochastic Analysis, J. BLATH, P. MÖRTERS & M. SCHEUTZOW (eds)
354 Groups and Analysis, K. TENT (ed.)
356 Elliptic Curves and Big Galois Representations, D. DELBOURGO

London Mathematical Society Lecture Note Series. 353

Trends in Stochastic Analysis

Festschrift in Honour of Heinrich von Weizsäcker

Edited by

JOCHEN BLATH, PETER MÖRTERS AND
MICHAEL SCHEUTZOW

CAMBRIDGE
UNIVERSITY PRESS

CAMBRIDGE
UNIVERSITY PRESS

University Printing House, Cambridge CB2 8BS, United Kingdom

One Liberty Plaza, 20th Floor, New York, NY 10006, USA

477 Williamstown Road, Port Melbourne, VIC 3207, Australia

314-321, 3rd Floor, Plot 3, Splendor Forum, Jasola District Centre, New Delhi - 110025, India

103 Penang Road, #05-06/07, Visioncrest Commercial, Singapore 238467

Cambridge University Press is part of the University of Cambridge.

It furthers the University's mission by disseminating knowledge in the pursuit of education, learning and research at the highest international levels of excellence.

www.cambridge.org
Information on this title: www.cambridge.org/9780521718219

© Cambridge University Press 2009

First published 2009

A catalogue record for this publication is available from the British Library

ISBN 978-0-521-71821-9 Paperback

Contents

Preface   *page* 1
Heinrich von Weizsäcker's students   3
Heinrich von Weizsäcker's publications   5

I. Foundations and techniques in stochastic analysis   11

1 Random variables – without basic space
Götz Kersting   13

2 Chaining techniques and their application to stochastic flows
Michael Scheutzow   35

3 Ergodic properties of a class of non-Markovian processes
Martin Hairer   65

4 Why study multifractal spectra?
Peter Mörters   99

II. Construction, simulation, discretization of stochastic processes 121

5 Construction of surface measures for Brownian motion
Nadia Sidorova, Olaf Wittich   123

6 Sampling conditioned diffusions
Martin Hairer, Andrew Stuart, Jochen Voß   159

7 Asymptotic formulae for coding problems: a review
Steffen Dereich   187

III. Stochastic analysis in mathematical physics   233

8 Intermittency on catalysts
Jürgen Gärtner, Frank den Hollander, Grégory Maillard   235

v

9 Stochastic dynamical systems in infinite dimensions
Salah-Eldin A. Mohammed   249

10 Feynman formulae for evolutionary equations
Oleg G. Smolyanov   283

11 Deformation quantization in infinite dimensional analysis
Rémi Léandre   303

IV. Stochastic analysis in mathematical biology   327

12 Measure-valued diffusions, coalescents and genetic inference
Matthias Birkner, Jochen Blath   329

13 How often does the ratchet click? Facts, heuristics, asymptotics
Alison M. Etheridge, Peter Pfaffelhuber, Anton Wakolbinger   365

Preface

This collection of papers on stochastic analysis is dedicated to Professor Heinrich von Weizsäcker on the occasion of his 60th birthday. The papers, written by a group of his students, coauthors, friends and colleagues, capture various important trends in the field, providing overviews of recent developments and often new results. They also give a hint of many of Heinrich's interests, and the profound influence he has, both within the field and on his collaborators. All papers have been peer-reviewed.

Heinrich von Weizsäcker began his research in mathematics as a graduate student in the early seventies. At the time, his focus was on real analysis and measure theory. He obtained his Doctorate at the Ludwig-Maximilian-Universtät München in 1973 under the supervision of Professor Hans Richter, for a thesis entitled 'Vektorverbände und meßbare Funktionen' (Vector lattices and measurable functions). In 1977 he defended his habilitation with a thesis entitled 'Einige maßtheoretische Formen der Sätze von Krein-Milman und Choquet' (Some measure theoretic variants of the theorems of Krein-Milman and Choquet) and after brief spells at the universities of Regensburg and Marburg, he moved to a chair at Universität Kaiserslautern.

In Kaiserslautern he built a strong research group, focusing more and more on stochastic analysis. He supervised a total of 11 PhDs and two habilitations; six of his former students remain in academia today. His current PhD students are Richard Kiefer, Martin Kolb and Yang Zou.

On the occasion of his 60th birthday, in April 2007, many of his former students, collaborators and friends met for an international conference on Real and Stochastic Analysis at the University of Kaiserslautern. Some of the contributions to this volume were inspired by presentations given at this conference, but most of them are specially written survey papers, chosen to reflect some of the important modern trends in stochastic analysis and neighbouring fields.

We hope that this book is not only a valuable guide to some modern trends in stochastic analysis for the research community, but also a showcase for the diversity of research activities in which Heinrich's former students and colleagues endulge. As such it is the best evidence of the lasting mathematical and personal influence of Heinrich von Weizsäcker.

Jochen Blath
Peter Mörters
Michael Scheutzow

Heinrich von Weizsäcker's students

Doctorates

1983 W. Krüger
Approximation von Integraldarstellungen und verwandte Fragen.

1983 M. Scheutzow
*Qualitatives Verhalten der Lösungen von eindimensionalen nicht-
linearen stochastischen Differentialgleichungen mit Gedächtnis.*

1990 D. Zimmermann
*Über die Integraldarstellung stochastischer Felder unter
Verwendung von Methoden der nonstandard-Analysis.*

1992 J. Krob
Kapazität statistischer Experimente.

1995 W. Doster
*Zur Berechnung des Hausdorffmaßes von Familien von
Wahrscheinlichkeitsmaßen.*

1997 P. Scheffel
Exponential Risk Rates in Discrete Markov Models.

1998 H. Scholl
Optimal Prior Distributions for Statistical Experiments.

2002 J. Blath
*Refined Multifractal Analysis of Super-Brownian Motion: the
Dimension Spectrum of Thick Points.*

2003 N. Sidorova
*Surface Measures on Paths in an Embedded Riemannian
Manifold.*

2004 M.-F. Ong
*Die Feynman-Kac-Formel für unbeschränkte Potentiale und
allgemeine Anfangsbedingungen.*

2004 J. Voß
Some Large Deviation Results for Diffusion Processes.

Habilitations

1988 M. Scheutzow
    Stationary and Periodic Stochastic Differential Systems: A Study of Qualitative Changes with Respect to the Noise Level and Asymptotics.

2001 P. Mörters
    Contributions to Fractal Geometry and Stochastic Processes.

Heinrich von Weizsäcker's publications

H.v. Weizsäcker. Zur Gleichwertigkeit zweier Arten der Randomisierung. *Manuscr. Math.*, 11:91–94, 1974.

H.v. Weizsäcker. Sublineare Abbildungen und ein Konvergenzsatz von Banach. *Math. Ann.*, 212:165–171, 1974.

H.v. Weizsäcker. Der Satz von Choquet-Bishop-de Leeuw für konvexe nicht kompakte Mengen straffer Maße über beliebigen Grundräumen. *Math. Z.*, 142:161–165, 1975.

H.v. Weizsäcker. Some negative results in the theory of liftings. In: *Measure Theory.* Springer Lecture Notes in Mathematics 541, 159–172, 1976.

S. Graf and H.v. Weizsäcker. On the existence of lower densities in noncomplete measure spaces. In: *Measure Theory.* Springer Lecture Notes in Mathematics 541, 155–158, 1976.

H.v. Weizsäcker. A note on infinite dimensional convex sets. *Math. Scand.*, 38:321–324, 1976.

H.v. Weizsäcker. Eine notwendige Bedingung für die Existenz invarianter maßtheoretischer Liftings. *Arch. Math.*, 28:91–97, 1977.

H.v. Weizsäcker. Strong measurability, liftings and the Choquet-Edgar Theorem. In: *Vector Measures.* Springer Lecture Notes in Mathematics 645, 209–218, 1978.

H.v. Weizsäcker and G. Winkler. Integral representations in the set of solutions of a generalized moment problem. *Math. Ann.*, 246:23–32, 1979.

H.v. Weizsäcker and G. Winkler. On the extremal decomposition in noncompact sets of measures. *Colloq. Math. Soc. Janos Bolyai*, 27:1105–1111, 1979.

'K. Blizzard' (joint pseudonym), W. Schachermayer, E. Thomas, and H.v. Weizsäcker. A Krein-Milman set without the Integral Representation Property. In Z. Frolik, editor, *Eighth Winter School in Abstract Analysis.* Math. Inst. Czech. Acad. Sci, 1980.

H.v. Weizsäcker and G. Winkler. Non compact extremal integral representations: Some Probabilistic aspects. In: *Functional Analysis: Surveys and Recent Results II*, North Holland, 115–148, 1980.

H.v. Weizsäcker. Remark on extremal measure extensions. In: *Measure Theory*, Springer Lecture Notes in Mathematics 794, 79–80, 1980.

H.v. Weizsäcker. A simple example concerning the global Markov property of lattice random fields. In Z. Frolik, editor, *Eighth Winter School in Abstract Analysis*, Math. Inst. Czech. Acad. Sci., 1980.

R.D. Mauldin, D. Preiss, and H.v. Weizsäcker. A survey of problems and results concerning orthogonal transition kernels. In: *Measure Theory*. Springer Lecture Notes in Mathematics 945, 419–426, 1981.

H.v. Weizsäcker. Nonexistence of liftings for arithmetic density. *Proc. Am. Math. Soc.*, 86:692–693, 1982.

H.v. Weizsäcker. Exchanging the order of taking suprema and countable intersections of σ-algebras. *Ann. Inst. Henri Poincaré*, Sect. B, 19:91–100, 1983.

R.D. Mauldin, D. Preiss, and H.v. Weizsäcker. Orthogonal transition kernels. *Ann. Probab.*, 11:970–988, 1983.

S.E.A. Mohammed, M. Scheutzow, and H.v. Weizsäcker. Hyperbolic state space decomposition for a linear stochastic delay equation. *SIAM J. Contr. Opt.*, 24:543–551, 1986.

H.v. Weizsäcker and G. Winkler. *Stochastic Integrals: An Introduction*. Friedrich Vieweg, Advanced Lectures in Math., 1990.

R.D. Mauldin and H.v. Weizsäcker. Some orthogonality preserving kernels which are not completely orthogonal. *Ann. Probab.*, 19:396–400, 1991.

O.G. Smolyanov and H.v. Weizsäcker. Differentiable families of measures. *J. Funct. Anal.*, 118:454–476, 1993.

O.G. Smolyanov and H.v. Weizsäcker. Smooth curves in spaces of measures and shifts of diffferentiable measures along vector fields. *Dokl. Akad. Nauk, Mathematics*, 339:584–587, 1994.

O.G. Smolyanov and H.v. Weizsäcker. Change of measures and their logarithmic derivatives under smooth transformations. *C. R. Acad. Sci., Paris*, 321:103–108, 1995.

O.G. Smolyanov and H.v. Weizsäcker. Formulae with logarithmic derivatives of measures related to the quantization of infinite-dimensional Hamiltonian systems. *Russ. Math. Surv.*, 51:357–358, 1996.

H.v. Weizsäcker. Some reflections on and experiences with SPLIFs. In T. Ferguson, MacQueen, and Shapley, editors, *Festschrift for David Blackwell*, IMS Lecture Notes. IMS, 1996.

R.D. Mauldin, M. Monticino, and H.v. Weizsäcker. A directionally reinforced random walk. *Adv. Math.*, 117:239–252, 1996.

H.v. Weizsäcker. Sudakov's typical marginals, random linear functionals and a conditional central limit theorem. *Probab. Theory Relat. Fields*, 107:313–324, 1997.

J. Krob and H.v. Weizsäcker. On the rate of information gain in experiments with a finite parameter set. *Stat. Decis.*, 15:281–294, 1997.

P. Scheffel and H.v. Weizsäcker. On risk rates and large deviations in finite Markov chain experiments. *Math. Methods Stat.*, 3:293–312, 1997.

M. Scheutzow and H.v. Weizsäcker. Which Moments of a Logarithmic Derivative Imply Quasiinvariance? *Documenta Mathematica*, 3:261–272, 1998.

V.N. Sudakov and H.v. Weizsäcker. Typical distributions: the infinite-dimensional approach. In: *Asymptotic Methods in Probability and Mathematical Statistics*. Conference 1998, 277–280. St Petersburg University, 1998.

H.v. Weizsäcker. *Differenzierbare Maße und Stochastische Analysis*. 6 Vorlesungen. Graduiertenkolleg Berlin, 1998.

H.v. Weizsäcker. Identifying a finite graph by its random walk. In: *Asymptotic Methods in Probability and Mathematical Statistics*. Conference 1998, 303–304. St. Petersburg University, 1998.

O.G. Smolyanov and H.v. Weizsäcker. Noether Theorem for infinite-dimensional variational problems. *Russ. Math. Dokl.*, 361:91–94 (583–586), 1998.

O.G. Smolyanov and H.v. Weizsäcker. Smooth probability measures and associated differential operators. *Infin. Dimens. Anal. Quantum Probab. Relat. Top.*, 2:51–78, 1999.

H.v. Weizsäcker and O.G. Smolyanov. Differential forms on infinite-dimensional spaces and an axiomatic approach to the Stokes formula. *Dokl. Math.*, 60:22–25, 1999.

H.v. Weizsäcker and O.G. Smolyanov. Relations between smooth measures and their logarithmic gradients and derivatives. *Dokl. Math.*, 60:337–340, 1999.

H.v. Weizsäcker, R. Léandre, and O.G. Smolyanov. Algebraic properties of infinite-dimensional differential forms of finite codegree. *Dokl. Math.*, 60:412–415, 1999.

O.G. Smolyanov, H.v. Weizsäcker, and O. Wittich. Brownian motion on a manifold as limit of stepwise conditioned standard Brownian motions. In: *Stochastic Processes, Physics and Geometry: New Interplays.* II: A Volume in Honour of S. Albeverio, vol. 29 of Can. Math. Soc. Conf. Proc., 589–602. AMS, 2000.

H.v. Weizsäcker, O.G. Smolyanov, and O. Wittich. Diffusions on compact Riemannian manifolds and surface measures. *Dokl. Math.*, 61:230–234, 2000.

P.A. Loeb and H.v. Weizsäcker. Appendix to the paper P. Loeb and J. Bliedtner: 'Sturdy Harmonic Functions'. *Positivity*, 7:384–387, 2003.

O.G. Smolyanov, H.v. Weizsäcker, and O. Wittich. Chernoff's theorem and the construction of semigroups. In: *Evolution Equations: Applications to Physics, Industry, Life Sciences and Economics* EVEQ 2000, M. Ianelli, G. Lumer (eds.), 355–364. Birkhäuser, 2003.

N. Sidorova, O.G. Smolyanov, H.v. Weizsäcker, and O. Wittich. Brownian motion close to a submanifold of a Riemannian manifold. *Proceedings of the First Sino-German Conference on Stochastic Analysis*, Beijing, 2003.

N. Sidorova, O.G. Smolyanov, H.v. Weizsäcker, and O. Wittich. The surface limit of Brownian motion in tubular neighbourhoods of an embedded Riemannian manifold. *J. Funct. Anal.*, 206:391–413, 2004.

H.v. Weizsäcker. Can one drop L^1-boundedness in Komlos' subsequence theorem? *Am. Math. Monthly*, 111:900–903, 2004.

O.G. Smolyanov, Kh. fon Vaĭtszekker, O. Vittikh. Feynman formulas for the Cauchy problem in domains with a boundary. (Russian) *Dokl. Akad. Nauk*, 395:596–600, 2004.

N. Sidorova, O.G. Smolyanov, H.v. Weizsäcker, and O. Wittich. Brownian motion close to submanifolds of Riemannian manifolds. *Recent Developments in Stochastic Analysis and Related Topics*, 439–452, World Sci. Publ., Hackensack, NJ, 2004.

O.G. Smolyanov, Kh. fon Vaĭtszekker, and O. Vittikh. Construction of diffusions on the set of mappings from an interval to a compact Riemannian manifold. (Russian) *Dokl. Akad. Nauk* 402:316–320, 2005.

O.G. Smolyanov, H.v. Weizsäcker, and O. Wittich. Chernoff's theorem and discrete time approximations of Brownian motion on manifolds. *Potential Anal.* 26:1–29, 2007.

R. Siegmund-Schultze and H.v. Weizsäcker. Level crossing probabilities. I. One-dimensional random walks and symmetrization. *Adv. Math.* 208:672–679, 2007.

R. Siegmund-Schultze and H.v. Weizsäcker. Level crossing probabilities. II. Polygonal recurrence of multidimensional random walks. *Adv. Math.* 208:680–698, 2007.

I. Foundations and techniques in stochastic analysis

1

Random variables – without basic space

Götz Kersting

Fachbereich Informatik und Mathematik
Universität Frankfurt
D-60054 Frankfurt am Main, Germany

Abstract

The common definition of a random variable as a measurable function
works well 'in practice', but has conceptual shortcomings, as was pointed
out by several authors. Here we treat random variables not as derived
quantities but as mathematical objects, whose basic properties are given
by intuitive axioms. This requires that their target spaces fulfil a minimal regularity condition saying that the diagonal in the product space is
measurable. From the axioms we deduce the basic properties of random
variables and events.

1.1 Introduction

In this paper we define the concept of a *stochastic ensemble*. It is our
intention thereby to give an intuitive axiomatic approach to the concept of a random variable. The primary ingredient is a sufficiently rich
collection of random variables (with 'good' target spaces). The set of
observable events will be derived from it.

Among the notions of probability it is the *random variable* which in
our view constitutes the fundamental object of modern probability theory. Albeit in the history of mathematical probability *events* came first,
random variables are closer to the roots of understanding nondetermin-
istic phenomena. Nowadays events typically refer to random variables
and are no longer studied for their own sake, and for *distributions* the situation is not much different. Moreover, random variables turn out to be
flexible mathematical objects. They can be handled in other ways than
events or distributions (think of couplings), and these ways often con-
form to intuition. 'Probabilistic', 'pathwise' methods gain importance
and combinatorial constructions with random variables can substitute
(or nicely prepare) analytic methods. It was a common belief that first

Trends in Stochastic Analysis, ed. J. Blath, P. Mörters and M. Scheutzow.
Published by Cambridge University Press. ©Cambridge University Press 2008.

of all the distributions of random variables matter in probability, but this belief is outdated.

Today it is customary to adapt random variables to a context from measure theory. Yet the feeling has persisted that random variables are objects in their own right. This was manifest, when measure theory took over in probability: According to J. Doob (interviewed by Snell [9]) 'it was a shock for probabilists to realize that a function is glorified into a random variable as soon as its domain is assigned a probability distribution with respect to which the function is measurable'. Later the experts insisted that it is the idea of random variables which conforms to intuition. Legendary is L. Breiman's [2] statement: 'Probability theory has a right and a left hand. On the right is the rigorous foundational work using the tools of measure theory. The left hand "thinks probabilistically," reduces problems to gambling situations, coin-tossing, and motions of a physical particle'. In applications of probability the concept of a random variable never lost its appeal. We may quote D. Mumford [8]: 'There are two approaches to developing the basic theory of probability. One is to use wherever possible the reduction to measure theory, eliminating the probabilistical language ... The other is to put the concept of "random variable" on center stage and work with manipulations of random variables wherever possible'. And, 'for my part, I find the second way ... infinitely clearer'.

Example To illustrate this assertion let us consider different proofs of the central limit theorem saying that $(X_1 + \ldots + X_n)/\sqrt{n}$ is asymptotically normal for iid random variables $X_1, X_2, \ldots$ with mean 0 and variance 1. There is the established analytic approach via characteristic functions. In contrast let us recall a coupling method taken from [2], which essentially consists in replacing $X_1, \ldots, X_n$ one after the other by independent standard normal random variables $Y_1, \ldots, Y_n$. In more detail this looks as follows: Let $f : \mathbb{R} \to \mathbb{R}$ be thrice differentiable, bounded and with bounded derivatives. Then it is sufficient to show that

$$\mathbf{E}\Big[f\Big(\frac{X_1 + \cdots + X_n}{\sqrt{n}}\Big) - f\Big(\frac{Y_1 + \cdots + Y_n}{\sqrt{n}}\Big)\Big]$$

converges to zero. The integrand may be expanded into

$$\sum_{i=1}^{n} \Big[f\Big(\frac{Z_i}{\sqrt{n}}\Big) - f\Big(\frac{Z_{i-1}}{\sqrt{n}}\Big)\Big]$$

with $Z_i := X_1 + \cdots + X_i + Y_{i+1} + \cdots + Y_n$. By means of two Taylor expansions around $U_i := X_1 + \cdots + X_{i-1} + Y_{i+1} + \cdots + Y_n$ the summands turn into

$$\frac{X_i - Y_i}{\sqrt{n}} f'(U_i) + \frac{X_i^2 - Y_i^2}{2n} f''(U_i) + O_P(n^{-3/2})$$

Taking expectations the first two terms vanish because of independence, and a closer look at the remainder gives the assertion (see [2], page 168). □

From an architectural point of view these considerations and statements suggest that we try to start from random variables in the presentation of probability theory and therewith to bring intuition and methods closer together – rather than to gain random variables as derived quantities in the accustomed measure-theoretic manner. We like to show that this can be accomplished without much technical effort. For this purpose we may leave aside distributions in this paper.

Let us comment on the difference of our approach to the customary one of choosing a certain σ-field $\mathcal{E}$ on some basic set Ω as a starting point and then indentifying events and random variables with measurable sets and measurable functions. In our view this is a set-theoretic *model* of the probabilistic notions.

To explain this first by analogy let us recall how natural numbers are treated in mathematics. There are two ways: Either one starts from the well-known Peano axioms. Then the *set* of natural numbers is the object of study, and a natural number is nothing more than an element of this structured set. Or natural numbers are introduced by a set-theoretic construction, e.g. $0 := \emptyset, 1 := \{0\}, \ldots, n + 1 := n \cup \{n\}, \ldots$ (see [7]). This setting exhibits aspects which are completely irrelevant for natural numbers (such as $n \in n + 1$ and $n \subset n + 1$) and which stress that we are dealing with a *model* of the natural numbers. Different models may each have additional properties and structures which are actually irrelevant (and to some extent misleading) for the study of natural numbers. It is not important that they are 'isomorphic' in any respect.

Analoguous observations can be made in our context, if events and random variables are represented by a measure space $(\Omega, \mathcal{E})$ and associated measurable functions. Note the following: There are subsets of Ω not belonging to $\mathcal{E}$, which are totally irrelevant. To some extent this is also true for the elements ω of Ω (as also Mumford [8] pointed out;

already Caratheodory considered integration on spaces without points in his theory of soma [3]). The 'small omegas' do not show up in any relevant result of probability theory, and one could do without them, if they were not needed to *define* measurable functions. Next the notion of a random variable is ambiguous: There are random variables and a.s. defined random variables, represented by measurable functions and equivalence classes of measurable functions. This distinction, though unavoidable in the traditional setting, is somewhat annoying. Finally note that probabilists leave aside the question of isomorphy of measurable spaces.

All these observations indicate that measurable spaces and mappings indeed make up a model of events and random variables. This is not to say that such models should be avoided, but one should not overlook that they might mislead. Aspects like the construction of non-Borel-sets are of no relevance in probability and may distract beginners. Also one should be cautious in giving the elements of Ω some undue relevance ('state of the world'), which may create misconceptions.

Example This example of possible misconception is taken from the textbook [1] (Example 4.6 and 33.11). Let $\Omega = [0,1]$, endowed with the Borel-σ-field and Lebesgue-measure λ. Let $\mathcal{F}$ be the sub-σ-field of sets B with $\lambda(B) = 0$ or 1. Then $\mathcal{F}$ presents an observer, who lacks information. It is mistaken to argue that $\mathcal{F}$ presents full information, because it contains all one-point sets such that the observer can recognize which event $\{\omega\}$ takes place and which 'state' ω is valid. Therefore for any Borel-set $E \subset \Omega$ the conditional probability $\lambda(E|\mathcal{F})$ is $\lambda(E)$ a.s., and in general not 1_E a.s.   □

The eminent geometer H. Coxeter pinpoints such delusion due to models in stating: 'When using models, it is desirable to have two rather than one, so as to avoid the temptation to give either of them undue prominence. Our ... reasoning should all depend on the axioms. The models, having served their purpose of establishing relative consistency, are no more essential than diagrams' (see Section 16.2 in [4]). Coxeter has the circle and halfplane models of hyperbolic geometry in mind, but certainly his remark applies more generally.

An axiomatic concept of random variables should avoid the asserted flaws. The reader may judge our approach from this viewpoint. This paper owes a lot to discussions with Hermann Dinges, who put forward related ideas already in [5] (jointly with H. Rost). For further discussion we refer to H. Dinges [6] and D. Mumford [8].

The paper is organized as follows. In Section 1.2 we have a look at those properties of events and random variables independent of a measure-theoretic representation (this section may be skipped). In Section 1.3 we discuss the class of measurable spaces which are suitable to serve as target spaces of random variables. Section 1.4 contains the axioms for general systems of random variables, which we call stochastic ensembles. In Section 1.5 we derive events and deduce their properties from these axioms. In Section 1.6 we discuss equality and a.s. equality of random variables. In Section 1.7 we address convergence of random variables in order to exemplify how to work within our framework of axioms.

1.2 Events and random variables – an outline

Random variables and events rely on each other. Random variables can be examined from the perspective of events, and vice versa. In this introductionary section we describe this interplay in an non-systematic manner and detached from the measure-theoretic model.

The field $\mathcal{E}$ of events is a σ-complete Boolean lattice. In particular:

- Each event E possesses a complementary event E^c.
- For any finite or infinite sequence $E_1, E_2, \ldots$ of events there exists its union $\bigcup_n E_n$ and its intersection $\bigcap_n E_n$.
- There are the sure and the impossible events E_{sure} and E_{imp}.

Also $E_1 \subset E_2$, iff $E_1 \cap E_2 = E_1$. Since events are no longer considered as subsets of some space, unions and intersections have to be interpreted here in the lattice-theoretic manner.

A random variable X first of all has a *target space* S equipped with a σ-field $\mathcal{B}$. Intuitively S is the set, where X may take its values. Collections of random variables obey the following simple rules:

- To each random variable X with target space S and to each measurable $\varphi : S \to S'$ a random variable with target space S' is uniquely associated, denoted by $\varphi(X)$.
- To each sequence $X_1, X_2, \ldots$ of random variables with target spaces $S_1, S_2, \ldots$ a random variable with target space $S_1 \times S_2 \times \cdots$ equipped with the product-σ-field is uniquely associated, denoted by $(X_1, X_2, \ldots)$.

The corresponding calculation rules are obvious; we will come back to them. We point out that not every measurable space is suitable as a target space – a minimal condition will be given in the next section. Uncountable products $\otimes_{i \in I}(S_i, \mathcal{B}_i)$ of measurable spaces are in general not admissible target spaces. This conforms to the fact that in probability an uncountable family of random variables $(X_i)_{i \in I}$ is at most provisionally considered as a single random variable with values in the product space, before proceeding to a better suited target space.

The connection between random variables and events is established by the remark that to any random variable X and to any measurable subset B of its target space S an event

$$\{X \in B\}$$

is uniquely associated. The events $\{X \in B\}$ uniquely determine X, where B runs through the measurable subsets of S. The calculation rules are

$$\left\{X \in \bigcup_n B_n\right\} = \bigcup_n \{X \in B_n\}, \quad \left\{X \in \bigcap_n B_n\right\} = \bigcap_n \{X \in B_n\},$$

$$\{X \in B^c\} = \{X \in B\}^c, \quad \{X \in S\} = E_{\text{sure}}, \quad \{X \in \emptyset\} = E_{\text{imp}},$$

where $B, B_1, B_2, \ldots$ are measurable subsets of the target space of X. If these properties hold, the mapping $B \mapsto \{X \in B\}$ is called a σ-*homomorphism*. Moreover

$$\{\varphi(X) \in B'\} = \{X \in B\}, \quad \text{where } B = \varphi^{-1}(B')$$

$$\{(X_1, X_2, \ldots) \in B_1 \times B_2 \times \cdots\} = \bigcap_n \{X_n \in B_n\}.$$

From the perspective of events the connection to random variables is as follows: For any event E there is a random variable I_E with values in $\{0, 1\}$, the indicator variable of E, fulfilling

$$\{I_E = 1\} = E, \quad \{I_E = 0\} = E^c.$$

For any infinite sequence $E_1, E_2, \ldots$ of disjoint events there is a random variable $N = \min\{n : E_n \text{ occurs}\}$ with values in $\{1, 2, \ldots, \infty\}$ such that

$$\{N = n\} = E_n, \quad \{N = \infty\} = \bigcap_n E_n^c.$$

For any infinite sequence $E_1, E_2, \ldots$ of events (disjoint or not) there is a

random variable X and measurable subsets $B_1, B_2, \ldots$ of its target space
such that

$$\{X \in B_n\} = E_n$$

for all n (see Section 1.5).

This is roughly all that mathematically can be stated about events and
random variables. A systematic treatment requires an axiomatic approach. There are two possibilities, namely to start from events or to
start from random variables.

Either the starting point is the field of events, which is assumed to
be a σ-complete Boolean lattice $\mathcal{E}$. Then a random variable X with
target space $(S, \mathcal{B})$ is nothing else but a σ-homomorphism from $\mathcal{B}$ to
$\mathcal{E}$. It is convenient to denote it as $B \mapsto \{X \in B\}$ again. In this
approach some technical efforts are required to show that any sequence
$X_1, X_2, \ldots$ of random variables may be combined to a single random
variable $(X_1, X_2, \ldots)$.

Starting from random variables instead is closer to intuition in our
view. Also it circumvents the technical efforts just mentioned. This
approach will be put forward in the following sections.

1.3 Spaces with denumerable separation

Not every measurable space qualifies as a possible target space. We
require that there exists a denumerable system of measurable sets separating points.

Definition *A measurable space* $(S, \mathcal{B})$ *is called a* measurable space with
denumerable separation (mSdS), *if there is a denumerable* $\mathcal{C} \subset \mathcal{B}$ *such
that for any pair* $x \neq y$ *of elements in* S *there is a* $C \in \mathcal{C}$ *such that
$x \in C$ and $y \notin C$.*

Examples

(i) Any separable metric space together with its Borel-σ-algebra is
an mSdS. This includes the case of denumerable S and in fact
any relevant target space of random variables considered in probability.

(ii) If $(S_1, \mathcal{B}_1), (S_2, \mathcal{B}_2), \ldots$ is a sequence of mSdS, then also the product space $\otimes_n (S_n, \mathcal{B}_n)$ is an mSdS. Indeed, if $\mathcal{C}_1, \mathcal{C}_2, \ldots$ are the

separating systems, then

$$\mathcal{C} := \bigcup_n \{S_1 \times \cdots \times S_{n-1} \times C_n \times S_{n+1} \times \cdots : C_n \in \mathcal{C}_n\}$$

is denumerable and separating in the product space.

(iii) An uncountable product of measurable spaces is not an mSdS (up to trivial cases). The reason is that this product-σ-field does not contain one-point sets (see below). □

An mSdS $(S, \mathcal{B})$ has two important properties. Firstly one point subsets $\{x\}$ are measurable, since

$$\{x\} = \bigcap_{C \in \mathcal{C}, x \in C} C$$

for all $x \in S$. Secondly the 'diagonal'

$$D := \{(x, y) \in S^2 : x = y\}$$

is measurable in the product space $(S^2, \mathcal{B}^2)$, since

$$D = \bigcap_{C \in \mathcal{C}} C \times C \cup C^c \times C^c. \tag{1.1}$$

These properties are crucial for target spaces of random variables. Remarkably the second one is characteristic for mSdS.

Proposition 1.1 *A measurable space $(S, \mathcal{B})$ is an mSdS, if and only if $D \in \mathcal{B}^2$.*

Proof. It remains to prove that $D \in \mathcal{B}^2$ implies the existence of a denumerable separating system $\mathcal{C}$. Let

$$\mathcal{F} := \bigcup_\mathcal{C} \sigma(\mathcal{C}) \otimes \sigma(\mathcal{C}),$$

where $\sigma(\mathcal{C})$ is the σ-field generated by $\mathcal{C}$ and the union is taken over all denumerable $\mathcal{C} \subset \mathcal{B}$. $\mathcal{F}$ is a sub-σ-field of $\mathcal{B} \otimes \mathcal{B}$ containing all $B_1 \times B_2$ with $B_1, B_2 \in \mathcal{B}$, thus

$$\mathcal{B} \otimes \mathcal{B} = \bigcup_\mathcal{C} \sigma(\mathcal{C}) \otimes \sigma(\mathcal{C}).$$

By assumption it follows that $D \in \sigma(\mathcal{C}) \otimes \sigma(\mathcal{C})$ for some denumerable $\mathcal{C} \subset \mathcal{B}$. We show that $\mathcal{C} \cup \{C^c : C \in \mathcal{C}\}$ is a separating system. Let $x, y \in S, x \neq y$. Then D does not belong to the σ-field

$$\mathcal{G} := \{B \in \sigma(\mathcal{C}) \otimes \sigma(\mathcal{C}) : \{(x, x), (y, x)\} \subset B \text{ or } \{(x, x), (y, x)\} \subset B^c\}.$$

It follows that $\mathcal{G} \neq \sigma(\mathcal{C}) \otimes \sigma(\mathcal{C})$, thus there are $B_1, B_2 \in \sigma(\mathcal{C})$ such that $B_1 \times B_2 \notin \mathcal{G}$. Thus B_1 contains x or y, but not both, and consequently is not an element of the σ-field

$$\mathcal{H} := \{B \in \sigma(\mathcal{C}) : \{x, y\} \subset B \text{ or } \{x, y\} \subset B^c\}.$$

Thus $\mathcal{H} \neq \sigma(\mathcal{C})$, therefore there is a $C \in \mathcal{C}$ such that x or y are elements of C, but not both. This finishes the proof. $\qquad\square$

The property of denumerable separation proves useful also in the study of σ-homomorphisms between measurable spaces.

Proposition 1.2 *Let* $(S, \mathcal{B})$ *be a mSdS, let* $(\Omega, \mathcal{E})$ *be a measurable space and let* $h : \mathcal{B} \to \mathcal{E}$ *be a* σ-*homomorphism. Then there is a unique measurable function* $\eta : \Omega \to S$ *such that* $\eta^{-1}(B) = h(B)$ *for all* $B \in \mathcal{B}$.

Proof. First we prove that h is not only a σ-homomorphism but a τ-homomorphism, that is

$$h(B) = \bigcup_{x \in B} h(\{x\}) \tag{1.2}$$

for all $B \in \mathcal{B}$. For the proof let $\{C_1, C_2, \ldots\}$ be a separating system of $\mathcal{B}$. Because h is a σ-homomorphism,

$$h(B) = \bigcap_n h(B \cap C_n) \cup h(B \cap C_n^c).$$

Since we consider sets here, this expression may be further transformed by general distributivity: Denoting $C_n^+ := C_n$ and $C_n^- := C_n^c$,

$$h(B) = \bigcup_\chi \bigcap_n h(B \cap C_n^{\chi(n)}) = \bigcup_\chi h\left(B \cap \bigcap_n C_n^{\chi(n)}\right),$$

where the union is taken over all mappings $\chi : \mathbb{N} \to \{+, -\}$. Since $\{C_1, C_2, \ldots\}$ is a separating system, $\bigcap_n C_n^{\chi(n)}$ contains at most one element, and for each $x \in S$ there is exactly one χ such that $\{x\} = \bigcap_n C_n^{\chi(n)}$. Therefore (1.2) follows.

In particular $\Omega = \bigcup_{x \in S} h(\{x\})$. This enables us to define η by means of

$$\eta(\omega) = x \quad :\Leftrightarrow \quad \omega \in h(\{x\}),$$

that is $\eta^{-1}(\{x\}) = h(\{x\})$. From (1.2)

$$h(B) = \bigcup_{x \in B} \eta^{-1}(\{x\}) = \eta^{-1}(B).$$

In particular η is measurable.   □

1.4 The axioms for random variables

In this section we introduce the concept of a *stochastic ensemble* $\mathcal{R}_S$, $S \in \mathcal{T}$. We require the following properties:

$\mathcal{T}$ is a collection of elements, which we call *target spaces*. They are assumed to be measurable spaces $(S, \mathcal{B})$ with denumerable separation. Since in concrete cases it is always clear which σ-field $\mathcal{B}$ is used within S, we often take the liberty to call S the target space and to suppress $\mathcal{B}$.

$\mathcal{R}_S$ is a set for each $S \in \mathcal{T}$. Its elements are called *random variables*, more precisely *random variables with target space* S. For $X \in \mathcal{R}_S$ we also say 'X takes values in S' and write

$$X \curvearrowright S.$$

$X, Y \curvearrowright S$ means $X, Y \in \mathcal{R}_S$.

Four axioms are needed to make this concept work. The first two ensure that stochastic ensembles are sufficiently rich. (Of course products of target spaces are always endowed with the product-σ-field.)

Axiom 1 $\{0,1\}^n \in \mathcal{T}$ *for* $n = 1, 2, \ldots, \infty$. *Moreover, if* $S_1, \ldots, S_n \in \mathcal{T}$, *then also* $S_1 \times \cdots \times S_n \in \mathcal{T}$.

This axiom contains the minimal assumptions needed for our purposes. In view of Axiom 3 below one might prefer to require that also countably infinite products of target spaces always belong to $\mathcal{T}$. This is a matter of taste – then $\mathcal{T}$ will be enlarged dramatically.

Axiom 2 *There are* $S \in \mathcal{T}$ *and* $X, Y \curvearrowright S$ *such that* $X \neq Y$.

The other two axioms describe how to build new random variables from given ones. The next axiom states that random variables transform like ordinary variables.

Axiom 3 *To each random variable* $X \curvearrowright S$ *and to each measurable mapping* $\varphi : S \to S'$ *with* $S' \in \mathcal{T}$ *a random variable* $X' \curvearrowright S'$ *is uniquely associated denoted* $X' = \varphi(X)$. *These random variables fulfil*

$$\mathrm{id}(X) = X$$

and

$$(\psi \circ \varphi)(X) = \psi(\varphi(X)),$$

whenever such expressions may be formed.

In the next axiom $\pi_i : S_1 \times S_2 \times \cdots \to S_i$ denotes the projection mapping to the ith coordinate,

$$\pi_i(x_1, x_2, \ldots) := x_i.$$

π_i is measurable.

Axiom 4 *Let $S_1, S_2, \ldots$ be a finite or infinite sequence of target spaces such that also $S_1 \times S_2 \times \ldots$ belongs to $\mathcal{T}$. Then to any $X_1 \curvearrowright S_1, X_2 \curvearrowright S_2, \ldots$ a random variable $X \curvearrowright S_1 \times S_2 \times \cdots$ is uniquely associated characterized by the property*

$$\pi_i(X) = X_i$$

for all i. It is denoted $X = (X_1, X_2, \ldots)$ and called the product variable *of $X_1, X_2, \ldots$.*

Axioms 3 and 4 can be summarized as follows: If $X_1 \curvearrowright S_1, X_2 \curvearrowright S_2, \ldots$ and if $\varphi : S_1 \times S_2 \cdots \times \to S$ is measurable, then we may form the random variable

$$\varphi(X_1, X_2, \ldots) := \phi(X)$$

with $X = (X_1, X_2, \ldots)$, provided the product space belongs to $\mathcal{T}$. Also

$$\psi\big(\varphi_1(X_1, X_2, \ldots), \varphi_2(X_1, X_2, \ldots), \ldots\big) = \big(\psi \circ (\varphi_1, \varphi_2, \ldots)\big)(X_1, X_2, \ldots) \tag{1.3}$$

for suitable measurable mappings $\psi, \varphi_1, \varphi_2, \ldots$ Indeed: From Axiom 3 we obtain

$$\pi_i\big((\varphi_1, \varphi_2, \ldots)(X)\big) = \varphi_i(X),$$

thus $(\varphi_1, \varphi_2, \ldots)(X) = (\varphi_1(X), \varphi_2(X), \ldots)$ from Axiom 4 and (1.3) follows from Axiom 3.

Examples

(i) **Real-valued random variables.** As to a concrete example let us look at real-valued random variables. Then $\mathcal{T}$ has to contain $\mathbb{R}^d$ for $d = 1, 2, \ldots$. The ordinary operations within $\mathbb{R}$ transfer to random variables without difficulties. For example, for $X_1, X_2 \curvearrowright$ $\mathbb{R}$ we may define $X_1 + X_2 := \varphi(X_1, X_2)$ using the measurable mapping $\varphi(x_1, x_2) := x_1 + x_2$. The calculation rules transfer from the real numbers to random variables by means of (1.3), i.e.

$$
\begin{aligned}
(X_1 + X_2) + X_3 &= \big(\varphi \circ (\varphi \circ (\pi_1, \pi_2), \pi_3)\big)(X_1, X_2, X_3) \\
&= \big(\varphi \circ (\pi_1, \varphi \circ (\pi_2, \pi_3))\big)(X_1, X_2, X_3) = X_1 + (X_2 + X_3)
\end{aligned}
$$

Other operations such as $|X|$ and $\max(X_1, X_2)$ are introduced in much the same way.

(ii) **Random variables with constant value.** Each element c of some target space S may be considered as a random variable: Let $\varphi' : S' \to S$ be any mapping taking only the value c and choose any random variable $X' \curvearrowright S'$. φ' is measurable. It is easy to show that $\varphi'(X')$ only depends on c: If $\varphi''(X'')$ is another choice, then by Axioms 3 and 4 $\varphi'(X') = (\varphi' \circ \pi')(X', X'') = (\varphi'' \circ \pi'')(X', X'') = \varphi''(X'')$. It is consistent to denote this random variable by c again. Indeed a measurable mapping φ may now be applied to c in two different meanings, but the result is the same because of $\varphi(\varphi'(X')) = (\varphi \circ \varphi')(X')$ and the observation that $\varphi \circ \varphi'$ takes the constant value $\varphi(c)$.

(iii) **Measure-theoretic models.** For any collection $\mathcal{T}$ of target spaces we get examples of stochastic ensembles by choosing some basic measurable space $(\Omega, \mathcal{E})$ and then letting $\mathcal{R}_S := \{X : \Omega \to S \mid X \text{ is measurable}\}$, $S \in \mathcal{T}$, $\varphi(X) := \varphi \circ X$ and $(X_1, X_2, \ldots)$ the product mapping. These are the canonical models. If a probability measure is given on the basic space, then also the sets $\mathcal{R}'_S := \{[X] \mid X \in \mathcal{R}_S\}$ make up a stochastic ensemble, where $[X]$ denotes the equivalence class of measurable functions, which are a.e. equal to X. Here $\varphi([X_1], [X_2], \ldots) := [\varphi(X_1, X_2, \ldots)]$, $([X_1], [X_2], \ldots) := [(X_1, X_2, \ldots)]$. The axioms are trivially fulfilled. Note that such stochastic ensembles do not consist of measurable functions in general. □

(1.3) says in short that ordinary variables may be replaced by random variables in functional relations. The following proposition substantiates this statement.

Proposition 1.3 *Let* S, S', S_1, S_2 *be target spaces and let* $\varphi_i : S_1 \times S_2 \rightarrow$ S, $\psi_i : S_1 \times S_2 \rightarrow S'$, $i = 1, 2$ *be measurable mappings fulfilling*

$$\varphi_1(x, y) = \varphi_2(x, y) \quad \Rightarrow \quad \psi_1(x, y) = \psi_2(x, y)$$

for all $x \in S_1, y \in S_2$. *Let also* $X \curvearrowright S_1$, $Y \curvearrowright S_2$. *Then*

$$\varphi_1(X, Y) = \varphi_2(X, Y) \quad \Rightarrow \quad \psi_1(X, Y) = \psi_2(X, Y).$$

The proposition comprises the possibility that φ_i and ψ_i do not depend on x or y. Note that e.g. $\varphi(x, y) = \varphi'(x)$ for all x, y implies $\varphi(X, Y) = (\varphi' \circ \pi)(X, Y) = \varphi'(X)$ by Axioms 3 and 4.

Proof. Let $z = (x, y)$ and $Z = (X, Y)$. Consider $\theta : S \times S \times S' \times S' \rightarrow S'$, given by

$$\theta(u, v, u', v') := \begin{cases} u' & \text{if } u = v, \\ v' & \text{if } u \neq v. \end{cases}$$

θ is measurable, owing to the fact that the diagonal in $S \times S$ is measurable. Then

$$\theta\big(\varphi_1(z), \varphi_1(z), \psi_1(z), \psi_2(z)\big) = \psi_1(z),$$

whereas by assumption

$$\theta\big(\varphi_1(z), \varphi_2(z), \psi_1(z), \psi_2(z)\big) = \psi_2(z).$$

By means of (1.3) replace the variable z by the random variable Z in these equations. Then by assumption the left-hand sides coincide, and our claim follows. $\qquad\square$

1.5 Events

To each random variable $X \curvearrowright S$ and to each measurable subset $B \subset S$ we associate now an *event*, written as

$$\{X \in B\}.$$

In particular, since target spaces contain one point sets, we may form the events

$$\{X = x\} := \{X \in \{x\}\}, \quad x \in S.$$

In order to carry out calculations, we have to define equality of events. Here we use that the characteristic function $\mathbf{1}_B(\cdot)$ of measurable $B \subset S$

is a measurable mapping from S to $\{0,1\}$, which allows us to apply Axioms 1 and 3.

Definition. *Two events* $\{X \in B\}$ *and* $\{X' \in B'\}$ *are said to be equal,*

$$\{X \in B\} = \{X' \in B'\},$$

if $\mathbf{1}_B(X) = \mathbf{1}_{B'}(X')$.

In other words: In our approach an event is an equivalence class of pairs (X, B). To each event E we may associate its *indicator variable* I_E, a random variable with values in $\{0,1\}$, given by

$$I_E := \mathbf{1}_B(X), \quad \text{if } E = \{X \in B\}.$$

Two events with the same indicator variable are equal. The set of events is denoted by $\mathcal{E}$.

Examples

(i) The equality

$$\{\varphi(X) \in B\} = \{X \in \varphi^{-1}(B)\}$$

holds, since $\mathbf{1}_B(\varphi(X)) = \mathbf{1}_B \circ \varphi(X) = \mathbf{1}_{\varphi^{-1}(B)}(X)$ in view of Axiom 3.

(ii) For any event E the equality

$$\{I_E = 1\} = E$$

holds, since $\mathbf{1}_{\{1\}} = \text{id}$ on $\{0,1\}$, thus $\mathbf{1}_{\{1\}}(I_E) = \text{id}(I_E) = I_E$. $\square$

Next we introduce the basic operations with events.

Proposition 1.4 *For any event E there exists a unique event, denoted E^c, such that*

$$\{X \in B\}^c = \{X \in B^c\}$$

for any $X \curvearrowright S$ and any measurable $B \subset S$.

Proof. For $E = \{X \in B\}$ we define $E^c := \{X \in B^c\}$. We only have to show that this definition is unambiguous. Note that $\mathbf{1}_{B^c} = \eta \circ \mathbf{1}_B$ with $\eta(0) = 1, \eta(1) = 0$. Thus $\{X \in B\} = \{X' \in B'\}$ and Axiom 3 imply $\mathbf{1}_{B^c}(X) = \eta(\mathbf{1}_B(X)) = \eta(\mathbf{1}_{B'}(X')) = \mathbf{1}_{(B')^c}(X')$. Therefore

$\{X \in B^c\} = \{X' \in (B')^c\}.$ □

Proposition 1.5 *For any finite or infinite sequence of events $E_1, E_2, \ldots$ there exist two unique events, denoted as $\bigcup_n E_n$ and $\bigcap_n E_n$, such that*

$$\bigcup_n \{X \in B_n\} = \left\{X \in \bigcup_n B_n\right\}, \quad \bigcap_n \{X \in B_n\} = \left\{X \in \bigcap_n B_n\right\}$$

for any $X \curvearrowright S$ and any measurable $B_1, B_2, \ldots \subset S$.

Proof. We proceed as in the last proof and define $\bigcup_n E_n := \{X \in \bigcup_n B_n\}$, if $E_n = \{X \in B_n\}$. Note that $1_{\bigcup_n B_n} = \max \circ (1_{B_1}, 1_{B_2}, \ldots)$ with the measurable function $\max(x_1, x_2, \ldots) := \max_n x_n$ from $\{0,1\}^\ell$ to $\{0,1\}$ (ℓ being the length of the sequence). Thus $\{X \in B_n\} = \{X' \in B'_n\}$ implies

$$1_{\bigcup_n B_n}(X) = \max(1_{B_1}(X), 1_{B_2}(X), \ldots)$$
$$= \max(1_{B'_1}(X'), 1_{B'_2}(X'), \ldots) = 1_{\bigcup_n B'_n}(X')$$

in view of Axioms 3 and 4. Therefore $\{X \in \bigcup_n B_n\} = \{X' \in \bigcup_n B'_n\}$ such that $\bigcup_n E_n$ is well-defined. $\bigcap_n E_n$ is obtained similarly.

It remains to show that each sequence $E_1, E_2, \ldots$ can be represented as $E_n = \{X \in B_n\}$ with *one* random variable X. Letting $X := (I_{E_1}, I_{E_2}, \ldots)$ with target space $S = \{0,1\}^\ell$ and $B_n := \{0,1\}^{n-1} \times \{1\} \times \{0,1\}^{\ell-n}$ we indeed obtain

$$\{X \in B_n\} = \{X \in \pi_n^{-1}(\{1\})\} = \{\pi_n(X) = 1\} = \{I_{E_n} = 1\} = E_n.$$
$$(1.4)$$

This is the claim. □

Proposition 1.6 *There are two unique events $E_{\text{sure}} \neq E_{\text{imp}}$ such that*

$$E_{\text{sure}} = \{X \in S\}, \quad E_{\text{imp}} = \{X \in \emptyset\}$$

for any $X \curvearrowright S$.

Proof. Again define $E_{\text{sure}} := \{X \in S\}$. Let also $X' \curvearrowright S'$. Then $1_S \circ \pi = 1_{S'} \circ \pi'$ on $S \times S'$ with projections π, π'. Axiom 3 implies $1_S(X) = 1_S \circ \pi(X, X') = 1_{S'} \circ \pi'(X, X') = 1_{S'}(X')$, thus $\{X \in S\} = \{X' \in S'\}$. Similarly $\{X \in \emptyset\} = \{X' \in \emptyset\}$, therefore E_{sure} and E_{imp} are well-defined.

Now suppose that $E_{\text{sure}} = E_{\text{imp}}$. Then all events are equal, as follows from Proposition 5:

$$
\begin{aligned}
E &= \{X \in B\} = \{X \in B\} \cap \{X \in S\} \\
&= \{X \in B\} \cap \{X \in \emptyset\} = \{X \in \emptyset\} = E_{\text{imp}}.
\end{aligned}
$$

This implies that any two random variables X, Y with the same target space S are equal. Indeed, let $C_1, C_2, \ldots$ be a separating system in S and let $\varphi := (\mathbf{1}_{C_1}, \mathbf{1}_{C_2}, \ldots)$. Then $\varphi(X) = (I_{\{X \in C_1\}}, I_{\{X \in C_2\}}, \ldots) = (I_{\{Y \in C_1\}}, I_{\{Y \in C_2\}}, \ldots) = \varphi(Y)$. φ is injective, thus we may conclude from Proposition 1.3 that $X = Y$. This contradicts Axiom 2, and our claim follows.   □

Thus we have introduced complementary events, unions and intersections of events as well as the sure and the impossible event. In view of the above characterizations and (1.4) it is obvious that properties of sets carry over to properties of events. Altogether we end up with a Boolean σ-lattice, equipped with the order relation

$$
E \subset E' \quad :\Leftrightarrow \quad E = E \cap E',
$$

with maximal element E_{sure} and minimal element E_{imp}. It is a standard procedure to obtain the other properties of fields of events.

Examples

(i) For $X = (X_1, X_2, \ldots)$ we have from Proposition 1.5

$$
\{X \in B_1 \times B_2 \times \cdots\} = \left\{X \in \bigcap_n \pi_n^{-1}(B_n)\right\} = \bigcap_n \{X \in \pi_n^{-1}(B_n)\}
$$

and from Axiom 4

$$
\{(X_1, X_2, \ldots) \in B_1 \times B_2 \times \cdots\} = \bigcap_n \{X_n \in B_n\}.
$$

(ii) Let $\sim$ be any relation on S such that $R := \{(x, y) \in S^2 : x \sim y\}$ is a measurable subset of S^2. Then it is natural to define $\{X \sim Y\} := \{(X, Y) \in R\}$ and to write $X \sim Y$, if $\{X \sim Y\} = E_{\text{sure}}$. This is in accordance with Proposition 1.7 below.   □

Remarks: σ-homomorphisms in stochastic ensembles. From Propositions 1.4 to 1.6 it is immediate that each random variable X induces a σ-homomorphism from $\mathcal{B}$ to $\mathcal{E}$ given by $B \mapsto h(B) := \{X \in$

B}. From Proposition 1.7 (see below) it follows that h determines X uniquely. Therefore one may ask whether in a stochastic ensemble every σ-homomorphism h from the σ-field $\mathcal{B}$ of some target space S into $\mathcal{E}$ comes from a random variable. This is true in two cases.

The first case is the classical one that the random variables with target space S are given as above by the system of measurable mappings from some basic measurable space $(\Omega, \mathcal{E})$ into S. Then Proposition 1.2 applies, stating that there are no other σ-homomorphisms.

In the other case we assume that S is a Polish space endowed with its Borel-σ-field. This case is more profound: Choose a separating sequence $C_1, C_2, \ldots$ Define the measurable function $\varphi := (\mathbf{1}_{C_1}, \mathbf{1}_{C_2}, \ldots)$ from S into $\{0,1\}^\infty$ and the random variable $Y := (I_{h(C_1)}, I_{h(C_2)}, \ldots)$. Then

$$\{Y \in B'\} = h(\varphi^{-1}(B'))$$

for each Borel-set $B' \subset \{0,1\}^\infty$. For the proof note that the system of B' fulfilling our claim is a σ-field containing $\{0,1\}^{m-1} \times \{1\} \times \{0,1\}^\infty$. Now φ is an injective measurable mapping from the Polish space S into the Polish space $\{0,1\}^\infty$. A celebrated theorem of Kuratowski says that then the image of each Borel-set is a Borel-set again. An immediate consequence is that there is a measurable mapping $\psi : \{0,1\}^\infty \to S$ such that $\psi \circ \varphi$ is the identity on S. Letting $X := \psi(Y)$ one obtains the claim: For each Borel-set $B \subset S$

$$\{X \in B\} = \{Y \in \psi^{-1}(B)\} = h(B).$$

1.6 Equality and a.s. equality

Recall from (1.1) that the diagonal $D \subset S^2$ is measurable in the case of target spaces. Thus we may define for $X, Y \curvearrowright S$

$$\{X = Y\} := \{(X,Y) \in D\}, \quad \{X \neq Y\} := \{(X,Y) \in D^c\}.$$

Proposition 1.7 *For $X, Y \curvearrowright S$ the following statements are equivalent:*

(i) $X = Y$,

(ii) $\{X = Y\} = E_{\text{sure}}$,

(iii) $\{X \in B\} = \{Y \in B\}$ *for all measurable $B \subset S$.*

Proof. Assume that $C_1, C_2, \ldots$ separate the elements of S. Then

$\varphi(x) := (\mathbf{1}_{C_1}, \mathbf{1}_{C_2}, \ldots)$ is an injective measurable function from S to $\{0,1\}^\infty$. Thus for $x, y \in S$

$$x = y \quad \Leftrightarrow \quad \mathbf{1}_D(x,y) = \mathbf{1}_{S \times S}(x,y) \quad \Leftrightarrow \quad \varphi(x) = \varphi(y) \, .$$

Proposition 1.3 implies

$$X = Y \quad \Leftrightarrow \quad \mathbf{1}_D(X,Y) = \mathbf{1}_{S \times S}(X,Y) \quad \Leftrightarrow \quad \varphi(X) = \varphi(Y) \, .$$

Thus $X = Y$ is equivalent to $\{X = Y\} = \{(X,Y) \in S \times S\} = E_{\text{sure}}$ as well as to $\{X \in C_i\} = \{Y \in C_i\}$ for all i. Since any measurable $B \subset S$ may be included into the sequence $C_1, C_2, \ldots$, our claim follows. $\qquad \square$

Next we discuss the notion of almost sure equality in stochastic ensembles. Shortly speaking this is any equivalence relation compatible with our axioms. As we shall see, this conforms to the traditional definition of a.s. equality. On the other hand it will become apparent that in our setting it is no longer necessary to distinguish between random variables and a.s. defined random variables as in the traditional approach. Both give rise to stochastic ensembles.

Definition *An equivalence relation $\sim$ on the collection of random variables of a stochastic ensemble is called an* a.s. equality, *if*

 (i) *$X \sim Y$ implies that X, Y have the same target space.*
 (ii) *There exist $X, Y \curvearrowright S$ such that $X \nsim Y$.*
 (iii) *$X \sim Y \quad \Rightarrow \quad \varphi(X) \sim \varphi(Y)$.*
 (iv) *$X_1 \sim Y_1, X_2 \sim Y_2, \ldots \quad \Rightarrow \quad (X_1, X_2, \ldots) \sim (Y_1, Y_2, \ldots)$.*

Let $X^\sim$ denote the equivalence class of the a.s. equality $\sim$ containing X. Then we may associate to $X^\sim$ a target space, namely that of X. Also we may define

$$\varphi(X^\sim) := \varphi(X)^\sim \, , \quad (X_1^\sim, X_2^\sim, \ldots) := (X_1, X_2, \ldots)^\sim \, .$$

Let

$$\mathcal{R}_S^\sim := \{X^\sim : X \in \mathcal{R}_S\}.$$

With these conventions the following result is obvious.

Proposition 1.8 $\mathcal{R}_S^\sim$, $S \in \mathcal{T}$ *is a stochastic ensemble.*

In what follows we show that our definition is intimately connected with the usual definition of a.s. equality. Let us recall the notion of a *null-system* (a *σ-ideal*). It is a system $\mathcal{N} \subset \mathcal{E}$ of events fulfilling

$$E_1, E_2, \ldots \in \mathcal{N} \quad \Rightarrow \quad \bigcup_n E_n \in \mathcal{N}$$

$$E \in \mathcal{N}, E' \subset E \quad \Rightarrow \quad E' \in \mathcal{N}$$

$$E_{\text{imp}} \in \mathcal{N} \qquad E_{\text{sure}} \notin \mathcal{N}$$

An important example of a null-system is the system of events of probability zero in case $\mathcal{E}$ is endowed with a probability measure.

Proposition 1.9 *To each a.s. equality $\sim$ the system of events*

$$\mathcal{N}^\sim := \{\{X \neq Y\} : X \sim Y\}$$

is a null-system. It fulfils

$$X \sim Y \quad \Leftrightarrow \quad \{X \neq Y\} \in \mathcal{N}^\sim$$

for any $X, Y \curvearrowright S$. The mapping $\sim \mapsto \mathcal{N}^\sim$ establishes a one-to-one correspondence between a.s. equality relations and null-systems.

Proof. From

$$\bigcup_n \{X_n \neq Y_n\} = \{(X_1, X_2, \ldots) \neq (Y_1, Y_2, \ldots)\}$$

and from condition (iv) of the definition it follows that $\mathcal{N}^\sim$ fulfils the first requirement of a null-system. Next let $X, Y \curvearrowright S$ and

$$E \subset \{X \neq Y\}.$$

We have to show $E \in \mathcal{N}^\sim$. Suppose $E = \{Z \in B'\}$ for a suitable $Z \curvearrowright S'$. Define the measurable function $\varphi : S \times S' \to S$ as

$$\varphi(x, z) := \begin{cases} x & \text{if } z \in B', \\ x_0 & \text{if } z \notin B', \end{cases}$$

for some $x_0 \in S$. It follows that

$$\{\varphi(X, Z) \neq \varphi(Y, Z)\} = \{Z \in B'\} \cap \{X \neq Y\} = E.$$

From $X \sim Y$ and conditions (iii) and (iv) of the definition we obtain $\varphi(X, Z) \sim \varphi(Y, Z)$. Thus $E \in \mathcal{N}^\sim$, which means that $\mathcal{N}^\sim$ fulfils the second requirement of a null-system. In particular: If $E_{\text{sure}} \in \mathcal{N}^\sim$, then $\{X \neq Y\} \in \mathcal{N}^\sim$ for all $X, Y \curvearrowright S$. This contradicts condition (ii) of the definition, thus we may conclude $E_{\text{sure}} \notin \mathcal{N}^{\sim}$. Finally $E_{\text{imp}} = \{X \neq X\} \in \mathcal{N}^{\sim}$. Therefore $\mathcal{N}^{\sim}$ is a null-system.

We come to the second claim of the proposition. The implication $\Rightarrow$ is obvious, thus let us assume $\{X \neq Y\} \in \mathcal{N}^{\sim}$. Then there exist $X', Y' \curvearrowright S$ such that $\{X \neq Y\} = \{X' \neq Y'\}$ and $X' \sim Y'$. We have to show that $X \sim Y$, too. For this purpose we use the measurable mapping θ, defined in the proof of Proposition 1.3. It fulfils

$$\mathbf{1}_D(x, y) = \mathbf{1}_{D'}(x', y') \quad \Rightarrow \quad \theta(x', y', x, y) = y,$$

where D and D' are the diagonals in S^2 and $(S')^2$. By assumption $\mathbf{1}_D(X, Y) = \mathbf{1}_{D'}(X', Y')$, therefore Proposition 3 implies $\theta(X', Y', X, Y) = Y$. Moreover $\theta(x', x', x, y) = x$ and consequently $\theta(X', X', X, Y) = X$. Since $X' \sim Y'$, we obtain $X \sim Y$ in view of (iii) and (iv) of the definition. Thus also the second statement is proved.

In particular this implies that $\sim \mapsto \mathcal{N}^{\sim}$ is an injective mapping. It remains to prove surjectivity. Thus let $\mathcal{N}$ be any null-system and define

$$X \sim Y \quad :\Leftrightarrow \quad \{X \neq Y\} \in \mathcal{N},$$

whenever X and Y have the same target space. We have to show that $\sim$ is an a.s. equality. Since $\{I_{E_{\text{sure}}} \neq I_{E_{\text{imp}}}\} = E_{\text{sure}}$, $I_{E_{\text{sure}}} \not\sim I_{E_{\text{imp}}}$. Thus condition (ii) of the definition holds. Condition (iii) follows from

$$\{\varphi(X) \neq \varphi(Y)\} \subset \{X \neq Y\}$$

and condition (iv) follows from

$$\{(X_1, X_2, \ldots) \neq (Y_1, Y_2, \ldots)\} = \bigcup_n \{X_n \neq Y_n\}.$$

This finishes the proof. □

Random variables $X \curvearrowright S$, $X' \curvearrowright S'$ with different target spaces are always unequal in our approach. It might be convenient to call them *indistinguishable*, if $S \cap S'$ is a measurable subset of S and of S' and if

$$\{X \in B\} = \{X' \in B\} \quad \text{for all measurable } B \subset S \cap S',$$
$$\{X \in S \cap S'\} = \{X' \in S \cap S'\} = E_{\text{sure}}.$$

In the same manner *a.s. indistinguishability* may be introduced.

1.7 Convergence of random variables

'In practice' the small omegas prove convenient in operating with random variables within the traditional setting. In our approach such manipulations may be reproduced without difficulties within the target spaces. This has been indicated already; here we like to exemplify this briefly in the context of convergence of random variables.

Let d be a metric on the target space S. We require

- The mapping $d : S^2 \to \mathbb{R}_+$ is measurable.
- Let $\varphi_1, \varphi_2, \ldots$ be mappings from some target space S' to S converging pointwise to $\varphi : S' \to S$ with respect to d. If the φ_n are measurable, then also φ is measurable.

These assumptions are satisfied if (S, d) is a separable metric space and $\mathcal{B}$ the corresponding Borel-σ-field on S, as follows from

$$d^{-1}([0, a)) = \bigcup_{\substack{x, y \in Q, \epsilon, \eta \in Q_+ \\ \epsilon + \eta + d(x, y) < a}} U_\epsilon(x) \times U_\eta(y) \, , \quad \varphi^{-1}(F) = \bigcap_{\epsilon \in Q_+} \bigcup_m \bigcap_{n \geq m} \varphi_n^{-1}(F^\epsilon) \, .$$

Here Q denotes a dense and countable subset of S, $U_\epsilon(x)$ the open ϵ-neighbourhood of x and F a closed subset of S with open ϵ-neighbourhood F^ϵ. An example of a non-separable metric space of importance fulfilling both requirements is the space of càdlàg functions endowed with the metric of locally uniform convergence (this metric is used in the theory of stochastic integration).

Now we assume that besides S also S^∞ is a target space. Let us consider the set of convergent sequences with given limit and the set of Cauchy-sequences,

$$B_{\lim} \quad := \quad \{(x, x_1, x_2, \ldots) \in S \times S^\infty : x = \lim_n x_n\},$$

$$B_{\text{Cauchy}} \quad := \quad \{(x_1, x_2, \ldots) \in S^\infty : (x_n) \text{ is Cauchy}\}.$$

Since $B_{\lim} = \bigcap_{k=1}^\infty \bigcup_{m=1}^\infty \bigcap_{n=m}^\infty \{(x, x_1, x_2, \ldots) : d(x_n, x) \leq \epsilon_k\}$ and $B_{\text{Cauchy}} = \bigcap_{k=1}^\infty \bigcup_{m=1}^\infty \bigcap_{n=m}^\infty \{(x_1, x_2, \ldots) : d(x_m, x_n) \leq \epsilon_k\}$ for any sequence $\epsilon_k \downarrow 0$, these are measurable subsets of $S \times S^\infty$ and S^∞. Thus for any random variables $X, X_1, X_2, \ldots \curvearrowright S$ we may define the events

$$\{X_n \to X\} \quad := \quad \{(X, X_1, X_2, \ldots) \in B_{\lim}\},$$

$$\{X_n \text{ is Cauchy}\} \quad := \quad \{(X_1, X_2, \ldots) \in B_{\text{Cauchy}}\}.$$

Proposition 1.10. *For any* $X, X_1, X_2, \ldots \curvearrowright S$

$$\{X_n \to X\} \subset \{X_n \text{ is Cauchy}\}.$$

Moreover, if (S, d) is a complete metric space, then for any $X_1, X_2, \ldots \curvearrowright S$ there is a $X \curvearrowright S$ such that

$$\{X_n \to X\} = \{X_n \text{ is Cauchy}\}.$$

Proof. Since $B_{\lim} \subset S \times B_{\text{Cauchy}}$

$$\{X_n \to X\} \subset \{(X, X_1, X_2, \ldots) \in S \times B_{\text{Cauchy}}\} = \{X_n \text{ is Cauchy}\}.$$

Moreover in the case of a complete metric space let $\varphi_n, \varphi : S^\infty \to S$ be given by

$$\varphi_n(x_1, x_2, \ldots) := \begin{cases} x_n & \text{if } (x_1, x_2, \ldots) \in B_{\text{Cauchy}}, \\ z & \text{otherwise,} \end{cases}$$

$$\varphi(x_1, x_2, \ldots) := \begin{cases} \lim_n x_n & \text{if } (x_1, x_2, \ldots) \in B_{\text{Cauchy}}, \\ z & \text{otherwise} \end{cases}$$

with some given $z \in S$. φ_n is measurable and

$$\varphi(x_1, x_2, \ldots) = \lim_n \varphi_n(x_1, x_2, \ldots)$$

for all $(x_1, x_2, \ldots)$, thus φ is measurable too. Define $X := \varphi(X_1, X_2, \ldots)$. Because of completeness $\psi^{-1}(B_{\lim}) = B_{\text{Cauchy}}$ for the measurable mapping $\psi : S^\infty \to S \times S^\infty$, given by

$$\psi(x_1, x_2, \ldots) := (\varphi(x_1, x_2, \ldots), x_1, x_2, \ldots).$$

Thus

$$
\begin{aligned}
\{X_n \to X\} &= \{\psi(X_1, X_2, \ldots) \in B_{\lim}\} \\
&= \{(X_1, X_2, \ldots) \in B_{\text{Cauchy}}\} = \{X_n \text{ is Cauchy}\},
\end{aligned}
$$

which is the claim.   □

Bibliography

[1] Billingsley, P. *Probability and Measure*, New York (1979).
[2] Breiman, L. *Probability*, New York (1968).
[3] Caratheodory, C. *Mass und Integral und ihre Algebraisierung*, Basel (1956).
[4] Coxeter, H.S.M. *Introduction to Geometry*, New York (1961).
[5] Dinges, H.; Rost, H. *Prinzipien der Stochastik*, Stuttgart (1982).
[6] Dinges, H. Variables, in particular random variables. *In Activity and Sign, Grounding Mathematics Education*, M. Hoffmann, J. Lenhard, F. Seeger (eds.), 305–311 (2005).
[7] Halmos, P.R. *Naive Set Theory*, Princeton (1960).
[8] Mumford, D. The dawning of the age of stochasticity. *Mathematics: Frontiers and Perspectives*, 197–218 (2000).
[9] Snell, L. A conversation with Joe Doob. *Stat. Science* **12**, 301–311. http://www.dartmouth.edu/~chance/Doob/conversation.html (1997).

2

Chaining techniques and their application to stochastic flows

Michael Scheutzow

Institut für Mathematik
Technische Universität Berlin
Straße des 17. Juni 136, D-10623 Berlin, Germany

Abstract

We review several competing *chaining* methods to estimate the supremum, the diameter of the range or the modulus of continuity of a stochastic process in terms of tail bounds of their two-dimensional distributions. Then we show how they can be applied to obtain upper bounds for the growth of bounded sets under the action of a stochastic flow.

2.1 Introduction

Upper and lower bounds for the (linear) growth rates of the diameter of the image of a bounded set in $\mathbf{R}^d$ under the action of a stochastic flow under various conditions have been shown in [4, 5, 6, 16, 17, 20]. In this survey, we will discuss upper bounds only. A well-established class of methods to obtain probability bounds for the supremum of a process are *chaining* techniques. Typically they transform bounds for the one- and two-dimensional distributions of the process into upper bounds of the supremum (for a real-valued process) or the diameter of the range of the process (for a process taking values in a metric space). In the next section, we will present some of these techniques, the best-known being Kolmogorov's continuity theorem, which not only states the existence of a continuous modification, but also provides explicit probabilistic upper bounds for the modulus of continuity and the diameter of the range of the process. We will also state a result which we call *basic chaining*. Further we will briefly review some of the results from Ledoux and Tala-

Trends in Stochastic Analysis, ed. J. Blath, P. Mörters and M. Scheutzow.
Published by Cambridge University Press. ©Cambridge University Press 2008.

grand [15] and a rather general version of the *GRR-Lemma* named after
Garsia, Rodemich and Rumsey [12]. Except for the result of Ledoux
and Talagrand, we will provide proofs for the chaining lemmas in the
appendix (in order to keep the article reasonably self-contained but also
because we chose to formulate the chaining results slightly differently
compared with the literature). We wish to point out, however, that
nothing in that section is essentially new and that it is not meant to
be a complete survey about chaining. The reader who is interested in
learning more about chaining should consult the literature, for example
the monograph by Talagrand [21].

In order to obtain good upper bounds on the diameter of the image
of a bounded set $\mathcal{X}$ under a stochastic flow ϕ which is generated (say)
by a stochastic differential equation on $\mathbf{R}^d$ with coefficients which are
bounded and Lipschitz continuous, one can try to apply the chaining
techniques directly to the process $\phi_{0,T}(x)$, $x \in \mathcal{X}$. This is what we
did in [6] using basic chaining. It worked, but it was a nightmare (for
the reader, the referee and us). The reason was that the two-point
motion of such a flow behaves quite differently depending on whether
the two points are very close (then the Lipschitz constants determine
the dynamics) or not (then the bounds on the coefficients do). This
requires a rather sophisticated choice of the parameters or functions in
the chaining lemmas. The papers [16, 17] provided somewhat simpler
proofs using the chaining methods of Ledoux-Talagrand and the GRR-
Lemma respectively. The approach presented here is (in our opinion)
much simpler and transparent than the previous ones. The reason is that
we strictly separate the local and the global behaviour in the following
sense: for a given (large) time T and a positive number γ, we cover
the set $\mathcal{X}$ with balls (or cubes) of radius $\exp\{-\gamma T\}$. For each center
of such a ball, we estimate the probability that it leaves a ball with
radius κT around zero up to time T using large deviations estimates
(Proposition 2.8). This probability bound depends only on the bounds
on the coefficients and not on the Lipschitz constants. In addition, we
provide an upper bound for the probablity that a particular one of the
small balls achieves a diameter of 1 (or some other fixed positive number)
up to time T (Theorem 2.1). This bound only involves the Lipschitz
constants and not the bounds on the coefficients. To obtain such a
bound, we use chaining. We will allow ourselves the luxury of five proofs
of this result using each of the chaining methods – with two proofs even
using the GRR-Lemma. Since we are only interested in the behaviour of
the image of a very small ball up to the time its radius becomes 1, things

become much easier compared with the approach in [6] mentioned above. In fact, we can use a polynomial function Ψ when applying the GRR-Lemma or the LT-Lemma and this is why Kolmogorov's Theorem, which also uses polynomial moment bounds, turns out to be just as efficient as the other (more sophisticated) methods. The proof of Theorem 2.3, which provides an explicit upper bound for the linear growth rate, now becomes almost straightforward: the probability that the diameter of the image of $\mathcal{X}$ under the flow up to time T exceeds κT is bounded from above by the number of small balls multiplied by the (maximal) probability that a center reaches a modulus of $\kappa T - 1$ or the diameter of a small ball exceeds 1. This bound – which is still a function of the parameter γ – turns out to be exponentially small in T provided κ is large enough and γ is chosen appropriately. An application of the first Borel-Cantelli Lemma then completes the proof.

We talk about stochastic flows above, but the results are true under less restrictive conditions. For the upper bound of the growth of a small ball (Theorem 2.1), it suffices that the underlying motion $\phi_t(x)$ is jointly continuous and that (roughly speaking) the distance of two trajectories does not grow faster than a geometric Brownian motion (this is hypothesis (H) in Section 2.3). In the special case of a (spatially) differentiable and translation invariant Brownian flow, Theorem 2.1 can be improved slightly. This is shown in Theorem 2.2. Its proof is completely different from that of Theorem 2.1: it does not use any chaining whatsoever.

2.2 The competitors

In the following, we will always assume that $(\hat{E}, \hat{\rho})$ is a complete, separable, metric space. Further, $\Psi : [0, \infty) \to [0, \infty)$ will always be a strictly increasing function which satisfies $\Psi(0) = 0$. If – in addition – Ψ is convex, then it is called a *Young function*. For a Young function Ψ, one defines the corresponding *Orlicz norm* of a real-valued random variable Z by

$$\|Z\|_\Psi := \inf\{c > 0 : \mathbf{E}\Psi(|Z|/c) \leq 1\}.$$

We will also need a totally bounded metric space (Θ, d) with diameter $D > 0$. The minimal number of closed balls of radius ε needed to cover Θ will be denoted by $N(\Theta, d; \varepsilon)$ and will be called *covering numbers*. A finite subset Θ_0 of Θ is called an *ε-net*, if $d(x, \Theta_0) \leq \varepsilon$ for each $x \in \Theta$ (we use x rather than t, because in our application Θ will be a subset of the space $\mathbf{R}^d$). We will abbreviate

$$J := \int_0^D \Psi^{-1}(N(\Theta, d; \varepsilon)) \, d\varepsilon.$$

Further, let Z_x, $x \in \Theta$ be an $\hat{E}$-valued process on some probability space $(\Omega, \mathcal{F}, \mathbf{P})$. We will denote the Euclidean norm, the l_1-norm and the maximum norm on $\mathbf{R}^d$ by $|.|$, $|.|_1$ and $|.|_\infty$ respectively. Whenever a constant is denoted by c with some index, then its value can change from line to line. We start with the well-known continuity theorem of Kolmogorov.

Lemma 2.1 (Kolmogorov) *Let $\Theta = [0,1]^d$ and assume that there exist $a, b, c > 0$ such that, for all $x, y \in [0,1]^d$, we have*

$$\mathbf{E}\left((\hat{\rho}(Z_x, Z_y))^a\right) \leq c|x - y|_1^{d+b}.$$

Then Z has a continuous modification (which we denote by the same symbol). For each $\kappa \in (0, b/a)$, there exists a random variable S such that $\mathbf{E}(S^a) \leq \frac{cd2^{a\kappa - b}}{1 - 2^{a\kappa - b}}$ and

$$\sup\left\{\hat{\rho}(Z_x(\omega), Z_y(\omega)) : x, y \in [0,1]^d, |x - y|_\infty \leq r\right\} \leq \frac{2d}{1 - 2^{-\kappa}} S(\omega) r^\kappa$$

for each $r \in [0,1]$. In particular, for all $u > 0$, we have

$$\mathbf{P}\left\{\sup_{x,y \in [0,1]^d} \hat{\rho}(Z_x, Z_y) \geq u\right\} \leq \left(\frac{2d}{1 - 2^{-\kappa}}\right)^a \frac{cd2^{a\kappa - b}}{1 - 2^{a\kappa - b}} u^{-a}. \qquad (2.1)$$

Lemma 2.2 (Basic chaining) *Let Z have continuous paths. Further, let δ_j, $j = 0, 1, \ldots$ be a sequence of positive real numbers such that $\sum_{j=0}^\infty \delta_j < \infty$ and let Θ_j be a δ_j-net in (Θ, d), $j = 0, 1, 2, \ldots$, such that $\Theta_0 = \{x_0\}$ is a singleton.*

Then, for any $u > 0$ and any sequence of positive ε_j with $\sum_{j=0}^\infty \varepsilon_j \leq 1$,

$$\mathbf{P}\left\{\sup_{x,y \in \Theta} \hat{\rho}(Z_x, Z_y) \geq u\right\} \leq \sum_{j=0}^\infty |\Theta_{j+1}| \sup_{d(x,y) \leq \delta_j} \mathbf{P}\left\{\hat{\rho}(Z_x, Z_y) \geq \varepsilon_j u/2\right\}.$$

The following lemma combines Theorems 11.1, 11.2, 11.6, and (11.3) in [15] (observe the obvious typo in (11.3) of [15]: ψ^{-1} should be replaced by ψ).

Lemma 2.3 (LT-chaining) *Let Ψ be a Young function such that $J < \infty$. Assume that there exists a constant $c > 0$ such that for all $x, y \in \Theta$*

$$\|\hat{\rho}(Z_x, Z_y)\|_\Psi \leq cd(x, y).$$

Then Z has a continuous modification (which we denote by the same symbol). Further, for each set $A \in \mathcal{F}$, we have

$$\int_A \sup_{x, y \in \Theta} \hat{\rho}(Z_x, Z_y)\, \mathrm{d}\mathbf{P} \leq 8\mathbf{P}(A)c \int_0^D \Psi^{-1}\left(\frac{N(\Theta, d; \varepsilon)}{\mathbf{P}(A)}\right) \mathrm{d}\varepsilon.$$

If moreover there exists $c_\Psi \geq 0$ satisfying $\Psi^{-1}(\alpha\beta) \leq c_\Psi \Psi^{-1}(\alpha)\Psi^{-1}(\beta)$ for all $\alpha, \beta \geq 1$, then for all $u > 0$, we have

$$\mathbf{P}\left\{ \sup_{x, y \in \Theta} \hat{\rho}(Z_x, Z_y) \geq u \right\} \leq \left(\Psi\left(\frac{u}{8cc_\Psi J}\right)\right)^{-1}.$$

The following version of the GRR-Lemma seems to be new. It is a joint upgrade (up to constants) of [8], Theorem B.1.1 and [1], Theorem 1. Even though the version in [8] meets our demands, we present a more general version below and prove it in the appendix.

Lemma 2.4 (GRR) *Let (Θ, d) be an arbitrary metric space (not necessarily totally bounded), m a measure on the Borel sets of Θ which is finite on bounded subsets and let $p : [0, \infty) \to [0, \infty)$ be continuous and strictly increasing and $p(0) = 0$. If $f : \Theta \to \hat{E}$ is continuous such that*

$$V := \int_\Theta \int_\Theta \Psi\left(\frac{\hat{\rho}(f(x), f(y))}{p(d(x, y))}\right) \mathrm{d}m(x)\, \mathrm{d}m(y) < \infty,$$

then we have

(i) $\hat{\rho}(f(x), f(y)) \leq 8 \max_{z \in \{x, y\}} \int_0^{4d(x, y)} \Psi^{-1}\left(\frac{4V}{m(K_{s/2}(z))^2}\right) \mathrm{d}p(s)$,

(ii) $\hat{\rho}(f(x), f(y)) \leq 8N \max_{z \in \{x, y\}} \int_0^{4d(x, y)} \Psi^{-1}\left(\frac{4}{m(K_{s/2}(z))^2}\right) \mathrm{d}p(s)$,

where

$$N := \inf\{\kappa > 0 : \int \int \Psi\left(\frac{1}{\kappa}\frac{\hat{\rho}(f(x), f(y))}{p(d(x, y))}\right) \mathrm{d}m(x)\, \mathrm{d}m(y) \leq 1\}$$

and $K_s(z)$ denotes the closed ball with center z and radius s. In the definition of V and N, $0/0$ is interpreted as zero, while in the conclusions $V/0$ and $N \times \infty$ are interpreted as ∞ even if $V = 0$ or $N = 0$.

Remark   When applying one of the chaining methods above, one is forced to choose the function Ψ (for LT-chaining and GRR) or other parameters (in basic chaining and Kolmogorov's Theorem). One might suspect that it is wise to choose Ψ in such a way that it increases as quickly as possible subject to the constraint that $J < \infty$ (in LT-chaining) because this will guarantee sharper tail estimates for the suprema in question. It may therefore come as a surprise that we will be able to obtain optimal estimates by choosing polynomial functions Ψ and that Kolmogorov's Theorem, which only allows for polynomial functions, will be just as good as the much more sophisticated LT-chaining (for example). The reason for this is that we will use chaining only to estimate the probability that the diameter of the image of a small ball under a flow (for example) exceeds a fixed value (for example 1) up to a given time T and we do not care how large the diameter is if it exceeds this value.

Remarks about the chaining literature The GRR-Lemma was first published in [12] in the special case $\Theta = [0,1]$. A version where Θ is an open bounded set in $\mathbf{R}^d$ can be found in [8], Appendix B (with $m = $ Lebesgue measure). Walsh ([22], Theorem 1.1) requires $\Theta = [0,1]^d$, $m = $ Lebesgue measure, Ψ convex and f real-valued but does not assume that f is continuous. The GRR–Lemma in [1] is similar to ours but they assume that $p = $ identity. Dalang et al. [7] prove a version which is also similar to ours. They assume that the function Ψ is convex (which we don't) and in turn obtain a smaller multiplicative constant. Like Walsh [22], they do not need to assume that the function f is continuous.

Lemma 2.2 appeared (in a slightly different form) in [6], but even at that time it was adequate to call it *essentially well-known*. Indeed, the idea of choosing a sequence of finite δ-nets with $\delta \to 0$ is at the heart of the chaining method (see e.g. [18]).

One can find more general *anisotropic* versions of Kolmogorov's continuity theorem (Lemma 2.1) in which the right-hand side $c|x - y|_1^{d+b}$ is replaced by $c\sum_{i=1}^{d} |x_i - y_i|^{\alpha_i}$ where $\sum_{i=1}^{d} \alpha_1^{-1} < 1$; see e.g. [14] or [7]. We point out that Kolmogorov's Theorem can be regarded as a corollary (possibly up to multiplicative constants) of both LT-chaining (Lemma 2.3) and certain variants of the GRR–Lemma; see [15] and [22] respectively.

2.3 Chaining at work

Let $(t, x) \mapsto \phi_t(x)$ be a continuous random field, $(t, x) \in [0, \infty) \times \mathbf{R}^d$ taking values in a separable complete metric space (E, ρ). We will always assume that ϕ satisfies the following condition:

(H): There exist $\Lambda \geq 0$, $\sigma > 0$ and $\bar{c} > 0$ such that, for each $x, y \in \mathbf{R}^d$, $T > 0$, and $q \geq 1$, we have

$$\left(\mathbf{E} \sup_{0 \leq t \leq T} (\rho(\phi_t(x), \phi_t(y)))^q \right)^{1/q} \leq \bar{c} |x - y| \exp\{(\Lambda + \frac{1}{2}q\sigma^2)T\}.$$

A sufficient condition for (H) to hold (with $\bar{c} = 2$) is the following condition (H').

(H'): There exist $\Lambda \geq 0$, $\sigma > 0$ such that, for each $x, y \in \mathbf{R}^d$, there exists a standard Brownian motion W such that

$$\rho(\phi_t(x), \phi_t(y)) \leq |x - y| \exp\{\Lambda t + \sigma W_t^*\}, \tag{2.2}$$

where $W_t^* := \sup_{0 \leq s \leq t} W(s)$.

We will verify in Lemma 2.6 that (H') and hence (H) is satisfied for the solution flow of a stochastic differential equation on $\mathbf{R}^d$ with global Lipschitz coefficients.

If there exists some $\nu > 0$ such that (2.2) holds only for $t \leq \inf\{s \geq 0 : \rho(\phi_s(x), \phi_s(y)) \geq \nu\}$, then (H') holds provided that ρ is replaced by the metric $\bar{\rho}(x_1, x_2) := \rho(x_1, x_2) \wedge \nu$. Choosing ν small allows us in some cases to use smaller values of Λ and/or σ and thus to improve the asymptotic bounds in the following theorem.

In fact the application of Lemmas 2.1 or 2.3 below shows that the existence of a continuous modification of ϕ with respect to x follows from (H).

In the following theorem, we will provide an upper bound for the probability that the image of a ball which is exponentially small in T attains diameter 1 (say) up to time T.

Theorem 2.1 *Assume* (H) *and let* $\gamma > 0$. *Define*

$$I(\gamma) := \begin{cases} \frac{(\gamma - \Lambda)^2}{2\sigma^2} & \text{if } \gamma \geq \Lambda + \sigma^2 d, \\ d(\gamma - \Lambda - \frac{1}{2}\sigma^2 d) & \text{if } \Lambda + \frac{1}{2}\sigma^2 d \leq \gamma \leq \Lambda + \sigma^2 d, \\ 0 & \text{if } \gamma \leq \Lambda + \frac{1}{2}\sigma^2 d. \end{cases}$$

Then, for each $u > 0$, *we have*

$$\limsup_{T \to \infty} \frac{1}{T} \sup_{\mathcal{X}_T} \log \mathbf{P}\{ \sup_{x, y \in \mathcal{X}_T} \sup_{0 \leq t \leq T} \rho(\phi_t(x), \phi_t(y)) \geq u \} \leq -I(\gamma),$$

where $\sup_{\mathcal{X}_T}$ *means that we take the supremum over all cubes* $\mathcal{X}_T$ *in* $\mathbf{R}^d$ *with side length* $\exp\{-\gamma T\}$.

We will first provide five different proofs of Theorem 2.1 by using Lemmas 2.1, 2.2, 2.3, and 2.4 respectively. We will always use the space $\hat{E} = C([0,T], E)$ equipped with the sup-norm $\hat{\rho}$, where (E, ρ) is a complete separable metric space as above.

Proof of Theorem 2.1 using Lemma 2.1. Let $T > 0$. Without loss of generality, we assume that $\mathcal{X} := \mathcal{X}_T = [0, e^{-\gamma T}]^d$. Define $Z_x(t) := \phi_t(e^{-\gamma T} x)$, $x \in \mathbf{R}^d$. For $q \geq 1$, (H) implies

$$\left(\mathbf{E} \sup_{0 \leq t \leq T} \rho(Z_x(t), Z_y(t))^q \right)^{1/q} \leq \bar{c} e^{-\gamma T} |x - y| e^{(\Lambda + \frac{1}{2} q \sigma^2)T},$$

i.e. the assumptions of Lemma 2.1 are satisfied with $a = q$, $c = \bar{c}^q \exp\{(\Lambda - \gamma + \frac{1}{2} q \sigma^2) q T\}$ and $b = q - d$ for any $q \cdot> d$. Therefore we get for $\kappa \in (0, b/a)$:

$$\mathbf{P}\{ \sup_{x,y \in \mathcal{X}} \sup_{0 \leq t \leq T} \rho(\phi_t(x), \phi_t(y)) \geq u \}$$

$$\leq \left(\frac{2d}{1 - 2^{-\kappa}} \right)^q \frac{\bar{c}^q d 2^{a\kappa - b}}{1 - 2^{a\kappa - b}} \exp\{(\Lambda - \gamma + \frac{1}{2} q \sigma^2) q T\} u^{-q}.$$

Taking logs, dividing by T, letting $T \to \infty$ and optimizing over $q > d$ yields Theorem 2.1. $\qquad\qquad\qquad\qquad\qquad\qquad\qquad\qquad\qquad$ $\square$

Proof of Theorem 2.1 using Lemma 2.2. Let $\gamma > \Lambda$ and take a cube $\Theta = \mathcal{X}_T$ of side length $\exp\{-\gamma T\}$. Then we apply the Chaining Lemma 2.2 to Θ with $\delta_j = \exp\{-\gamma T\} \sqrt{d} 2^{-j-1}$ and $\varepsilon_j = C/(j+1)^2$, $j = 0, 1, \ldots$, where the constant C is chosen such that the ε_j sum up to 1. Then there exist subsets Θ_j of B with cardinality $|\Theta_j| = 2^{jd}$ such that the assumptions of Lemma 2.2 are satisfied. In particular, x_0 is the center of the cube Θ. For $q > d$, we get

$$\mathbf{P}\Big\{ \sup_{x,y\in\Theta} \sup_{t\in[0,T]} \rho(\phi_t(x),\phi_t(y)) \ge u \Big\}$$

$$\le 2^d \sum_{j=0}^{\infty} 2^{dj} \sup_{|x-y|\le\delta_j} \mathbf{P}\Big\{ \sup_{t\in[0,T]} \rho(\phi_t(x),\phi_t(y)) \ge \varepsilon_j u/2 \Big\}$$

$$\le 2^d \sum_{j=0}^{\infty} 2^{dj} (\varepsilon_j u/2)^{-q} \sup_{|x-y|\le\delta_j} \mathbf{E}\Big(\sup_{0\le t\le T} \rho(\phi_t(x),\phi_t(y)) \Big)^q$$

$$\le 2^d e^{((\Lambda-\gamma)q+\frac{1}{2}\sigma^2 q^2)T} \bar{c}^q d^{q/2} u^{-q} \sum_{j=0}^{\infty} 2^{(d-q)j} \varepsilon_j^{-q}. \qquad (2.3)$$

The sum converges since $q > d$ and the ε_j decay polynomially. Taking logs in (2.3), dividing by T, letting $T \to \infty$ and optimizing over $q > d$ yields Theorem 2.1. □

Proof of Theorem 2.1 using Lemma 2.3. Fix $T > 0$ and $q > d$. We apply Lemma 2.3 with $\Psi(x) = x^q$ (then $c_\Psi = 1$). Inequality (H) shows that the assumptions are satisfied with $c = \bar{c}\exp\{(\Lambda + \frac{1}{2}q\sigma^2)T\}$. Further, we have

$$J := \int_0^{\sqrt{d}e^{-\gamma T}} N([0, e^{-\gamma T}]^d, |.|; \varepsilon)^{1/q} \, d\varepsilon \le c_{d,q} e^{-\gamma T}.$$

Therefore, we obtain

$$\mathbf{P}\{ \sup_{x,y\in\mathcal{X}} \sup_{0\le t\le T} \rho(\phi_t(x),\phi_t(y)) \ge u \} \le (8Jc)^q u^{-q}$$

$$\le \bar{c}^q \tilde{c}_{d,q} \exp\{(\Lambda q - \gamma q + \frac{1}{2}q^2\sigma^2)T\} u^{-q}$$

Taking logarithms, dividing by T, letting $T \to \infty$ and optimizing over $q > d$ yields the claim in Theorem 2.1. □

Proof of Theorem 2.1 using Lemma 2.4. Let $p(s) := s^{(2d+\varepsilon)/q}$, where $\varepsilon \in (0,1)$ and $q > d + \varepsilon$. Define

$$V := \int_{\mathcal{X}_T} \int_{\mathcal{X}_T} \sup_{0\le t\le T} \frac{\rho(\phi_t(x),\phi_t(y))^q}{p(|y-x|)^q} \, dx \, dy.$$

Let m be Lebesgue measure restricted to $\mathcal{X}_T$ and $\Psi(x) = x^q$. By (H),

$$
\begin{aligned}
\mathbf{EV} &\leq \bar{c}^q e^{(\Lambda+\frac{1}{2}\sigma^2 q)qT} \int_{\mathcal{X}_T} \int_{\mathcal{X}_T} |y-x|^{q-2d-\varepsilon}\, dx\, dy \\
&\leq \bar{c}^q e^{(\Lambda+\frac{1}{2}\sigma^2 q)qT} e^{-\gamma dT} \int_{\{|y|\leq\sqrt{d}e^{-\gamma T}\}} |y|^{q-2d-\varepsilon}\, dy \\
&= \bar{c}^q c_d e^{(\Lambda+\frac{1}{2}\sigma^2 q)qT} e^{-\gamma dT} \int_0^{\sqrt{d}e^{-\gamma T}} r^{q-d-1-\varepsilon}\, dr \\
&= \bar{c}^q c_{d,q,\varepsilon} e^{(\Lambda-\gamma+\frac{1}{2}\sigma^2 q)qT} e^{\gamma\varepsilon T}.
\end{aligned}
$$

Therefore, the assumptions of Lemma 2.4 are satisfied for almost all $\omega \in \Omega$ and we obtain

$$
\begin{aligned}
\mathbf{P}\{&\sup_{x,y\in\mathcal{X}_T} \sup_{0\leq t\leq T} \rho(\phi_t(x),\phi_t(y)) \geq u\} \\
&\leq \mathbf{P}\left\{ V^{1/q}(\omega) \int_0^{4\sqrt{d}e^{-\gamma T}} s^{\frac{\varepsilon}{q}-1}\, ds \geq c_{d,q,\varepsilon} u \right\} \\
&\leq \mathbf{EV} e^{-\gamma\varepsilon T} c_{d,q,\varepsilon} u^{-q} \\
&\leq \bar{c}^q c_{d,q,\varepsilon} e^{(\Lambda-\gamma+\frac{1}{2}\sigma^2 q)qT} u^{-q}.
\end{aligned}
$$

Taking logarithms, dividing by T, letting $T \to \infty$ and then $\varepsilon \to 0$ and optimizing over $q > d$ yields the claim in Theorem 2.1. $\square$

Occasionally, the GRR-Lemma is formulated only for p being the identity (e.g. in [1]). The following proof shows that we don't lose anything in this case but a few modifications are necessary.

Proof of Theorem 2.1 using Lemma 2.4 with $p =$ id. Fix $u > 0$. We start as in the previous proof except that we choose $p(s) = s$, $q > 2d$, $Q \in (0,1)$, $qQ \geq 1$, and

$$
\begin{aligned}
V(x,y) &:= \sup_{0\leq t\leq T} \left(\frac{\rho(\phi_t(x),\phi_t(y)) \wedge u}{|x-y|} \right)^q \\
V &:= \int_{\mathcal{X}_T} \int_{\mathcal{X}_T} V(x,y)\, dx\, dy.
\end{aligned}
$$

Using Chebychev's inequality and (H), we get

$$
\begin{aligned}
\mathbf{E}V(x,y) &= \int_0^{(u/|x-y|)^q} \mathbf{P}\{V(x,y) \geq s\}\, \mathrm{d}s \\
&\leq \mathbf{E}\sup_{0 \leq t \leq T}\left(\frac{\rho(\phi_t(x),\phi_t(y))}{|x-y|}\right)^{qQ} \int_0^{(u/|x-y|)^q} s^{-Q}\, \mathrm{d}s \\
&\leq \bar{c}^{qQ}(1-Q)^{-1}\exp\{(\Lambda+\tfrac{1}{2}\sigma^2 qQ)qQT\}\left(\frac{u}{|x-y|}\right)^{q(1-Q)}.
\end{aligned}
$$

Hence Lemma 2.4 with $p = \mathrm{id}$ implies

$$
\mathbf{P}\{\sup_{x,y \in \mathcal{X}_T}\sup_{0 \leq t \leq T}\rho(\phi_t(x),\phi_t(y)) \geq u\} \tag{2.4}
$$
$$
\leq \mathbf{P}\left\{V^{1/q} \geq c_{q,d}\, u\mathrm{e}^{\gamma(-\frac{2d}{q}+1)T}\right\}
$$
$$
\leq \mathbf{E}V c_{d,q}\, u^{-q}\mathrm{e}^{(2d-q)\gamma T}
$$
$$
\leq c_{q,Q,d}\bar{c}^{qQ}\mathrm{e}^{(\Lambda+\frac{1}{2}\sigma^2 qQ)qQT}\mathrm{e}^{(2d-q)\gamma T}u^{-qQ}\iint\limits_{\mathcal{X}_T\ \mathcal{X}_T}|x-y|^{-q(1-Q)}\,\mathrm{d}x\,\mathrm{d}y.
$$

The double integral is finite if $q(1-Q) < d$. Observe that for any $\kappa > d$ we can find $q > 2d$ and $Q \in (0,1)$ such that $qQ = \kappa$ and $q(1-Q) < d$. Therefore we obtain the same asymptotics for (2.4) as in the previous proof. $\qquad\square$

Theorem 2.1 can be improved in the case where ϕ_t is a homeomorphism on $\mathbf{R}^d$ for each $t \geq 0$ and each $\omega \in \Omega$.

Corollary 2.5 *Let ϕ_t be a homeomorphism on $\mathbf{R}^d$, $d \geq 1$ for each $t \geq 0$ and $\omega \in \Omega$. If ϕ satisfies (H) with respect to the Euclidean norm ρ, then the conclusion of Theorem 2.1 holds when, in the definition of I, d is replaced by $d-1$.*

Proof. Owing to the homeomorphic property, the sup over $\mathcal{X}_T$ in Theorem 2.1 is attained on one of the faces of $\mathcal{X}$. Applying Theorem 2.1 to each of the faces (which have dimension $d-1$), the assertion in the corollary follows. $\qquad\square$

2.4 Examples and complements

Let us first show that a solution flow of a stochastic differential equation on $\mathbf{R}^d$ with Lipschitz coefficients satisfies hypothesis (H') and therefore also (H).

For each $x \in \mathbf{R}^d$, let $t \mapsto M(t,x)$ be an $\mathbf{R}^d$-valued continuous martingale with $M(0,x) = 0$ such that the joint quadratic variation can be represented as

$$\langle M(.,x.\omega), M(.,y,\omega)\rangle_t = \int_0^t a(s,x,y,\omega)\,\mathrm{d}s,$$

for a jointly measurable matrix-valued function a which is continuous in (x,y) and predictable in (s,ω). Defining

$$\mathcal{A}(s,x,y,\omega) := a(s,x,x,\omega) - a(s,y,x,\omega) - a(s,x,y,\omega) + a(s,y,y,\omega),$$

we will require that a satisfies the following Lipschitz property: there exists some constant $a \geq 0$ such that, for all $x,y \in \mathbf{R}^d$, all $s \geq 0$ and almost all ω, we have

$$\|\mathcal{A}(s,x,y,\omega)\| \leq a^2|x-y|^2,$$

where $\|.\|$ denotes the operator norm. Note that

$$\mathcal{A}(t,x,y,\omega) = \frac{\mathrm{d}}{\mathrm{d}t}\langle M(.,x) - M(.,y)\rangle_t.$$

Further, we assume that $b : [0,\infty) \times \mathbf{R}^d \times \Omega \to \mathbf{R}^d$ is a vector field which is jointly measurable, predictable in (t,ω) and Lipschitz continuous with constant b in the spatial variable uniformly in (t,ω). In addition, we require that the functions $a(.)$ and $b(.)$ are bounded on each compact subset of $[0,\infty) \times \mathbf{R}^{d \times d}$ resp. $[0,\infty) \times \mathbf{R}^d$ uniformly w.r.t. $\omega \in \Omega$. Under these assumptions, it is well known that the *Kunita type* stochastic differential equation

$$\mathrm{d}X(t) = b(t,X(t))\,\mathrm{d}t + M(\mathrm{d}t,X(t)) \tag{2.5}$$

generates a stochastic flow of homeomorphisms ϕ (see [14], Theorem 4.5.1), i.e.

(i) $t \mapsto \phi_{s,t}(x)$, $t \geq s$ solves (2.5) with initial condition $X(s) = x$ for all $x \in \mathbf{R}^d$, $s \geq 0$.

(ii) $\phi_{s,t}(\omega)$ is a homeomorphism on $\mathbf{R}^d$ for all $0 \leq s \leq t$ and all $\omega \in \Omega$.

(iii) $\phi_{s,u} = \phi_{t,u} \circ \phi_{s,t}$ for all $0 \leq s \leq t \leq u$ and all $\omega \in \Omega$.

(iv) $(s,t,x) \mapsto \phi_{s,t}(x)$ is continuous.

We will write $\phi_t(x)$ instead of $\phi_{0,t}(x)$.

For readers who are unfamiliar with Kunita type stochastic differential equations, we point out that if one replaces the term $M(\mathrm{d}t, X(t))$ in equation (2.5) by $\sum_{i=1}^m \sigma_i(X(t)) \,\mathrm{d}W_i(t)$, where W_i are independent scalar standard Brownian motions and the functions $\sigma_i : \mathbf{R}^d \to \mathbf{R}^d$ are Lipschitz continuous, then the Lipschitz condition imposed above holds. In fact

$$\mathcal{A}(t, x, y, \omega) = \sum_{i=1}^m (\sigma_i(x) - \sigma_i(y))(\sigma_i(x) - \sigma_i(y))^T.$$

The following lemma is identical with Lemma 5.1 in [6]. The proof below is slightly more elementary since it avoids the use of a comparison theorem by Ikeda and Watanabe.

Lemma 2.6 *Under the assumptions above,* (H') *holds with* $\sigma = a$ *and* $\Lambda = b + (d-1)a^2/2$.

Proof. Fix $x, y \in \mathbf{R}^d$, $x \neq y$ and define

$$D_t := \phi_t(x) - \phi_t(y), \quad Z_t := \frac{1}{2}\log(|D_t|^2).$$

Therefore, $Z_t = f(D_t)$ where $f(z) := \frac{1}{2}\log(|z|^2)$. Note that $D_t \neq 0$ for all $t \geq 0$ by the homeomorphic property. Using Itô's formula, we get

$$
\begin{aligned}
\mathrm{d}Z_t \;=\; & \frac{D_t \cdot (M(\mathrm{d}t, \phi_t(x)) - M(\mathrm{d}t, \phi_t(y)))}{|D_t|^2} \\
& + \frac{D_t \cdot (b(t, \phi_t(x)) - b(t, \phi_t(y)))}{|D_t|^2} \,\mathrm{d}t \\
& + \frac{1}{2}\frac{1}{|D_t|^2}\mathrm{Tr}\left(\mathcal{A}(t, \phi_t(x), \phi_t(y), \omega)\right)\mathrm{d}t \\
& - \sum_{i,j} \frac{D_t^i D_t^j}{(|D_t|^2)^2}\mathcal{A}_{i,j}\left(t, \phi_t(x), \phi_t(y), \omega\right)\mathrm{d}t.
\end{aligned}
$$

We define the local martingale $N_t, t \geq 0$, by

$$N_t = \int_0^t \frac{D_s}{|D_s|^2} \cdot (M(\mathrm{d}s, \phi_s(x)) - M(\mathrm{d}s, \phi_s(y)))$$

and obtain

$$Z_t = Z_0 + N_t + \int_0^t \alpha(s, \omega)\,\mathrm{d}s,$$

where

$$\sup_{x,y} \sup_s \operatorname{esssup}_\omega |\alpha(s,\omega)| \leq b + (d-1)a^2/2 =: \Lambda$$

and

$$d\langle N\rangle_t = \sum_{i,j} \frac{D_t^i D_t^j}{(|D_t|^2)^2} \mathcal{A}_{i,j}(t, \phi_t(x), \phi_t(y), \omega)\, dt \leq a^2\, dt. \qquad (2.6)$$

Since N is a continuous local martingale with $N_0 = 0$, there exists a standard Brownian motion W (possibly on an enlarged probability space) such that $N_t = aW_{\tau(t)}$, $t \geq 0$ and (2.6) implies $\tau(t) \leq t$ for all $t \geq 0$. Hence

$$Z_t \leq \log|x - y| + a\, W_t^* + \Lambda t. \qquad (2.7)$$

Exponentiating the last inequality completes the proof of the lemma. $\square$

The following simple example shows that the upper bound in Theorem 2.1 is sharp for $\gamma \geq \Lambda + \sigma^2 d$.

Example Consider the linear stochastic differential equation

$$dX(t) = \left(\Lambda + \frac{1}{2}\sigma^2\right) X(t)\, dt + \sigma X(t)\, dW(t), \quad X(0) = x \in \mathbf{R}^d,$$

where $W(t)$, $t \geq 0$, is a one-dimensional Brownian motion, $\Lambda \geq 0$ and $\sigma > 0$. The solution (flow) $\phi_t(x)$ is given by

$$\phi_t(x) = xe^{\Lambda t + \sigma W(t)},$$

which satisfies (H′) and hence (H). If $\mathcal{X}$ is a cube of side length $e^{-\gamma T}$ in $\mathbf{R}^d$ for some $\gamma \geq \Lambda$ and $u > 0$, then

$$\sup_{x,y \in \mathcal{X}} |\phi_t(x) - \phi_t(y)| = \sqrt{d}e^{-\gamma T} e^{\Lambda t + \sigma W(t)}$$

and

$$\limsup_{T \to \infty} \frac{1}{T} \log \mathbf{P}\{\sup_{x,y \in \mathcal{X}} \sup_{0 \leq t \leq T} |\phi_t(x) - \phi_t(y)| \geq u\} = -\frac{1}{2\sigma^2}(\gamma - \Lambda)^2,$$

for $\gamma \geq \Lambda$ and $u > 0$. $\square$

Next, we provide an example which shows that the conclusion in Theorem 2.1 is sharp also for $\gamma < \Lambda + \sigma^2 d$.

Example Let $h : \mathbf{R}^d \to [0, \infty)$ be Lipschitz continuous with Lipschitz constant 2 and support contained in $[-1/2, 1/2]^d$. Further suppose that $h(0) = 1$. Let W^i, $i \in \mathbf{Z}^d$ be independent standard Brownian motions and let $\Lambda \geq 0$ and $\sigma > 0$ be constants. For $\delta > 0$, define

$$\phi_t(x) := \sum_{i \in \mathbf{Z}^d} \delta h\left(\frac{x}{\delta} - i\right) e^{\Lambda t + \sigma W_t^i}, \quad x \in \mathbf{R}^d, \, t \geq 0.$$

Note that at most one term in the sum is nonzero. Therefore

$$\left(\mathbf{E} \sup_{0 \leq t \leq T} |\phi_t(x) - \phi_t(y)|^q\right)^{1/q} \leq 2|x - y| e^{(\Lambda + \frac{1}{2} q \sigma^2)T}$$

for each $q \geq 1$, so (H) is satisfied with $\bar{c} = 2$. Let $T > 0$, $\gamma > \Lambda$, $\mathcal{X} = [0, e^{-\gamma T}]^d$ and $\delta = e^{-\xi T}$, where $\xi > \gamma$ will be optimized later. Since the processes $\phi_t(i\delta)$, $i \in \mathbf{Z}^d$, are independent and identically distributed, we conclude

$$\mathbf{P}\{\sup_{x \in \mathcal{X}} \phi_T(x) \leq \delta e^{\Lambda T} + 1\}$$

$$\leq \mathbf{P}\{\max_{i \in \mathbf{Z}^d, i\delta \in \mathcal{X}} \phi_T(i\delta) \leq \delta e^{\Lambda T} + 1\}$$

$$\leq \left(\mathbf{P}\{\phi_T(0) \leq \delta e^{\Lambda T} + 1\}\right)^{\exp\{(\xi - \gamma)dT\}}$$

$$= \left(1 - \mathbf{P}\left\{W_1 > \frac{1}{\sigma \sqrt{T}} \log(1 + e^{(\xi - \Lambda)T})\right\}\right)^{\exp\{(\xi - \gamma)dT\}}.$$

From this and the asymptotic behaviour of the last probability, it follows that the last term will converge to 0 as $T \to \infty$ provided that

$$2\sigma^2 d(\xi - \gamma) > (\xi - \Lambda)^2,$$

which holds true in case $\xi = \Lambda + \sigma^2 d$ and $\gamma \in (\Lambda, \Lambda + \sigma^2 d/2)$. Since the probability that the infimum of $\phi_T(x)$, $x \in \mathcal{X}$, is at most $\delta \exp\{\Lambda T\}$ converges to 1 as $T \to \infty$, we obtain for $\gamma < \Lambda + \sigma^2 d/2$

$$\limsup_{T \to \infty} \frac{1}{T} \log \mathbf{P}\{\sup_{x,y \in \mathcal{X}} \sup_{0 \leq t \leq T} |\tilde{\phi}_t(x) - \phi_t(y)| \geq 1\} = 0$$

(in fact we just showed that this is true even if the sup over t is replaced by T). Similarly, we obtain

$$\limsup_{T \to \infty} \frac{1}{T} \log \mathbf{P}\{\sup_{x \in \mathcal{X}} \phi_T(x) \geq \delta e^{\Lambda T} + 1\} = (\xi - \gamma)d - \frac{1}{2\sigma^2}(\xi - \Lambda)^2 \quad (2.8)$$

in case the last expression is strictly negative, which holds true in case $\gamma \in (\Lambda + \frac{1}{2}\sigma^2 d, \Lambda + \sigma^2 d)$ and $\xi = \Lambda + \sigma^2 d$. Inserting this value for ξ in

(2.8) yields

$$\limsup_{T \to \infty} \frac{1}{T} \log \mathbf{P}\{\sup_{x \in \mathcal{X}} \phi_T(x) \geq \delta e^{\Lambda T} + 1\} = -I(\gamma)$$

for all $\gamma < \Lambda + \sigma^2 d$ with $I(\gamma)$ defined as in Theorem 2.1.

The reader may complain that in this example the field ϕ actually depends on T (via $\delta = \delta(T)$), i.e. as we let $T \to \infty$, we keep changing ϕ. It is easy to see, however, that we can define a single field ϕ by spatially piecing together fields as above for an appropriate sequence $T_i \to \infty$.

Remark　The previous example(s) show that the conclusion in Theorem 2.1 is sharp, but in the last example ϕ is not a stochastic flow of homeomorphisms. Can we do better in that case? The following theorem shows that we can, provided the flow is C^1. More precisely, we consider a stochastic flow of homeomorphisms ϕ as introduced at the beginning of this section and require that it has – in addition – independent and stationary increments and that its law is invariant under shifts in $\mathbf{R}^d$. We will call such a flow a *translation invariant Brownian flow*.

Theorem 2.2 *Let ϕ be a translation invariant Brownian flow on $\mathbf{R}^d$ such that the map $(t, x) \mapsto D\phi_t(x)$ is continuous (for all $\omega \in \Omega$). In addition, we assume that there exist $\bar{c} \geq 1, \Lambda \geq 0$ and $\sigma > 0$ and a standard Wiener process such that, for each $T \geq 0$, we have*

$$\|D\phi_T(0)\| \leq \bar{c} \exp\{ \sup_{0 \leq s \leq T} (\Lambda s + \sigma W_s)\}. \tag{2.9}$$

Then, for each $u > 0$ and $\xi \geq 0$, we have

$$\limsup_{T \to \infty} \frac{1}{T} \log \mathbf{P} \left\{ \sup_{|x| \leq \exp\{-(\Lambda+\xi)T\}} \sup_{0 \leq t \leq T} |\phi_t(x) - \phi_t(0)| \geq u \right\} \leq -\frac{\xi^2}{2\sigma^2}.$$

Note that, owing to the fact that the flow is translation invariant and stationary, the statement is invariant under a shift in space and time as well. We mention that the hypotheses of the theorem are for example fulfilled for isotropic Brownian flows; see [2].

Proof of Theorem 2.2. Fix $\xi > 0$, $\varepsilon \in (0, 1)$, $z > \varepsilon + \bar{c} \exp\{-\frac{\sigma^2}{\xi} \log(1 - \varepsilon)\}$ and $u > 0$. We abbreviate $D_t := \|D\phi_t(0)\|$. Let

$$\tau_z := \inf \left\{ t \geq 0 : \Lambda t + \sigma W_t \geq \log \frac{z}{\bar{c}} \right\}.$$

Using the formula for the Laplace transform of the hitting time of Brownian motion with drift ([3], page 223, formula 2.2.0.1), we get for $\lambda > 0$

$$\mathbf{E}e^{-\lambda \tau_z} = \exp\left\{\frac{1}{\sigma^2}(\Lambda - \sqrt{2\lambda\sigma^2 + \Lambda^2})\log(z/\bar{c})\right\}. \qquad (2.10)$$

For $\delta > 0$, we define

$$\tilde{T}^{(\delta)} := \inf\{t > 0 : \sup_{|x| \leq \delta} |\phi_t(x) - \phi_t(0)| \geq \delta z\}.$$

Since the flow ϕ is C^1 and $\tau_{z-\varepsilon} < \infty$, there exists $\delta_0 = \delta_0(z, \varepsilon) > 0$ such that

$$\mathbf{P}\left\{\sup_{0 \leq t \leq \tau_{z-\varepsilon}} \sup_{|x| \leq \delta} \frac{|\phi_t(x) - \phi_t(0)|}{\delta} < \sup_{0 \leq t \leq \tau_{z-\varepsilon}} D_t + \varepsilon\right\} \geq 1 - \varepsilon$$

for all $\delta \in (0, \delta_0]$. Note that (2.9) implies $\sup_{0 \leq t \leq \tau_{z-\varepsilon}} D_t + \varepsilon \leq z$. Hence

$$\mathbf{E}e^{-\lambda \tilde{T}^{(\delta)}} \leq \mathbf{E}e^{-\lambda \tau_{z-\varepsilon}} + \varepsilon \qquad (2.11)$$

for $\delta \in (0, \delta_0]$ and all $\lambda > 0$. Define $\hat{u} := u \wedge \delta_0$.

Let $T > 0$ such that $\exp\{-(\Lambda + \xi)T\} < \hat{u}$. Further, let $T_1, T_2, \ldots$ be independent random variables such that the laws of T_j and $\tilde{T}^{(\delta_j)}$ coincide, where $\delta_j = \exp\{-(\Lambda + \xi)T\}z^{j-1}$. Define

$$m = \left\lfloor \frac{(\Lambda + \xi)T + \log \hat{u}}{\log z} \right\rfloor. \qquad (2.12)$$

Using the fact that ϕ has independent and stationary increments, and Markov's inequality, we obtain

$$\mathbf{P}\left\{\sup_{|x| \leq \exp\{-(\Lambda+\xi)T\}} \sup_{0 \leq t \leq T} |\phi_t(x) - \phi_t(0)| \geq u\right\} \leq \mathbf{P}\left\{\sum_{j=1}^{m} T_j \leq T\right\}$$

$$= \mathbf{P}\left\{\exp\{-\lambda \sum_{j=1}^{m} T_j\} \geq \exp\{-\lambda T\}\right\}$$

$$\leq \exp\{\lambda T\} \max_{j=1,\ldots m} (\mathbf{E}\exp\{-\lambda T_j\})^m$$

$$\leq \exp\{\lambda T\}(\mathbf{E}\exp\{-\lambda \tau_{z-\varepsilon}\} + \varepsilon)^m,$$

where we used (2.11) and $\hat{u} \leq \delta_0$ in the last step. Using (2.10) and (2.12) and inserting $\lambda := \frac{1}{2\sigma^2}((\Lambda + \xi)^2 - \Lambda^2)$, we get

$$\log \mathbf{P} \left\{ \sup_{|x| \leq \exp\{-(\Lambda+\xi)T\}} \sup_{0 \leq t \leq T} |\phi_t(x) - \phi_t(0)| \geq u \right\}$$

$$\leq \frac{(\Lambda + \xi)^2 - \Lambda^2}{2\sigma^2} T$$

$$+ \left\lfloor \frac{(\Lambda + \xi)T + \log \hat{u}}{\log z} \right\rfloor \log \left(\exp \left\{ -\frac{\xi}{\sigma^2} \log \frac{z-\varepsilon}{\bar{c}} \right\} + \varepsilon \right).$$

Dividing by T, and letting (in this order) $T \to \infty$, $\varepsilon \to 0$ and $z \to \infty$, the assertion follows. $\square$

2.5 Dispersion of sets: upper bounds

We will now formulate the dispersion result mentioned in the introduction and prove it using Theorem 2.1. In addition to hypothesis (H) we require a growth condition for the one-point motion. In Proposition 2.8 we will provide explicit conditions on the coefficients of a stochastic differential equation which guarantee that the associated stochastic flow fulfills that condition. The value of the linear bound K in Theorem 2.3 improves previous ones in [6, 16, 17] but the main improvement is its simpler proof.

Theorem 2.3 *Let* $\phi : [0, \infty) \times \mathbf{R}^d \times \Omega \to \mathbf{R}^d$ *be a continuous random field satisfying the following.*

(i) *(H).*

(ii) *There exist* $A > 0$ *and* $B \geq 0$ *such that, for each* $k > 0$ *and each bounded set* $S \subset \mathbf{R}^d$, *we have*

$$\limsup_{T \to \infty} \frac{1}{T} \log \sup_{x \in S} \mathbf{P} \left\{ \sup_{0 \leq t \leq T} |\phi_t(x)| \geq kT \right\} \leq -\frac{(k-B)_+^2}{2A^2},$$

where $r_+ = r \vee 0$ *denotes the positive part of* $r \in \mathbf{R}$.

Let $\mathcal{X}$ *be a compact subset of* $\mathbf{R}^d$ *with box (or upper entropy) dimension* $\Delta > 0$. *Then*

$$\limsup_{T \to \infty} \left(\sup_{t \in [0,T]} \sup_{x \in \mathcal{X}} \frac{1}{T} |\phi_t(x)| \right) \leq K \ a.s., \tag{2.13}$$

where

$$K = \begin{cases} B + A\sqrt{2\Delta \left(\Lambda + \sigma^2\Delta + \sqrt{\sigma^4\Delta^2 + 2\Delta\Lambda\sigma^2} \right)} & \text{if } \Lambda \geq \Lambda_0 \\ B + A\sqrt{2\Delta \frac{d}{d-\Delta} \left(\Lambda + \frac{1}{2}\sigma^2 d \right)} & \text{otherwise,} \end{cases} \tag{2.14}$$

where

$$\Lambda_0 := \frac{\sigma^2 d}{\Delta} \left(\frac{d}{2} - \Delta \right).$$

Proof. Let $N(\mathcal{X}, r)$, $r > 0$, denote the minimal number of subsets of $\mathbf{R}^d$ of diameter at most r which cover $\mathcal{X}$. By definition, we have

$$\Delta = \limsup_{r \downarrow 0} \frac{\log N(\mathcal{X}, r)}{\log \frac{1}{r}}.$$

Choose $\varepsilon > 0$ and $r_0 > 0$ such that $\log N(\mathcal{X}, r) \leq (\Delta + \varepsilon) \log \frac{1}{r}$ for all $0 < r \leq r_0$. Further, let $\gamma, T > 0$ satisfy $\mathrm{e}^{-\gamma T} \leq r_0$. Then $N(\mathcal{X}, \mathrm{e}^{-\gamma T}) \leq \exp\{\gamma T(\Delta + \varepsilon)\}$. Let $\mathcal{X}_i$, $i = 1, \dots, N(\mathcal{X}, \mathrm{e}^{-\gamma T})$, be compact sets of diameter at most $\mathrm{e}^{-\gamma T}$ which cover $\mathcal{X}$ and choose arbitrary points $x_i \in \mathcal{X}_i$. Define

$$\widetilde{\mathcal{X}} := \{x_i, \, i = 1, \dots, N(\mathcal{X}, \mathrm{e}^{-\gamma T})\}.$$

For $\kappa > 0$, we have

$$\mathbf{P}\{\sup_{x \in \mathcal{X}} \sup_{0 \leq t \leq T} |\phi_t(x)| \geq \kappa T\} \leq S_1 + S_2,$$

where

$$S_1 := \exp\{\gamma T(\Delta + \varepsilon)\} \max_{x \in \widetilde{\mathcal{X}}} \mathbf{P}\{\sup_{0 \leq t \leq T} |\phi_t(x)| \geq \kappa T - 1\}$$

and

$$S_2 := \exp\{\gamma T(\Delta + \varepsilon)\} \max_i \mathbf{P}\{\sup_{0 \leq t \leq T} \mathrm{diam}(\phi_t(\mathcal{X}_i)) \geq 1\}.$$

Using (ii) in the theorem, we get

$$\limsup_{T \to \infty} \frac{1}{T} \log S_1 \leq \gamma(\Delta + \varepsilon) - \frac{(\kappa - B)_+^2}{2A^2}.$$

Further, Theorem 2.1 implies

$$\limsup_{T \to \infty} \frac{1}{T} \log S_2 \leq \gamma(\Delta + \varepsilon) - I(\gamma). \tag{2.15}$$

Therefore,

$$\zeta(\gamma, \kappa) := \limsup_{T \to \infty} \frac{1}{T} \log \mathbf{P}\{\sup_{x \in \mathcal{X}} \sup_{0 \leq t \leq T} |\phi_t(x)| \geq \kappa T\}$$

$$\leq \gamma \Delta - \left(\frac{(\kappa - B)_+^2}{2A^2} \wedge I(\gamma) \right).$$

Let γ_0 be the unique positive solution of $I(\gamma) = \gamma\Delta$, where $I(\gamma)$ is defined in Theorem 2.1. Then

$$
\gamma_0 \;=\; \begin{cases} \dfrac{d}{d-\Delta}(\Lambda + \tfrac{1}{2}\sigma^2 d) & \text{if } \Lambda \le \frac{\sigma^2 d}{\Delta}\left(\frac{d}{2} - \Delta\right) \\[2mm] \Lambda + \sigma^2\Delta + \sqrt{2\Lambda\sigma^2\Delta + \sigma^4\Delta^2} & \text{otherwise} \end{cases}
$$

and $\zeta(\gamma,\kappa) < 0$ whenever $\gamma > \gamma_0$ and $\kappa > \kappa_0(\gamma)$, where

$$
\kappa_0(\gamma) := B + A\sqrt{2\gamma\Delta}.
$$

Therefore, for any $\gamma > \gamma_0$ we have

$$
\sum_{n=1}^{\infty} \mathbf{P}\left\{ \frac{1}{n}\sup_{x\in\mathcal{X}}\sup_{0\le t\le n} |\phi_t(x)| \ge B + A\sqrt{2\gamma\Delta} \right\} < \infty.
$$

Using the Borel–Cantelli Lemma, we obtain

$$
\limsup_{T\to\infty} \frac{1}{T}\sup_{x\in\mathcal{X}}\sup_{0\le t\le T} |\phi_t(x)| \le K := B + A\sqrt{2\gamma_0\Delta} \quad \text{a.s.,}
$$

which proves the theorem. $\qquad\qquad\square$

Corollary 2.7 *Assume in addition to the hypotheses in Theorem 2.3 that ϕ_t is a (random) homeomorphism on $\mathbf{R}^d$ for each $t \ge 0$. Then (2.13) holds with Δ replaced by $\Delta \wedge (d-1)$.*

Proof. Let $\mathcal{X}$ be compact and have box dimension $> (d-1)$ and let $\tilde{\mathcal{X}}$ be a compact set which contains $\mathcal{X}$ such that $\partial\tilde{\mathcal{X}}$ has box dimension $d-1$. We then apply Theorem 2.3 to $\partial\tilde{\mathcal{X}}$ instead of $\mathcal{X}$. By the homeomorphic property, we know that $\sup_{x\in\mathcal{X}} |\phi_t(x)| \le \sup_{x\in\partial\tilde{\mathcal{X}}} |\phi_t(x)|$ and the assertion of the corollary follows. $\qquad\square$

Remark   If ϕ is a flow which satisfies the assumptions of Theorem 2.2, then the upper bound for K in Theorem 2.3 can be improved by changing $I(\gamma)$ in (2.15) accordingly. In this case the upper formula for K in (2.14) holds for all values of Λ.

Now we provide a class of stochastic differential equations for which the assumptions of the previous theorem are satisfied. For simplicity we will assume that the drift b is autonomous and deterministic (if it is not, but the bound on b in the proposition is uniform with respect to (t,ω), then the proposition and its proof remain true without further change).

Proposition 2.8 *Let the assumptions of Lemma 2.6 be satisfied and assume in addition (for simplicity) that $a(.)$ and $b(.)$ are deterministic and autonomous. Further, we require that there exists $A > 0$ such that $\|a(x,x)\| \leq A^2$ for all x and that*

$$\limsup_{|x| \to \infty} \frac{x}{|x|} \cdot b(x) \leq B \in \mathbf{R}.$$

Then for each compact set S and each $k > 0$

$$\limsup_{T \to \infty} \frac{1}{T} \log \sup_{x \in S} \mathbf{P} \left\{ \sup_{0 \leq t \leq T} |\phi_t(x)| \geq kT \right\} \leq \begin{cases} -\frac{(k-B)_+^2}{2A^2} & \text{if } k \geq -B \\ 2Bk\frac{1}{A^2} & \text{otherwise.} \end{cases}$$

Proof. Let S be a compact subset of $\mathbf{R}^d$ and $k > B$ (otherwise there is nothing to show). Fix $0 < \varepsilon < k - B$ and let $r_0 > 1$ be such that

$$\frac{x}{|x|} \cdot b(x) + \frac{d-1}{2|x|} A^2 \leq B + \varepsilon \quad \text{for all } |x| \geq r_0$$

and such that S is contained in a ball around 0 of radius r_0. Let h be an even, smooth function from $\mathbf{R}$ to $\mathbf{R}$ such that $h(y) = |y|$ for $|y| \geq 1$ and $|h'(y)| \leq 1$ for all $y \in \mathbf{R}$ and define $\rho_t(x) = h(|\phi_t(x)|)$. Applying Itô's formula, we get

$$d\rho_t(x) = dN_t + f(\phi_t(x))\, dt,$$

where

$$N_t = \sum_{i=1}^{d} \int_0^t h'(\rho_s(x)) \frac{\phi_s^i(x)}{\rho_s(x)} M^i(ds, \phi_s(x)) \quad \text{and}$$

$$f(x) = \frac{x}{|x|} \cdot b(x) + \frac{1}{2|x|} \operatorname{Tr} a(x,x) - \frac{1}{2|x|^3} x^T a(x,x)x$$

$$\leq \frac{x}{|x|} \cdot b(x) + \frac{d-1}{2|x|} A^2 \leq B + \varepsilon \quad \text{on } \{|x| \geq r_0\}.$$

For the quadratic variation of N, we have the following bound:

$$\langle N \rangle_t - \langle N \rangle_s \leq \int_s^t \frac{1}{\rho_u^2(x)} \phi_u^T(x) a(\phi_u(x), \phi_u(x)) \phi_u(x)\, du \leq A^2(t-s).$$

The continuous local martingale N can be represented (possibly on an enriched probability space) in the form $N_t = A W_{\tau(t)}$, where W is a standard Brownian motion and the family of stopping times $\tau(s)$ satisfies

$\tau(t) - \tau(s) \leq t - s$ whenever $s \leq t$. For $|x| \leq r_0 < kT$ we get

$$\mathbf{P}\left\{\sup_{0 \leq t \leq T} |\phi_t(x)| \geq kT\right\}$$

$$\leq \mathbf{P}\left\{\exists 0 \leq s \leq t \leq T : \rho_t(x) - \rho_s(x) \geq kT - r_0, \inf_{s \leq u \leq t} \rho_u(x) \geq r_0\right\}$$

$$\leq \mathbf{P}\left\{\exists 0 \leq s \leq t \leq T : A(W_{\tau(t)} - W_{\tau(s)}) + (B + \varepsilon)(t - s) \geq kT - r_0\right\}$$

$$=: \bar{\mathbf{P}}$$

Now we distinguish between two cases:

Case 1: $B \geq 0$. Then

$$\bar{\mathbf{P}} \leq \mathbf{P}\left\{\max_{0 \leq s \leq 1} W_s - \min_{0 \leq s \leq 1} W_s \geq \frac{k - B - \varepsilon}{A}\sqrt{T} - \frac{r_0}{A\sqrt{T}}\right\}.$$

The density of the range $R := \max_{0 \leq s \leq 1} W_s - \min_{0 \leq s \leq 1} W_s$ equals

$$8 \sum_{j=1}^{\infty} (-1)^{j-1} j^2 \varphi(jr),$$

on $[0, \infty)$ (see [11]), where φ denotes the density of a standard normal law. Therefore, for all $u \geq 0$,

$$\mathbf{P}\{R \geq u\} \leq 8 \sum_{j=1}^{\infty} j\frac{1}{2}\exp\left\{-\frac{1}{2}j^2 u^2\right\} \sim 4\exp\left\{-\frac{u^2}{2}\right\}.$$

Hence

$$\limsup_{T \to \infty} \frac{1}{T} \log \sup_{x \in S} \mathbf{P}\left\{\sup_{0 \leq t \leq T} |\phi_t(x)| \geq kT\right\} \leq -\frac{(k - B)_+^2}{2A^2}.$$

Case 2: $B < 0$. We may assume that $\varepsilon > 0$ is so small that also $-\tilde{B} := (B + \varepsilon)/A < 0$. We have

$$\bar{\mathbf{P}} \leq \mathbf{P}\left\{\exists 0 \leq s \leq t \leq 1 : \frac{1}{\sqrt{T}}(W_t - W_s) - \tilde{B}(t - s) \geq k/A - \frac{r_0}{AT}\right\}. \tag{2.16}$$

To estimate this term, we use large deviations estimates for the standard Wiener process. Let

$$M := \{f \in C[0, 1] : \exists 0 \leq s \leq t \leq 1 : f_t - f_s - \tilde{B}(t - s) \geq k/A\}.$$

The set M is closed in $C[0, 1]$ and therefore Schilder's Theorem [9] implies

$$\limsup_{T \to \infty} \frac{1}{T} \log \mathbf{P}\{T^{-1/2} W \in M\} \leq -\inf_{f \in M} I(f), \tag{2.17}$$

where

$$I(f) \;:=\; \begin{cases} \frac{1}{2} \int_0^1 (f_u')^2 \, du & \text{if } f \text{ is abs. cont. with } L^2 \text{ derivative} \\ +\infty & \text{otherwise.} \end{cases}$$

The infimum in (2.17) can be computed explicitly. Let

$$I \;:=\; \begin{cases} \frac{1}{2} \left(\frac{k}{A} + \tilde{B} \right)^2 & \text{if } \tilde{B} \leq \frac{k}{A} \\ 2\tilde{B}\frac{k}{A} & \text{otherwise.} \end{cases}$$

For $f_t = (\tilde{B} + \tilde{B} \vee (k/A))t$ on $[0, k/((A\tilde{B}) \vee k)]$ and f constant on $[k/((A\tilde{B}) \vee k), 1]$, we have $f \in M$ and $I(f) = I$. On the other hand, if $f \in M$ with $I(f) < \infty$, then there exist $0 \leq s < t \leq 1$ such that $f_t - f_s - \tilde{B}(t - s) \geq k/A$. It follows that

$$I(f) \geq \frac{1}{2} \int_s^t (f_u')^2 \, du \geq \frac{1}{2} \frac{1}{t-s} \left(\int_s^t f_u' \, du \right)^2$$
$$= \frac{1}{2} \frac{1}{t-s} (f_t - f_s)^2 \geq \frac{1}{2} \frac{1}{t-s} \left(\frac{k}{A} + \tilde{B}(t-s) \right)^2 \geq I.$$

Therefore, using (2.16) and (2.17), we obtain

$$\limsup_{T \to \infty} \frac{1}{T} \log \sup_{x \in S} \mathbf{P} \left\{ \sup_{0 \leq t \leq T} |\phi_t(x)| \geq kT \right\} \leq -I$$

and the proof of the proposition is complete. □

Remark One can modify Theorem 2.3 in such a way that it also applies to solution flows generated by stochastic differential equations like in the previous proposition with negative B. In this case condition (ii) in Theorem 2.3 has to be changed accordingly. The corresponding linear upper bound will still be strictly positive no matter how small $B < 0$ is (namely $\gamma_0 \Delta A^2 / (-2B)$ as long as this number is at most $-B$). In reality however, the linear growth rate turns out to be zero when B is sufficiently small. This is shown in [10].

2.6 Appendix: Proofs of the chaining lemmas

In this section, we provide proofs of those chaining lemmas which are not available in the literature in the form presented here.

Proof of Lemma 2.1. We skip the proof of the existence of a continuous modification which can be found in many textbooks (e.g. [13]) and only show the estimates, assuming continuity of Z.

For $n \in \mathbf{N}$ define

$$D_n := \{(k_1, \ldots, k_d) \cdot 2^{-n}; \; k_1, \ldots k_d \in \{1, \ldots, 2^n\}\}$$
$$\xi_n(\omega) := \max\{\hat{\rho}(Z_x(\omega), Z_y(\omega)) : x, y \in D_n, |x - y| = 2^{-n}\}.$$

The $\xi_n, n \in \mathbf{N}$ are measurable since $(\hat{E}, \hat{\rho})$ is separable. Further,

$$|\{x, y \in D_n : |x - y| = 2^{-n}\}| \leq d \cdot 2^{dn}.$$

Hence, for $\kappa \in (0, \frac{b}{a})$,

$$\mathbf{E}\left(\sum_{n=1}^{\infty} (2^{\kappa n} \xi_n)^a\right) = \sum_{n=1}^{\infty} 2^{\kappa n a} \mathbf{E}(\xi_n^a)$$

$$\leq \sum_{n=1}^{\infty} 2^{\kappa n a} \mathbf{E}\left(\sum_{(x,y) \in D_n^2, |x-y|=2^{-n}} (\hat{\rho}(Z_x(\omega), Z_y(\omega))^a\right).$$

$$\leq \sum_{n=1}^{\infty} 2^{\kappa n a} \cdot d \cdot 2^{dn} \cdot c \cdot 2^{-n(d+b)}$$

$$= cd \sum_{n=1}^{\infty} 2^{-n(b-a\kappa)} = \frac{cd 2^{a\kappa - b}}{1 - 2^{a\kappa - b}} < \infty.$$

Hence, there exists $\Omega_0 \in \mathcal{F}$, $\mathbf{P}(\Omega_0) = 1$ such that

$$S(\omega) := \sup_{n \geq 1}(2^{\kappa n} \xi_n(\omega)) < \infty \quad \text{for all } \omega \in \Omega_0.$$

Further,

$$\mathbf{E}(S^a) \leq \mathbf{E}\left(\sum_{n=1}^{\infty} (2^{\kappa n} \xi_n)^a\right) \leq \frac{cd 2^{a\kappa - b}}{1 - 2^{a\kappa - b}}.$$

Let $x, y \in \bigcup_{n=1}^{\infty} D_n$ such that $|x - y|_\infty \leq r < 2^{-m}$, where $m \in \mathbf{N}_0$. There exists a sequence

$$x = x_1, x_2 \ldots, x_l = y$$

in $\bigcup_{n=m+1}^{\infty} D_n$, such that for each $i = 1, \ldots l - 1$ there exists $n(i) \geq m + 1$ which satisfies $x_i, x_{i+1} \in D_{n(i)}$ and $|x_i - x_{i+1}| = 2^{-n(i)}$ and

$$|\{i \in \{1, \ldots, l - 1\} : n(i) = k\}| \leq 2d \quad \text{for all } k \geq m + 1.$$

For $\omega \in \Omega_0$ and $0 < r < 1$ with $2^{-m-1} \le r < 2^{-m}$, we get

$$\sup\{\hat{\rho}(Z_x(\omega), Z_y(\omega)); x, y \in \bigcup_{n=1}^{\infty} D_n, |x - y|_\infty \le r\}$$

$$\le 2d \sum_{n=m+1}^{\infty} \xi_n(\omega) \le 2dS(\omega) \sum_{n=m+1}^{\infty} 2^{-\kappa n}$$

$$= 2^{-\kappa(m+1)} \frac{2d}{1 - 2^{-\kappa}} S(\omega) \le \frac{2d}{1 - 2^{-\kappa}} S(\omega) r^\kappa.$$

The statement in the lemma now follows by the continuity of Z. The final statement follows by an application of Chebychev's inequality. $\square$

Proof of Lemma 2.2. For each $j \in \mathbf{N}_0$ and each $x \in \Theta_{j+1}$, define $g_j(x) \in \Theta_j$ such that $d(x, g_j(x)) \le \delta_j$ (such a $g_j(x)$ exists due to the assumptions in the lemma). We will show that for each $x \in \Theta$ there exists a sequence $x_0, x_1, \ldots$ such that $x = \lim_{j \to \infty} x_j, x_j \in \Theta_j$ and $x_j = g_j(x_{j+1})$ for all $j \in \mathbf{N}_0$.

To see this, let $\delta_j^* = \sum_{i=j}^{\infty} \delta_i$ and $\tilde{\Theta}_j(x) := \{y \in \Theta_j : d(y, x) \le \delta_j^*\}$ for $x \in \Theta$. Then $\tilde{\Theta}_j(x) \ne \emptyset$ and $\tilde{x} \in \tilde{\Theta}_{j+1}(x)$ implies $g_j(\tilde{x}) \in \tilde{\Theta}_j(x)$. Therefore, there exists a sequence $x_0, x_1, x_2, \ldots$ which satisfies $x_j \in \tilde{\Theta}_j(x)$ and $x_j = g_j(x_{j+1})$ for all $j \in \mathbf{N}_0$. Since $\lim_{j \to \infty} \delta_j^* = 0$, we have $x = \lim_{j \to \infty} x_j$. We will write $x_j(x)$ instead of x_j.

Fix $x \in \Theta$. The continuity of Z implies

$$\hat{\rho}(Z_x, Z_{x_0}) \le \sum_{j=0}^{\infty} \hat{\rho}\left(Z_{x_{j+1}(x)}, Z_{x_j(x)}\right).$$

Therefore,

$$\sup_{x \in \Theta} \hat{\rho}(Z_x, Z_{x_0}) \le \sup_{x \in \Theta} \sum_{j=0}^{\infty} \hat{\rho}\left(Z_{x_{j+1}(x)}, Z_{x_j(x)}\right)$$

$$\le \sum_{j=0}^{\infty} \max_{x_{j+1} \in \Theta_{j+1}} \hat{\rho}\left(Z_{x_{j+1}}, Z_{g_j(x_{j+1})}\right).$$

Hence,

$$
\begin{aligned}
\mathbf{P}\{ \sup_{x,y \in \Theta} \hat{\rho}(Z_x, Z_y) \geq u \} &\leq \mathbf{P}\{ \sup_{x \in \Theta} \hat{\rho}(Z_x, Z_{x_0}) \geq u/2 \} \\
&\leq \mathbf{P}\{ \sum_{j=0}^{\infty} \max_{x \in \Theta_{j+1}} \hat{\rho}\left(Z_x, Z_{g_j(x)}\right) \geq \frac{u}{2} \sum_{j=0}^{\infty} \varepsilon_j \} \\
&\leq \sum_{j=0}^{\infty} \sum_{x \in \Theta_{j+1}} \mathbf{P}\{ \hat{\rho}\left(Z_x, Z_{g_j(x)}\right) \geq \varepsilon_j u/2 \} \\
&\leq \sum_{j=0}^{\infty} |\Theta_{j+1}| \sup_{d(x,y) \leq \delta_j} \mathbf{P}\{ \hat{\rho}\left(Z_x, Z_y\right) \geq \varepsilon_j u/2 \}.
\end{aligned}
$$

This completes the proof of the lemma.   □

Proof of Lemma 2.4. The proof is essentially a combination of those of [1] and [8]. The case $V = 0$ is clear (by our conventions about $V/0$), so we assume $V > 0$. We abbreviate

$$
\tilde{\Psi}(x,y) := \begin{cases} \Psi\left(\frac{\hat{\rho}(f(x),f(y))}{p(d(x,y))}\right) & \text{if } x \neq y, \\ 0 & \text{if } x = y. \end{cases}
$$

Fix $x \neq y$, $x, y \in \Theta$ and define $\rho := d(x,y)$ and

$$
I(u) := \int_{\Theta} \tilde{\Psi}(u,z) m(dz), \quad u \in \Theta.
$$

If either $m(K_\varepsilon(x)) = 0$ or $m(K_\varepsilon(y)) = 0$ for some $\varepsilon > 0$ or $V = 0$, then there is nothing to show, so we will assume $V > 0$, $m(K_\varepsilon(x)) > 0$ and $m(K_\varepsilon(y)) > 0$ for all $\varepsilon > 0$.

Let

$$
U := \{ z \in \Theta : d(x,z) \leq \rho \text{ and } d(y,z) \leq \rho \}.
$$

By the definition of I there exists $x_{-1} \in U$ such that

$$
I(x_{-1}) \leq \frac{V}{m(U)}. \tag{2.18}
$$

Let $\rho = r_{-1} \geq r_0 \geq r_1 \geq \ldots$ be a sequence of strictly positive reals which we will specify below. We will recursively define $x_n \in K_{r_n}(x), n \in \mathbf{N}_0$ such that

$$
I(x_n) \leq \frac{2V}{m(K_{r_n}(x))} \quad \text{and} \tag{2.19}
$$

$$\tilde{\Psi}(x_n, x_{n-1}) \leq \frac{2I(x_{n-1})}{m(K_{r_n}(x))}, \quad n \in \mathbf{N}_0. \tag{2.20}$$

For $n \in \mathbf{N}_0$ define

$$A_n := \left\{ z \in K_{r_n}(x) : I(z) > \frac{2V}{m(K_{r_n}(x))} \text{ or } \tilde{\Psi}(z, x_{n-1}) > \frac{2I(x_{n-1})}{m(K_{r_n}(x))} \right\}.$$

Then

$$m(A_n) \leq \frac{m(K_{r_n}(x))}{2V} \int_\Theta I(z) m(\mathrm{d}z) + \frac{m(K_{r_n}(x))}{2I(x_{n-1})} \int_\Theta \tilde{\Psi}(z, x_{n-1}) m(\mathrm{d}z)$$

$$\leq m(K_{r_n}(x))$$

and the first inequality is strict if $m(A_n) > 0$. In any case we have $m(A_n) < m(K_{r_n}(x))$. Now any $x_n \in K_{r_n}(x) \backslash A_n$ will satisfy (2.19) and (2.20). Using the fact that $K_\rho(x) \subseteq U$, it follows that

$$\hat{\rho}(f(x_{n+1}), f(x_n)) \leq \Psi^{-1}\left(\frac{2I(x_n)}{m(K_{r_{n+1}}(x))} \right) p(d(x_n, x_{n+1}))$$

$$\leq \Psi^{-1}\left(\frac{4V}{m(K_{r_{n+1}}(x)) m(K_{r_n}(x))} \right) p(d(x_n, x_{n+1})), n \geq -1.$$

Now, we choose the sequence r_n recursively as follows:

$$p(2r_{n+1}) = \frac{1}{2} p(r_n + r_{n+1}), \quad r_{-1} = 2\rho.$$

It is easy to check that this defines the sequence uniquely and that it decreases to zero as $n \to \infty$. If $n \geq -1$, then

$$p(d(x_n, x_{n+1})) \leq p(d(x_n, x) + d(x, x_{n+1})) \leq p(r_n + r_{n+1}) \leq 2p(2r_{n+1})$$
$$= 4p(2r_{n+1}) - 2p(2r_{n+1}) \leq 4p(2r_{n+1}) - 4p(2r_{n+2}).$$

Hence,

$$\hat{\rho}(f(x_{n+1}), f(x_n)) \leq 4 \int_{2r_{n+2}}^{2r_{n+1}} \Psi^{-1}\left(\frac{4V}{(m(K_{s/2}(x)))^2} \right) \mathrm{d}p(s).$$

The fact that f is continuous (at x) implies

$$\hat{\rho}(f(x), f(x_{-1})) \leq 4 \int_0^{2r_0} \Psi^{-1}\left(\frac{4V}{m(K_{s/2}(x))^2} \right) \mathrm{d}p(s)$$

$$\leq 4 \int_0^{4\rho} \Psi^{-1}\left(\frac{4V}{m(K_{s/2}(x))^2} \right) \mathrm{d}p(s).$$

The same estimate holds with x replaced by y (with $y_{-1} := x_{-1}$). Using the triangle inequality we get

$$\hat{\rho}(f(x), f(y)) \le 8 \max_{z \in \{x,y\}} \int_0^{4\rho} \Psi^{-1}\left(\frac{4V}{m(K_{s/2}(z))^2}\right) dp(s).$$

showing (i) of the lemma. If $N = 0$, then there is nothing to show. If $N = 1$, then $V \le 1$ and (ii) follows. The general case $N > 0$ can be reduced to the case $N = 1$ by considering the metric $\hat{\rho}'(x, y) := N^{-1}\hat{\rho}(x, y)$.   □

Bibliography

[1] L. Arnold and P. Imkeller, Stratonovich calculus with spatial parameters and anticipative problems in multiplicative ergodic theory, *Stoch. Proc. Appl.* **62**, 19–54 (1996).

[2] P. Baxendale and T. Harris, Isotropic stochastic flows, *Ann. Probab.* **14**, 1155–1179 (1986).

[3] A. Borodin and P. Salminen, *Handbook of Brownian Motion – Facts and Formulae*, Basel: Birkhäuser, 1996.

[4] M. Cranston and M. Scheutzow, Dispersion rates under finite mode Kolmogorov flows, *Ann. Appl. Probab.*, **12**, 511–532 (2002).

[5] M. Cranston, M. Scheutzow, and D. Steinsaltz, Linear expansion of isotropic Brownian flows, *Electron. Commun. Probab.* **4**, 91–101 (1999).

[6] M. Cranston, M. Scheutzow, and D. Steinsaltz, Linear bounds for stochastic dispersion, *Ann. Probab.* **28**, 1852–1869 (2000).

[7] R. Dalang, D. Khoshnevisan, and E. Nualart, Hitting probabilities for systems of non-linear stochastic heat equations with additive noise, *Alea* **3**, 231–271 (2007).

[8] G. Da Prato and J. Zabczyk, *Ergodicity for Infinite Dimensional Systems*, Cambridge University Press, 1996.

[9] A. Dembo and O. Zeitouni, *Large Deviations Techniques and Applications*, 2nd edition, New York: Springer, 1998.

[10] G. Dimitroff and M. Scheutzow, Attractors and expansion for Brownian flows, submitted.

[11] W. Feller, The asymptotic distribution of the range of sums of independent random variables, *Ann. Math. Stat.* **22**. 427–432 (1951).

[12] A.M. Garsia, E. Rodemich, and H. Rumsey, A real variable lemma and the continuity of paths of some Gaussian processes, *Indiana Univ. Math. Journal* **20**, 565–578 (1970).

[13] O. Kallenberg, *Foundations of Modern Probability*, 2nd edition, Berlin: Springer, 2002.

[14] H. Kunita, *Stochastic Flows and Stochastic Differential Equations*, Cambridge University Press, 1990.

[15] M. Ledoux and M. Talagrand, *Probability in Banach Spaces*, Berlin: Springer, 1991.

[16] H. Lisei and M. Scheutzow, Linear bounds and Gaussian tails in a stochastic dispersion model, *Stoch. Dynam.* **1**, 389–403 (2001).

[17] H. Lisei and M. Scheutzow, *On the dispersion of sets under the action of an*

isotropic Brownian flow, in: Proceedings of the Swansea 2002 Workshop *Probabilistic Methods in Fluids*, ed. I. Davies, 224–238, World Scientific, 2003.

[18] D. Pollard, *Empirical Processes: Theory and Applications*, IMS, 1990.

[19] M. Scheutzow, Attractors for Ergodic and Monotone Random Dynamical Systems, in: *Seminar on Stochastic Analysis, Random Fields and Applications V*, ed. R. Dalang, M. Dozzi, and F. Russo, 331–344, Basel: Birkhäuser, 2007.

[20] M. Scheutzow and D. Steinsaltz, Chasing balls through martingale fields, *Ann. Probab.* **30**, 2046–2080 (2002).

[21] M. Talagrand, *The Generic Chaining*, Berlin: Springer, 2005.

[22] J. Walsh, An introduction to stochastic partial differential equations, in: *École d'été de probabilités de Saint-Flour XIV – 1984*, Lect. Notes Math. 1180, 265–437, Berlin: Springer, 1986.

3

Ergodic properties of a class of non-Markovian processes

Martin Hairer

Warwick Mathematics Institute
University of Warwick
Coventry CV4 7AL, UK

Abstract

We study a fairly general class of time-homogeneous stochastic evolutions driven by noises that are not white in time. As a consequence, the resulting processes do not have the Markov property. In this setting, we obtain constructive criteria for the uniqueness of stationary solutions that are very close in spirit to the existing criteria for Markov processes.

In the case of discrete time, where the driving noise consists of a stationary sequence of Gaussian random variables, we give optimal conditions on the spectral measure for our criteria to be applicable. In particular, we show that, under a certain assumption on the spectral density, our assumptions can be checked in virtually the same way as one would check that the Markov process obtained by replacing the driving sequence by a sequence of independent identically distributed Gaussian random variables is strong Feller and topologically irreducible. The results of the present paper are based on those obtained previously in the continuous time context of diffusions driven by fractional Brownian motion.

3.1 Introduction

Stochastic processes have been used as a powerful modelling tool for decades in situations where the evolution of a system has some random component, be it intrinsic or to model the interaction with a complex environment. In its most general form, a stochastic process describes

Trends in Stochastic Analysis, ed. J. Blath, P. Mörters and M. Scheutzow.
Published by Cambridge University Press. ©Cambridge University Press 2008.

the evolution $X(t, \omega)$ of a system, where t denotes the time parameter and ω takes values in some probability space and abstracts the 'element of chance' describing the randomness of the process.

In many situations of interest, the evolution of the system can be described (at least informally) by the solutions of an evolution equation of the type

$$\frac{\mathrm{d}x}{\mathrm{d}t} = F(x, \xi), \tag{3.1a}$$

where ξ is the 'noise' responsible for the randomness in the evolution. In the present paper, we will not be interested in the technical subtleties arising from the fact that the time parameter t in (3.1a) takes continuous values. We will therefore consider its discrete analogue

$$x_{n+1} = F(x, \xi_n), \tag{3.1b}$$

were ξ_n describes the noise acting on the system between times n and $n + 1$. Note that (3.1a) can always be reduced to (3.1b) by allowing x_n to represent not just the state of the system at time n, but its evolution over the whole time interval $[n-1, n]$. We were intentionally vague about the precise meaning of the symbol x in the right-hand side of (3.1) in order to suggest that there are situations where it makes sense to let the right-hand side depend not only on the current state of the system, but on the whole collection of its past states as well.

The process x_n defined by a recursion of the type (3.1b) has the Markov property if both of the following properties hold:

a. The noises $\{\xi_n\}_{n \in \mathbf{Z}}$ are mutually independent.
b. For a fixed value of ξ, the function $x \mapsto F(x, \xi)$ depends only on the last state of the system.

In this paper, we will be interested in the study of recursion relations of the type (3.1b) when condition b still holds, but the Markov property is lost because condition a fails to hold. Our main focus will be on the ergodic properties of (3.1b), with the aim of providing concrete conditions that ensure the uniqueness (in law) of a stationary sequence of random variables x_n satisfying a given recursion of the type (3.1b).

Many such criteria exist for Markov processes and we refer to Meyn and Tweedie [17] for a comprehensive overview of the techniques developed in this regard over the past seven decades. The aim of the present paper will be to present a framework in which recursions of the type (3.1b) can be studied and such that several existing ergodicity results for

Markov processes have natural equivalents whose assumptions can also be checked in similar ways. This framework (which should be considered as nothing but a different way of looking at random dynamical systems, together with some slightly more restrictive topological assumptions) was developed in [9] and further studied in [11] in order to treat the ergodicity of stochastic differential equations driven by fractional Brownian motion. The main novelty of the present paper is to relax a number of assumptions from the previous works and to include a detailed study of the discrete-time case when the driving noise is Gaussian.

The remainder of this paper is organised as follows. After introducing our notations at the end of this section, we will introduce in Section 3.2 the framework studied in the present paper. We then proceed in Section 3.3 to a comparison of this framework with that of random dynamical systems. In Section 3.4 we recall a few general ergodicity criteria for Markov processes and give a very similar criterion that can be applied in framework. In Section 3.5 we finally study in detail the case of a system driven by a (time-discrete) stationary sequence of Gaussian random variables. We derive an explicit condition on the spectral measure of the sequence that ensures that such a system behaves qualitatively like the same system driven by an i.i.d. sequence of Gaussian random variables.

3.1.1 Notation

The following notation will be used throughout this paper. Unless stated otherwise, measures will always be Borel measures over Polish (i.e. metrizable, complete, separable) spaces and they will always be positive. We write $\mathscr{M}_+(\mathcal{X})$ for the set of all such measures on the space $\mathcal{X}$ and $\mathscr{M}_1(\mathcal{X})$ for the subset of all probability measures. We write $\mu \approx \nu$ to indicate that μ and ν are equivalent (i.e. they are mutually absolutely continuous, that is they have the same negligible sets) and $\mu \perp \nu$ to indicate that they are mutually singular.

Given a map $f: \mathcal{X} \to \mathcal{Y}$ and a measure μ on $\mathcal{X}$, we denote by $f^*\mu$ the push-forward measure $\mu \circ f^{-1}$ on $\mathcal{Y}$. Given a product space $\mathcal{X} \times \mathcal{Y}$, we will use the notation $\Pi_\mathcal{X}$ and $\Pi_\mathcal{Y}$ to denote the projections onto the two factors. For infinite products like $\mathcal{X}^{\mathbf{Z}}$, $\mathcal{X}^{\mathbf{Z}-}$ or $\mathcal{X}^{\mathbf{N}}$, we denote by Π_n the projection onto the nth factor (n can be negative in the first two cases).

We will also make use of the concatenation operator $\sqcup$ from $\mathcal{X}^{\mathbf{Z}-} \times \mathcal{X}^n$

to $\mathcal{X}^{\mathbf{Z}_-}$ defined in the natural way by

$$(w \sqcup w')_k = \left\{ \begin{array}{ll} w_{k+n} & \text{if } k \leq -n, \\ w'_{k+n} & \text{otherwise.} \end{array} \right.$$

Finally, given a Markov transition probability $\mathcal{P} \colon \mathcal{X} \to \mathcal{M}_1(\mathcal{X})$, we will use the same symbol for the associated Markov operator acting on observables $\phi \colon \mathcal{X} \to \mathbf{R}$ by $(\mathcal{P}\phi)(x) = \int_{\mathcal{X}} \phi(y)\,\mathcal{P}(x, \mathrm{d}y)$, and the dual operator acting on probability measures μ by $(\mathcal{P}\mu)(A) = \int_{\mathcal{X}} \mathcal{P}(x, A)\,\mu(\mathrm{d}x)$.

3.2 Skew-products

Whatever stochastic process X one may wish to consider, it is always possible to turn it into a Markov process by adding sufficiently many 'hidden' degrees of freedom to the state space. For example, one can take the state space large enough to contain all possible information about the past of X, as well as all possible information on the future of the driving noise ξ. The evolution (3.1) is then deterministic, with all randomness injected once and for all by drawing ξ initially according to the appropriate distribution. This is the point of view of random dynamical systems explained in more detail in Section 3.3 below.

On the other hand, one could take a somewhat smaller 'noise space' that contains only information about the *past* of the driving noise ξ. In this case, the evolution is no longer deterministic, but it becomes a skew-product between a Markovian evolution for the noise (with the transition probabilities given informally by the conditional distribution of the 'future' given the 'past') and a deterministic map that solves (3.1). This is the viewpoint that was developed in [9] and [11] and will be studied further in this paper.

The framework that will be considered here is the following. Let $\mathcal{W}$ and $\mathcal{X}$ be two Polish spaces that will be called the 'noise space $\mathcal{W}$' and the 'state space $\mathcal{X}$' respectively, let $\mathcal{P}$ be a Markov transition kernel on $\mathcal{W}$, and let $\Phi \colon \mathcal{W} \times \mathcal{X} \to \mathcal{X}$ be an 'evolution map'. Throughout this paper, we will make the following standing assumptions:

1. There exists a probability measure $\mathbf{P}$ on $\mathcal{W}$ which is invariant for $\mathcal{P}$ and such that the law of the corresponding stationary process is ergodic under the shift map.
2. The map $\Phi \colon \mathcal{W} \times \mathcal{X} \to \mathcal{X}$ is continuous in the product topology.

We will also occasionally impose some regularity of the kernel $\mathcal{P}(w, \cdot)$

as a function of w. We therefore state the following property which will not always be assumed to hold:

3. The transition kernel $\mathcal{P}$ is Feller, that is the function $\mathcal{P}\phi$ defined by $(\mathcal{P}\phi)(w) = \int_{\mathcal{W}} \phi(w')\mathcal{P}(w, dw')$ is continuous as soon as ϕ is continuous.

There are two objects that come with a construction such as the one above. First, we can define a Markov transition operator $\mathcal{Q}$ on $\mathcal{X} \times \mathcal{W}$ by

$$(\mathcal{Q}\phi)(x, w) = \int_{\mathcal{W}} \phi(\Phi(x, w'), w')\mathcal{P}(w, dw'). \tag{3.2}$$

In words, we first draw an element w' from the noise space according to the law $\mathcal{P}(w, \cdot)$ and we then update the state of the system with that noise according to Φ. We also introduce a 'solution map' $\mathcal{S} \colon \mathcal{X} \times \mathcal{W} \to \mathscr{M}_1(\mathcal{X}^{\mathbf{N}})$ that takes as arguments an initial condition $x \in \mathcal{X}$ and an 'initial noise' w and returns the law of the corresponding solution process, that is the marginal on $\mathcal{X}$ of the law of the Markov process starting at (x, w) with transition probabilities $\mathcal{Q}$.

The point of view that we take in this paper is that $\mathcal{S}$ encodes all the 'physically relevant' part of the evolution (3.1), and that the particular choice of noise space is just a mathematical tool. This motivates the introduction of an equivalence relation between probability measures on $\mathcal{X} \times \mathcal{W}$ by

$$\mu \simeq \nu \quad \Leftrightarrow \quad \mathcal{S}\mu = \mathcal{S}\nu. \tag{3.3}$$

(Here, we used the shorthand $\mathcal{S}\mu = \int \mathcal{S}(x, w)\, \mu(dx, dw)$.) In the remainder of this paper, when we will be looking for criteria that ensure the uniqueness of the invariant measure for $\mathcal{Q}$, this will always be understood to hold up to the equivalence relation (3.3).

Remark 3.1 *The word 'skew-product' is sometimes used in a slightly different way in the literature. In our framework, given a realization of the noise, that is a realization of a Markov process on $\mathcal{W}$ with transition probabilities $\mathcal{P}$, the evolution in $\mathcal{X}$ is purely deterministic. This is different from, for example, the skew-product decomposition of Brownian motion where, given one realization of the evolution of the radial part, the evolution of the angular part is still random.*

3.2.1 Admissible measures

We consider the invariant measure $\mathbf{P}$ for the noise process to be fixed. Therefore, we will usually consider measures on $\mathcal{X} \times \mathcal{W}$ such that their projections on $\mathcal{W}$ are equal to $\mathbf{P}$. Let us call such probability measures *admissible* and let us denote the set of admissible probability measures by $\mathcal{M}_{\mathbf{P}}(\mathcal{X})$. Obviously, the Markov operator $\mathcal{Q}$ maps the set of admissible probability measures into itself. Since we assumed that $\mathcal{W}$ is a Polish space, it is natural to endow $\mathcal{M}_{\mathbf{P}}(\mathcal{X})$ with the topology of weak convergence. This topology is preserved by $\mathcal{Q}$ if we assume that $\mathcal{P}$ is Feller:

Lemma 3.2 *If Φ is continuous and $\mathcal{P}$ is Feller, then the Markov transition operator $\mathcal{Q}$ is also Feller and therefore continuous in the topology of weak convergence on $\mathcal{X} \times \mathcal{W}$.*

The proof of this result is straightforward and of no particular interest, so we leave it as an exercise. $\qquad\square$

There are, however, cases of interest in which we do not wish to assume that $\mathcal{P}$ is Feller. In this case, a natural topology for the space $\mathcal{M}_{\mathbf{P}}(\mathcal{X})$ is given by the 'narrow topology'; see [26] and [4]. In order to define this topology, denote by $\mathcal{C}_{\mathbf{P}}(\mathcal{X})$ the set of functions $\phi \colon \mathcal{X} \times \mathcal{W} \to \mathbf{R}$ such that $x \mapsto \phi(x, w)$ is bounded and continuous for every $w \in \mathcal{W}$, $w \mapsto \phi(x, w)$ is measurable for every $x \in \mathcal{X}$, and $\int_{\mathcal{W}} \sup_{x \in \mathcal{X}} |\phi(x, w)| \, \mathbf{P}(\mathrm{d}w) < \infty$. The narrow topology on $\mathcal{M}_{\mathbf{P}}(\mathcal{X})$ is then the coarsest topology such that the map $\mu \mapsto \int \phi(x, w) \, \mu(\mathrm{d}x, \mathrm{d}w)$ is continuous for every $\phi \in \mathcal{C}_{\mathbf{P}}(\mathcal{X})$. Using Lebesgue's dominated convergence theorem, it is straightforward to show that $\mathcal{Q}$ is continuous in the narrow topology without requiring any assumption besides the continuity of Φ.

An admissible probability measure μ is now called an *invariant measure* for the skew-product $(\mathcal{W}, \mathbf{P}, \mathcal{P}, \mathcal{X}, \Phi)$ if it is an invariant measure for $\mathcal{Q}$, that is if $\mathcal{Q}\mu = \mu$. We call it a *stationary measure* if $\mathcal{Q}\mu \simeq \mu$, that is if the law of the $\mathcal{X}$-component of the Markov process with transition probabilities $\mathcal{Q}$ starting from μ is stationary. Using the standard Krylov-Bogoliubov argument, one shows that:

Lemma 3.3 *Given any stationary measure μ as defined above, there exists an invariant measure $\hat{\mu}$ such that $\hat{\mu} \simeq \mu$.*

Proof. Define a sequence of probability measures μ_N on $\mathcal{X} \times \mathcal{W}$ by $\mu_N = \frac{1}{N} \sum_{n=1}^{N} \mathcal{Q}^n \mu$. Since, for every N, the marginal of μ_N on $\mathcal{W}$ is equal to $\mathbf{P}$ and the marginal on $\mathcal{X}$ is equal to the marginal of μ on $\mathcal{X}$ (by stationarity), this sequence is tight in the narrow topology [4]. It therefore has at least one accumulation point $\hat{\mu}$ and the continuity of $\mathcal{Q}$ in the narrow topology ensures that $\hat{\mu}$ is indeed an invariant measure for $\mathcal{Q}$. $\qquad\qquad\qquad\qquad\qquad\qquad\qquad\qquad\qquad\qquad\qquad\qquad$ $\square$

The aim of this paper is to present some criteria that allow us to show the uniqueness *up to the equivalence relation (3.3)* of the invariant measure for a given skew-product. The philosophy that we will pursue is not to apply existing criteria to the Markov semigroup $\mathcal{Q}$. This is because, in typical situations like a random differential equation driven by some stationary Gaussian process, the noise space $\mathcal{W}$ is very 'large' and so the Markov operators $\mathcal{P}$ and $\mathcal{Q}$ typically do not have any of the 'nice' properties (strong Feller property, ψ-irreducibility, etc.) that are often required in the ergodic theory of Markov processes.

3.2.2 A simple example

In this section, we give a simple example that illustrates the fact that it is possible in some situations to have non-uniqueness of the invariant measure for $\mathcal{Q}$, even though $\mathbf{P}$ is ergodic and one has uniqueness up to the equivalence relation (3.3). Take $\mathcal{W} = \{0,1\}^{\mathbf{Z}_-}$ and define the 'concatenation and shift' map $\Theta \colon \mathcal{W} \times \{0,1\} \to \mathcal{W}$ by

$$\Theta(w,j)_n = \begin{cases} w_{n+1} & \text{for } n < 0, \\ j & \text{for } n = 0. \end{cases}$$

Fix $p \in (0,1)$, let ξ be a random variable that takes the values 0 and 1 with probabilities p and $1-p$ respectively, and define the transition probabilities $\mathcal{P}$ by

$$\mathcal{P}(w,\cdot) \overset{\text{law}}{=} \Theta(w,\xi).$$

We then take as our state space $\mathcal{X} = \{0,1\}$ and we define an evolution Φ by

$$\Phi(x,w) = \begin{cases} w_0 & \text{if } x = w_{-1}, \\ 1 - w_0 & \text{otherwise.} \end{cases} \tag{3.4}$$

It is clear that there are two extremal invariant measures for this evolution. One of them charges the set of pairs (w,x) such that $x = w_0$, the other one charges the set of pairs such that $x = 1 - w_0$. (The projection of these measures onto $\mathcal{W}$ is Bernoulli with parameter p in both cases.)

However, if $p = \frac{1}{2}$, both invariant measures give rise to the same stationary process in $\mathcal{X}$, which is just a sequence of independent Bernoulli random variables.

3.2.3 An important special case

In most of the remainder of this paper, we are going to focus on the following particular case of the setup described above. Suppose that there exists a Polish space $\mathcal{W}_0$ which carries all the information that is needed in order to reconstruct the dynamic of the system from one time-step to the next. We then take $\mathcal{W}$ of the form $\mathcal{W} = \mathcal{W}_0^{\mathbf{Z}_-}$ (with the product topology) and we assume that Φ is of the form $\Phi(x, w) = \Phi_0(x, w_0)$ for some jointly continuous function $\Phi_0 \colon \mathcal{X} \times \mathcal{W}_0 \to \mathcal{X}$. Here, we use the notation $w = (\ldots, w_{-1}, w_0)$ for elements of $\mathcal{W}$. Concerning the transition probabilities $\mathcal{P}$, we fix a Borel measure $\mathbf{P}$ on $\mathcal{W}$ which is invariant and ergodic for the shift map[†] $(\Theta w)_n = w_{n-1}$, and we define a measurable map $\bar{\mathcal{P}} \colon \mathcal{W} \to \mathcal{M}_1(\mathcal{W}_0)$ as the regular conditional probabilities of $\mathbf{P}$ under the splitting $\mathcal{W} \approx \mathcal{W} \times \mathcal{W}_0$.

The transition probabilities $\mathcal{P}(w, \cdot)$ are then constructed as the push-forward of $\bar{\mathcal{P}}(w, \cdot)$ under the concatenation map $f_w \colon \mathcal{W}_0 \to \mathcal{W}$ given by $f_w(w') = w \sqcup w'$. Since we assumed that $\mathbf{P}$ is shift-invariant, it follows from the construction that it is automatically invariant for $\mathcal{P}$.

Many natural situations fall under this setup, even if they do not look as if they do at first sight. For example, in the case of the example from the previous section, one would be tempted to take $\mathcal{W}_0 = \{0, 1\}$. This does not work since the function ϕ defined in (3.4) depends not only on w_0 but also on w_{-1}. However, one can choose $\mathcal{W}_0 = \{0, 1\}^2$ and identify $\mathcal{W}$ with the subset of all sequences $\{w_n\}_{n \leq 0}$ in $\mathcal{W}_0^{\mathbf{Z}_-}$ such that $w_n^2 = w_{n+1}^1$ for every $n < 0$.

3.3 Skew-products of Markov processes versus random dynamical systems

There already exists a mature theory which was developed precisely in order to study systems like (3.1). The theory in question is of course that of random dynamical systems (RDS hereafter), which was introduced under this name in the nineties by Arnold and then developed

---

† Recall that a probability measure μ is ergodic for a map T leaving μ invariant if all T-invariant measurable sets are of μ-measure 0 or 1.

further by a number of authors, in particular Caraballo, Crauel, De-
bussche, Flandoli, Robinson, Schmalfuß, and many others. Actually,
skew-products of flows had been considered by authors much earlier
(see for example [18], [19], or the monograph [14]), but previous authors usually made very restrictive assumptions on the structure of the
noise, either having independent noises at each step or some periodic-
ity or quasi-periodicity. We refer to the monograph by Arnold [1] for a
thorough exposition of the theory but, for the sake of completeness, we
briefly recall the main framework here. For simplicity, and in order to
facilitate comparison with the alternative framework presented in this
paper, we restrict ourselves to the case of discrete time. An RDS con-
sists of a dynamical system $(\Omega, \bar{\mathbf{P}}, \Theta)$ (here $\bar{\mathbf{P}}$ is a probability on the
measurable space Ω which is both invariant and ergodic for the map
$\Theta \colon \Omega \to \Omega$) together with a 'state space' $\mathcal{X}$ and a map $\bar{\Phi} \colon \Omega \times \mathcal{X} \to \mathcal{X}$.
For every initial condition $x \in \mathcal{X}$, this allows us to construct a stochastic
process X_n over $(\Omega, \bar{\mathbf{P}})$, viewed as a probability space, by

$$X_0(\omega) = x \,, \quad X_{n+1}(\omega) = \bar{\Phi}\bigl(\Theta^n \omega, X_n(\omega)\bigr) \,.$$

Note the similarity with (3.2). The main difference is that the evolution
Θ on the 'noise space' Ω is *deterministic*. This means that an element
of Ω must contain all possible information on the future of the noise
driving the system. In fact, one can consider an RDS as a dynamical
system $\hat{\Phi}$ over the product space $\Omega \times \mathcal{X}$ via

$$\hat{\Phi}(x, \omega) = \bigl(\Theta\omega, \bar{\Phi}(x, \omega)\bigr) \,.$$

The stochastic process X_n is then nothing but the projection on $\mathcal{X}$ of
a 'typical' orbit of $\hat{\Phi}$. An *invariant measure* for an RDS $(\Omega, \bar{\mathbf{P}}, \Theta, \mathcal{X}, \hat{\Phi})$
is a probability measure μ on $\Omega \times \mathcal{X}$ which is invariant under $\hat{\Phi}$ and
such that its marginal on Ω is equal to $\bar{\mathbf{P}}$. It can be shown [1] that such
measures can be described by their 'disintegration' over $(\Omega, \bar{\mathbf{P}})$, which is
a map $\omega \mapsto \mu_\omega$ from Ω into the set of probability measures on $\mathcal{X}$.

Consider the example of an elliptic diffusion X_t on a compact manifold
$\mathcal{M}$. Let us be even more concrete and take for X_t a simple Brownian
motion and for $\mathcal{M}$ the unit circle S^1. When considered from the point of
view of the Markov semigroup $\mathcal{P}_t$ generated by X_t, it is straightforward
to show that there exists a unique invariant probability measure for $\mathcal{P}_t$.
In our example, this invariant measure is of course simply the Lebesgue
measure on the circle.

Consider now X_t as generated from a random dynamical system (since
we focus on the discrete time case, choose t to take integer values). At

this stage, we realize that we have a huge freedom of choice when it comes to finding an underlying dynamical system $(\Omega, \bar{\mathbf{P}}, \Theta)$ and a map $\bar{\Phi} \colon \Omega \times S^1 \to S^1$ such that the corresponding stochastic process is equal to our Brownian motion X_t. The most immediate choice would be to take for Ω the space of real-valued continuous functions vanishing at the origin, $\bar{\mathbf{P}}$ equal to the Wiener measure, Θ the shift map $(\Theta B)(s) = B(s+1) - B(1)$, and $\bar{\Phi}(x, B) = x + B(1)$. In this case, it is possible to show that there exists a unique invariant measure for this random dynamical systems and that this invariant measure is equal to the product measure of $\bar{\mathbf{P}}$ with Lebesgue measure on S^1, so that μ_ω is equal to the Lebesgue measure for every ω.

However, we could also have considered X_t as the solution to the stochastic differential equation

$$\mathrm{d}X(t) = \sin(kX(t))\,\mathrm{d}B_1(t) + \cos(kX(t))\,\mathrm{d}B_2(t),$$

where B_1 and B_2 are two independent Wiener processes and k is an arbitrary integer. In this case, it turns out (see [15], [3]) that there are *two* invariant measures μ^+ and μ^- for the corresponding random dynamical system. Both of them are such that, for almost every ω, μ_ω is equal to a sum of k δ-measures of weights $1/k$. Furthermore, the map $\omega \mapsto \mu_\omega^-$ is measurable with respect to the filtration generated by the increments of $B_i(t)$ for negative t, whereas $\omega \mapsto \mu_\omega^+$ is measurable with respect to the filtration generated by the increments of $B_i(t)$ for positive t.

What this example makes clear is that, while the ergodic theory of X_t considered as a Markov process focuses on the long-time behaviour of *one instance* of X_t started at an arbitrary but fixed initial condition, the theory of random dynamical systems instead focuses on the (potentially much richer) simultaneous long-time behaviour of *several instances* of X_t driven by the same instance of the noise. Furthermore, it shows that a random dynamical system may have invariant measures that are 'unphysical' in the sense that they can be realized only by initializing the state of our system in some way that requires clairvoyant knowledge of the entire future of its driving noise.

In the framework presented in the previous section, such unphysical invariant measures never arise, since our noise space only contains information about the 'past' of the noise. Actually, given a skew-product of Markov processes as before, one can construct in a canonical way a random dynamical system by taking $\Omega = \mathcal{W}^{\mathbf{Z}}$, Θ the shift map, and $\bar{\mathbf{P}}$ the measure on $\Omega = \mathcal{W}^{\mathbf{Z}}$ corresponding to the law of the stationary

Markov process with transition probabilities $\mathcal{P}$ and one-point distribution **P**. The map $\bar{\Phi}$ is then given by $\bar{\Phi}(x,\omega) = \Phi(x,\omega_0)$. With this correspondence, an invariant measure for the skew-product yields an invariant measure for the corresponding random dynamical system but, as the example given above shows, the converse is not true in general.

3.4 Ergodicity criteria for Markov semigroups

Consider a Markov transition kernel $\mathcal{P}$ on some Polish space $\mathcal{X}$. Recall that an invariant measure μ for $\mathcal{P}$ is said to be ergodic if the law of the corresponding stationary process is ergodic for the shift map. It is a well-known fact that if a Markov transition kernel has more than one invariant measure then it must have at least two of them that are mutually singular. Therefore, the usual strategy for proving the uniqueness of the invariant measure for $\mathcal{P}$ is to assume that $\mathcal{P}$ has two mutually singular invariant measures μ and ν and to arrive at a contradiction.

This section is devoted to the presentation of some ergodicity criteria for a Markov process on a general state space $\mathcal{X}$ and to their extension to the framework presented in Section 3.2. If $\mathcal{X}$ happens to be countable (or finite), the transition probabilities are given by a transition matrix $P = (P_{ij})$ (P_{ij} being the probability of going from i to j in time 1) and there is a very simple characterization of those transition probabilities that can lead to at most one invariant probability measure. In a nutshell, ergodicity is implied by the existence of one point which cannot be avoided by the dynamic:

Proposition 3.4 *Let P be a transition matrix. If there exists a state j such that, for every i, $\sum_{n\geq 1}(P^n)_{ij} > 0$, then P can have at most one invariant probability measure. Conversely, if P has exactly one invariant probability measure, then there exists a state j with the above property.*

There is no such clean criterion available in the case of general state space, but the following comes relatively close. Recall that a Markov transition operator P over a Polish space $\mathcal{X}$ is said to be *strong Feller* if it maps the space of bounded measurable functions into the space of bounded continuous functions. This is equivalent to the continuity of transition probabilities in the topology of strong convergence of measures. With this definition, one has the following criterion, of which a proof can be found for example in [5]:

Proposition 3.5 *Let P be a strong Feller Markov transition operator on a Polish space $\mathcal{X}$. If μ and ν are two invariant measures for P that are mutually singular, then $\operatorname{supp} \mu \cap \operatorname{supp} \nu = \emptyset$.*

This is usually used together with some controllability argument in the following way:

Corollary 3.6 *Let P be strong Feller. If there exists x such that x belongs to the support of every invariant measure for P, then P can have at most one invariant measure.*

Remark 3.7 *The importance of the strong Feller property is that it allows us to replace a measure-theoretical statement (μ and ν are mutually singular) by a stronger topological statement (the topological supports of μ and ν are disjoint) which is then easier to invalidate by a controllability argument. If one further uses the fact that μ and ν are invariant measures, one can actually replace the strong Feller property by the weaker asymptotic strong Feller property [10], but we will not consider this generalization here.*

A version with slightly stronger assumptions that, however, leads to a substantially stronger conclusion is usually attributed to Doob and Khasminsk'ii:

Theorem 3.8 *Let P be a strong Feller Markov transition operator on a Polish space $\mathcal{X}$. If there exists $n \geq 1$ such that, for every open set $A \subset \mathcal{X}$ and every $x \in \mathcal{X}$, one has $P^n(x, A) > 0$, then the measures $P^m(x, \cdot)$ and $P^m(y, \cdot)$ are equivalent for every pair $(x, y) \in \mathcal{X}^2$ and for every $m > n$. In particular, P can have at most one invariant probability measure and, if it exists, it is equivalent to $P^{n+1}(x, \cdot)$ for every x.*

These criteria suggest that we should look for a version of the strong Feller property that is suitable for our context. Requiring Q to be strong Feller is a very strong requirement which will not be fulfilled in many cases of interest. On the one hand, since there is nothing like a semigroup on $\mathcal{X}$, it is not clear *a priori* how the strong Feller property should be translated to our framework. On the other hand, the *ultra Feller* property, that is the continuity of the transition probabilities in the total variation topology, is easier to generalize to our setting. Even though this property seems at first sight to be stronger than the strong Feller property (the topology on probability measures induced by the total variation distance is strictly stronger than the one induced by strong convergence), it turns out that the two are 'almost' equivalent. More precisely, if two Markov transition operators P and Q are strong Feller, then PQ is ultra Feller. Since this fact is not easy to find in the literature, we will give a self-contained proof in Appendix A (Section 3.6).

One possible generalization, and this is the one that we will retain here, is given by the following:

Definition 3.9 *A skew-product* $(\mathcal{W}, \mathbf{P}, \mathcal{P}, \mathcal{X}, \Phi)$ *is said to be strong Feller if there exists a measurable map* $\ell \colon \mathcal{W} \times \mathcal{X}^2 \to [0,1]$ *such that, for* $\mathbf{P}$*-almost every* w *one has* $\ell(w, x, x) = 0$ *for every* x*, and such that*

$$\|\mathcal{S}(x, w) - \mathcal{S}(y, w)\|_{\mathrm{TV}} \le \ell(w, x, y) \,, \tag{3.5}$$

for every $w \in \mathcal{W}$ *and every* $x, y \in \mathcal{X}$*.*

If we are furthermore in the setting of Section 3.2.3, we assume that, for $\mathbf{P}$*-almost every* w*, the map* $(w', x, y) \mapsto \ell(w \sqcup w', x, y)$ *with* $w' \in \mathcal{W}_0$ *is jointly continuous.*

If we are not in that setting, we impose the stronger condition that ℓ *is jointly continuous.*

A natural generalization of the topological irreducibility used in Theorem 3.8 is given by

Definition 3.10 *A skew-product* $(\mathcal{W}, \mathbf{P}, \mathcal{P}, \mathcal{X}, \Phi)$ *is said to be topologically irreducible if there exists* $n \ge 1$ *such that*

$$\mathcal{Q}^n(x, w; A \times \mathcal{W}) > 0 \,,$$

for every $x \in \mathcal{X}$*,* $\mathbf{P}$*-almost every* $w \in \mathcal{W}$*, and every open set* $A \subset \mathcal{X}$*.*

According to these definitions, the example given in Section 3.2.2 is both strong Feller and topologically irreducible. If $p \ne \frac{1}{2}$, it does, however, have two distinct (even up to the equivalence relation $\simeq$) invariant measures. The problem is that in the non-Markovian case it is of course perfectly possible to have two distinct ergodic invariant measures for $\mathcal{Q}$ that are such that their projections on $\mathcal{X}$ are *not* mutually singular. This shows that, if we are aiming for an extension of a statement along the lines of Theorem 3.8, we should impose some additional condition, which ideally should always be satisfied for Markovian systems.

3.4.1 Off-white noise systems

In order to proceed, we consider the measure $\hat{\mathbf{P}}$ on $\mathcal{W}^{\mathbf{Z}}$, which is the law of the stationary process with transition probabilities $\mathcal{P}$ and fixed-time law $\mathbf{P}$. We define the coordinate maps $\Pi_i \colon \mathcal{W}^{\mathbf{Z}} \to \mathcal{W}$ in the natural way and the shift map Θ satisfying $\Pi_i \Theta w = \Pi_{i+1} w$. We also define two natural σ-fields on $\mathcal{W}^{\mathbf{Z}}$. The past, $\mathscr{P}$, is defined as the σ-field generated by the coordinate maps Π_i for $i \leq 0$. The future, $\mathscr{F}$, is defined as the σ-field generated by all the maps of the form $w \mapsto \Phi(x, \Pi_i w)$ for $x \in \mathcal{X}$ and $i > 0$. With these definitions, we see that the process corresponding to our skew-product is Markov (which is sometimes expressed by saying that the system is a 'white noise system') if $\mathscr{P}$ and $\mathscr{F}$ are independent under $\hat{\mathbf{P}}$.

A natural weakening of the Markov property is therefore given by:

Definition 3.11 *A skew-product* $(\mathcal{W}, \mathbf{P}, \mathcal{P}, \mathcal{X}, \Phi)$ *is said to be an off-white noise system if there exists a probability measure* $\hat{\mathbf{P}}_0$ *that is equivalent to* $\hat{\mathbf{P}}$ *and such that* $\mathscr{P}$ *and* $\mathscr{F}$ *are independent under* $\hat{\mathbf{P}}_0$.

Remark 3.12 *The terminology 'off-white noise system' is used by analogy on the one hand with 'white noise systems' in the theory of random dynamical systems and on the other hand with 'off-white noise' (or 'slightly coloured noise') as studied by Tsirelson ([24], [25]).*

An off-white noise system behaves, as far as ergodic properties are concerned, pretty much like a white noise (Markovian) system. This is the content of the following proposition:

Theorem 3.13 *Let* $(\mathcal{W}, \mathbf{P}, \mathcal{P}, \mathcal{X}, \Phi)$ *be an off-white noise system and let* μ *and* ν *be two stationary measures for* $\mathcal{Q}$ *such that* $\mathcal{S}\mu \perp \mathcal{S}\nu$. *Then their projections* $\Pi_{\mathcal{X}}^* \mu$ *and* $\Pi_{\mathcal{X}}^* \nu$ *onto the state space* $\mathcal{X}$ *are also mutually singular.*

Proof. Denote by $\hat{\Phi} \colon \mathcal{X} \times \mathcal{W}^{\mathbf{Z}} \to \mathcal{X}^{\mathbf{N}}$ the solution map defined recursively by

$$\big(\hat{\Phi}(x, w)\big)_0 = x, \qquad \big(\hat{\Phi}(x, w)\big)_n = \Phi\big(\big(\hat{\Phi}(x, w)\big)_{n-1}, \Pi_n w\big). \qquad (3.6)$$

It follows from the construction that $\hat{\Phi}$ is $\mathcal{B}(\mathcal{X}) \otimes \mathscr{F}$-measurable, where $\mathcal{B}(\mathcal{X})$ denotes the Borel σ-algebra of $\mathcal{X}$. Denote now by μ_w and ν_w the disintegrations of μ and ν over $\mathcal{W}$, that is the only (up to $\mathbf{P}$-negligible sets) measurable functions from $\mathcal{W}$ to $\mathcal{M}_1(\mathcal{X})$ such that

$$\mu(A \times B) = \int_B \mu_w(A) \mathbf{P}(dw), \quad A \in \mathcal{B}(\mathcal{X}), \quad B \in \mathcal{B}(\mathcal{W}),$$

and similarly for ν. Using this, we construct measures $\hat{\mu}$ and $\hat{\nu}$ on $\mathcal{X} \times \mathcal{W}^{\mathbf{Z}}$ by

$$\hat{\mu}(A \times B) = \int_B \mu_{\Pi_0 w}(A) \hat{\mathbf{P}}(dw), \quad A \in \mathcal{B}(\mathcal{X}), \quad B \in \mathcal{B}(\mathcal{W}^{\mathbf{Z}}). \quad (3.7)$$

With these constructions, one has $\mathcal{S}\mu = \hat{\Phi}^* \hat{\mu}$ and similarly for ν. We also define $\hat{\mu}_0$ by the same expression as (3.7) with $\hat{\mathbf{P}}$ replaced by $\hat{\mathbf{P}}_0$. Since $\hat{\mathbf{P}}_0 \approx \hat{\mathbf{P}}$, one has $\hat{\mu} \approx \hat{\mu}_0$. However, since Π_0 is $\mathscr{P}$-measurable and since $\mathscr{P}$ and $\mathscr{F}$ are independent under $\hat{\mathbf{P}}_0$, one has $\hat{\mu}_0 \approx \Pi_{\mathcal{X}}^* \hat{\mu}_0 \otimes \hat{\mathbf{P}}_0$ when restricted to the σ-algebra $\mathcal{B}(\mathcal{X}) \otimes \mathscr{F}$. This implies in particular that

$$\mathcal{S}\mu = \hat{\Phi}^* \hat{\mu} \approx \hat{\Phi}^* \hat{\mu}_0 = \hat{\Phi}^* (\Pi_{\mathcal{X}}^* \hat{\mu}_0 \otimes \hat{\mathbf{P}}_0) \approx \hat{\Phi}^* (\Pi_{\mathcal{X}}^* \mu \otimes \hat{\mathbf{P}}),$$

which concludes the proof. $\qquad\square$

Remark 3.14 *Our definition of off-white noise is slightly more restrictive than the one in [25]. Translated to our present setting, Tsirelson defined $\mathscr{F}_n$ as the σ-field generated by all the maps of the form $w \mapsto \Phi(x, \Pi_i w)$ for $x \in \mathcal{X}$ and $i > n$ and a noise was called 'off-white' if there exists $n \geq 0$ and $\hat{\mathbf{P}}_0 \approx \hat{\mathbf{P}}$ such that $\mathscr{P}$ and $\mathscr{F}_n$ are independent under $\hat{\mathbf{P}}_0$.*

With this definition, one could expect to be able to obtain a statement similar to Theorem 3.13 with the projection on $\mathcal{X}$ of μ (and ν) replaced by the projection on the first $n + 1$ copies of $\mathcal{X}$ of the solution $\mathcal{S}\mu$. Such a statement is wrong, as can be seen again by the example from Section 3.2.2. It is, however, true if one defines $\mathscr{F}_n$ as the (larger) σ-algebra generated by all maps of the form $w \mapsto (\hat{\Phi}(x, w))_i$ for $x \in \mathcal{X}$ and $i > n$.

A consequence of this theorem is the following equivalent of Proposition 3.5:

Proposition 3.15 *Let $(\mathcal{W}, \mathbf{P}, \mathcal{P}, \mathcal{X}, \Phi)$ be an off-white noise system which is strong Feller in the sense of Definition 3.9. If there exists $x \in \mathcal{X}$ such that $x \in \operatorname{supp} \Pi_{\mathcal{X}}^* \mu$ for every stationary measure μ of $\mathcal{Q}$, then there can be at most one such measure, up to the equivalence relation (3.3).*

Proof. Assume by contradiction that there exist two distinct invariant measures μ and ν. For simplicity, denote $\mu_{\mathcal{X}} = \Pi_{\mathcal{X}}^* \mu$ and similarly for ν and let x be an element from the intersection of their supports (such an x exists by assumption). We can assume furthermore without any loss of generality that $\mathcal{S}\mu \perp \mathcal{S}\nu$.

Define, with the same notations as in the proof of Theorem 3.13,

$$\hat{\mathcal{S}}(x; \cdot) = \hat{\Phi}^*\big(\delta_x \otimes \hat{\mathbf{P}}\big) = \int_{\mathcal{W}} \mathcal{S}(x, w; \cdot)\,\mathbf{P}(dw)$$

and note that one has, as before, $\mathcal{S}\mu \approx \int \hat{\mathcal{S}}(x; \cdot)\,\mu_{\mathcal{X}}(dx)$ and similarly for ν. Since $\mathcal{S}\mu \perp \mathcal{S}\nu$, this shows that there exists a measurable set A such that $\mathcal{S}(x, w; A) = 0$ for $\mu_{\mathcal{X}} \otimes \mathbf{P}$-almost every (x, w) and $\mathcal{S}(x, w; A) = 1$ for $\nu_{\mathcal{X}} \otimes \mathbf{P}$-almost every (x, w). Define a function $\delta \colon \mathcal{W} \to \mathbf{R}_+$ by

$$\delta(w) = \inf\{\delta \,:\, \exists x_0 \text{ with } d(x_0, x) < \delta \text{ and } \mathcal{S}(x_0, w; A) = 0\}\,,$$

were d is any metric generating the topology of $\mathcal{X}$. Since x belongs to the support of $\mu_{\mathcal{X}}$ one must have $\delta(w) = 0$ for $\mathbf{P}$-almost every w. Since we assumed that $y \mapsto \mathcal{S}(y, w; \cdot)$ is continuous, this implies that $\mathcal{S}(x, w; A) = 0$ for $\mathbf{P}$-almost every w. Reversing the roles of μ and ν, one arrives at the fact that one also has $\mathcal{S}(x, w; A) = 1$ for $\mathbf{P}$-almost every w, which is the contradiction we were looking for. $\qquad\square$

Remark 3.16 *It follows from the proof that it is sufficient to assume that the map $x \mapsto \mathcal{S}(x, w)$ is continuous in the total variation norm for $\mathbf{P}$-almost every w.*

3.4.2 Another quasi-Markov property

While the result in the previous section is satisfactory in the sense that it shows a nice correspondence between results for Markov processes and results for off-white noise systems, it covers only a very restrictive class of systems. For example, in the case of continuous time, neither fractional noise (the derivative of fractional Brownian motion) nor the Ornstein–Uhlenbeck process fall into this class. It is therefore natural to look for weaker conditions that still allow us to obtain statements similar to Theorem 3.8. The key idea at this stage is to make use of the topology of $\mathcal{W}$ which has not been used in the previous section. This is also the main conceptual difference between the approach outlined in this paper and the approach used by the theory of random dynamical systems.

In the previous section, we made use of the fact that, for off-white noise systems, one has $\mathcal{S}(x, w; \cdot) \approx \mathcal{S}(x, w'; \cdot)$ for every x and every pair (w, w') in a set of full $\mathbf{P}$-measure. We now consider the set

$$\Lambda = \{(w, w') \in \mathcal{W}^2 : \mathcal{S}(x, w; \cdot) \approx \mathcal{S}(x, w'; \cdot)\}, \qquad (3.8)$$

and we require that the dynamic on $\mathcal{W}$ is such that one can construct couplings that hit Λ with positive probability. If we think of the driving noise to be some Gaussian process, the set Λ typically consists of pairs (w, w') such that the difference $w - w'$ is sufficiently 'smooth'.

Recall that a *coupling* between two probability measures μ and ν is a measure π on the product space such that its projections on the two factors are equal to μ and ν respectively (the typical example is $\pi = \mu \otimes \nu$ but there exist in general many different couplings for the same pair of measures). Given two positive measures μ and ν, we say that π is a *subcoupling* for μ and ν if the projections on the two factors are *smaller* than μ and ν respectively. With this definition at hand, we say that:

Definition 3.17 *A skew-product* $(\mathcal{W}, \mathbf{P}, \mathcal{P}, \mathcal{X}, \Phi)$ *is said to be quasi-Markovian if, for any two open sets* $U, V \subset \mathcal{W}$ *such that* $\min\{\mathbf{P}(U), \mathbf{P}(V)\} > 0$, *there exists a measurable map* $w \mapsto \mathcal{P}^{U,V}(w, \cdot) \in \mathcal{M}_+(\mathcal{W}^2)$ *such that:*

(i) For $\mathbf{P}$-*almost every* w, *the measure* $\mathcal{P}^{U,V}(w, \cdot)$ *is a subcoupling for* $\mathcal{P}(w, \cdot)|_U$ *and* $\mathcal{P}(w, \cdot)|_V$.

(ii) Given Λ *as in (3.8), one has* $\mathcal{P}^{U,V}(w, \Lambda) = \mathcal{P}^{U,V}(w, \mathcal{W}^2)$ *for* $\mathbf{P}$-*almost every* w.

Remark 3.18 *If we are in the setting of Section 3.2.3, this is equivalent to considering for* U *and* V *open sets in* $\mathcal{W}_0$ *and replacing every occurrence of* $\mathcal{P}$ *by* $\bar{\mathcal{P}}$. *The set* Λ *should then be replaced by the set*

$$\bar{\Lambda} = \{(w_0, w_0') \in \mathcal{W}_0^2 : \mathcal{S}(x, w \sqcup w_0; \cdot) \approx \mathcal{S}(x, w \sqcup w_0'; \cdot) \text{ for } \mathbf{P}\text{-a.e. } w\}$$

Remark 3.19 *In general, the transition probabilities* $\mathcal{P}$ *only need to be defined up to a* $\mathbf{P}$-*negligible set. In this case, the set* Λ *is defined up to a set which is negligible with respect to any coupling of* $\mathbf{P}$ *with itself. In particular, this shows that the 'quasi-Markov' property from Definition 3.17 does not depend on the particular choice of* $\mathcal{P}$.

With these definitions, we have the following result:

Theorem 3.20 *Let* $(\mathcal{W}, \mathbf{P}, \mathcal{P}, \mathcal{X}, \Phi)$ *be a quasi-Markovian skew-product which is strong Feller in the sense of Definition 3.9 and topologically* *irreducible in the sense of Definition 3.10. Then, it can have at most one invariant measure, up to the equivalence relation (3.3).*

Proof. Under slightly more restrictive assumptions, this is the content of [11, Theorem 3.10]. It is a tedious but rather straightforward task to go through the proof and to check that the arguments still hold under the weaker assumptions stated here.   □

3.4.3 Discussion

The insight that we would like to convey with the way of exposing the previous two subsections is the following. If one wishes to obtain a statement of the form 'strong Feller + irreducible + quasi-Markov ⇒ uniqueness of the stationary measure', one should balance the regularity of ℓ, defined in (3.5), as a function of w, with the class of sets U and V used in Definition 3.17. This in turn is closely related to the size of the set Λ from (3.8). The larger Λ is, the larger the admissible class of sets in Definition 3.17, and the lower the regularity requirements on ℓ.

The off-white noise case corresponds to the situation where $\Lambda = \mathcal{W}^2$. This in turn shows that one could take for U and V any two measurable sets and $\mathcal{P}^{U,V}(w, \cdot) = \mathcal{P}(w, \cdot)|_U \otimes \mathcal{P}(w, \cdot)|_V$. Accordingly, there is no regularity requirement (in w) on ℓ, except for it being measurable.

In the case of Section 3.2.3, the transition probabilities $\mathcal{P}$ have a special structure in the sense that $\mathcal{P}(w, A) = 1$ for $A = w \sqcup \mathcal{W}_0$. This implies that one can take for U and V any measurable set that is such that, if we decompose $\mathcal{W}$ according to $\mathcal{W} \approx \mathcal{W} \times \mathcal{W}_0$, the 'slices' of U and V in $\mathcal{W}_0$ are **P**-almost surely open sets. The corresponding regularity requirement on ℓ is that the map $(x, y, w') \mapsto \ell(w \sqcup w', x, y)$ is jointly continuous for **P**-almost every w.

Finally, if we do not assume any special structure on Λ or $\mathcal{P}$, we take for U and V arbitrary open sets in $\mathcal{W}$. In this case, the corresponding regularity requirement on ℓ is that it is jointly continuous in all of its arguments.

3.5 The Gaussian case

In this section, we study the important particular case of Gaussian noise. We place ourselves in the framework of Section 3.2.3 and we choose $\mathcal{W}_0 = \mathbf{R}$, so that $\mathcal{W} = \mathbf{R}^{\mathbf{Z}^-}$. We furthermore assume that the measure **P** is centred, stationary, and Gaussian with covariance C and spectral measure μ. In other words, we define μ as the (unique) finite Borel measure on $[-\pi, \pi]$ such that

$$C_n = \int_{\mathcal{W}} w_k w_{k+n} \, \mathbf{P}(\mathrm{d}w) = \frac{1}{2\pi} \int_{-\pi}^{\pi} \mathrm{e}^{inx} \mu(\mathrm{d}x), \qquad (3.9)$$

holds for every $n \geq 0$. A well-known result by Maruyama (see [16] or the textbook [7], Section 3.9) states that $\mathbf{P}$ is ergodic for the shift map if and only if the measure μ has no atoms. As in Section 3.2.3, denote by $\bar{\mathcal{P}}: \mathcal{W} \to \mathcal{M}_1(\mathbf{R})$ the corresponding regular conditional probabilities.

Since regular conditional probabilities of Gaussian measures are again Gaussian [2], one has

Lemma 3.21 *There exists $\sigma \geq 0$ and a $\mathbf{P}$-measurable linear functional* $\mathrm{m}: \mathcal{W} \to \mathbf{R}$ *such that, $\mathbf{P}$-almost surely, the measure $\bar{\mathcal{P}}(w, \cdot)$ is Gaussian with mean $\mathrm{m}(w)$ and variance σ^2.*

This, however, does not rule out the case where $\sigma = 0$. The answer to the question of when $\sigma \neq 0$ is given by the following classical result in linear prediction theory [22], [12]:

Theorem 3.22 *Decompose μ as $\mu(\mathrm{d}x) = f(x) \, \mathrm{d}x + \mu_s(\mathrm{d}x)$ with μ_s singular with respect to Lebesgue's measure. Then, one has*

$$\sigma^2 = \exp\left(\frac{1}{2\pi} \int \log f(x) \, \mathrm{d}x\right), \qquad (3.10)$$

if the expression on the right-hand side makes sense, and $\sigma = 0$ otherwise.

If $\sigma = 0$, all the randomness is contained in the remote past of the noise and no new randomness comes in as time evolves. We will therefore always assume that μ is non-atomic and that $\sigma^2 > 0$. Since in that case all elements of $\mathcal{W}$ with only finitely many non-zero entries belong to the reproducing kernel of $\mathbf{P}$ (see Section 3.7 below for the definition of the reproducing kernel of a Gaussian measure and for the notations that follow), the linear functional m can be chosen such that, for every $n > 0$, $\mathrm{m}(w)$ is jointly continuous in $(w_{-n}, \ldots, w_0)$ for $\mathbf{P}$-almost every $(\ldots, w_{-n-2}, w_{-n-1})$; see [2], Section 2.10.

We will denote by $\hat{\mathbf{P}}$ the Gaussian measure on $\hat{\mathcal{W}} = \mathbf{R}^{\mathbf{Z}}$ with correlations given by the C_n. We denote its covariance operator again by C. The measure $\hat{\mathbf{P}}$ is really the same as the measure $\hat{\mathbf{P}}$ defined in Section 3.4.1 if we make the necessary identification of $\hat{\mathcal{W}}$ with a subset of $\mathcal{W}^{\mathbf{N}}$, this is why we use the same notation without risking confusion.

We also introduce the equivalents to the two σ-algebras $\mathscr{P}$ and $\mathscr{F}$. We interpret them as σ-algebras on $\mathbf{R}^{\mathbf{Z}}$, so that $\mathscr{P}$ is the σ-algebra generated by the Π_n with $n \le 0$ and $\mathscr{F}$ is generated by the Π_n with $n > 0$. (Actually, $\mathscr{F}$ could be slightly smaller than that in general, but we do not want to restrict ourselves to one particular skew-product, and so we simply take for $\mathscr{F}$ the smallest choice which contains all 'futures' for all possible choices of Φ as in Section 3.2.3.)

It is natural to split $\hat{\mathcal{W}}$ as $\hat{\mathcal{W}} = \mathcal{W}_- \oplus \mathcal{W}_+$ where $\mathcal{W}_- \approx \mathcal{W}$ is the span of the images of the Π_n with $n \le 0$ and similarly for $\mathcal{W}_+$. We denote by $\mathcal{H}$ the reproducing kernel Hilbert space of $\hat{\mathbf{P}}$. Recall that via the map $\hat{\mathcal{W}}^* \ni \Pi_n \mapsto e^{inx}$ and the inclusion $\hat{\mathcal{W}}^* \subset \mathcal{H}$, one has the isomorphism $\mathcal{H} \approx L^2(\mu)$ with μ as in (3.9); see for example [7]. Following the construction of Section 3.7, we see that $\hat{\mathcal{H}}_-^p$ is given by the closure in $\mathcal{H}$ of the span of e^{inx} for $n \le 0$ and similarly for $\hat{\mathcal{H}}_-^p$. Denote by $P_\pm$ the orthogonal projection from $\hat{\mathcal{H}}_-^p$ to $\hat{\mathcal{H}}_+^p$ and by P the corresponding operator from $\mathcal{H}_-^p$ to $\mathcal{H}_+^p$.

With all these preliminaries in place, an immediate consequence of Proposition 3.42 is:

Proposition 3.23 *Let* $\hat{P}: \mathcal{W}_- \to \mathcal{W}_+$ *be the* **P**-*measurable extension of* P. *Then the set* Λ *is equal (up to a negligible set in the sense of Remark 3.19) to* $\{(w, w') : \hat{P}(w - w') \in \mathcal{H}_+^c\}$. *In particular, it always contains the set* $\{(w, w') : w - w' \in \hat{\mathcal{H}} \cap \mathcal{W}_-\}$.

Proof. The first statement follows from the fact that, by (3.17), $\mathcal{H}_+^c$ is the reproducing kernel space of the conditional probability of $\hat{P}$, given the past $\mathscr{P}$, and $P(w - w')$ yields the shift between the conditional probability given w and the conditional probability given w'. The second statement follows from the fact that P extends to a bounded operator from $\mathcal{H}_-^c$ to $\mathcal{H}_+^c$. □

3.5.1 The quasi-Markov property

We assume as above that $\mathcal{W}_0 = \mathbf{R}$ and that $\mathbf{P}$ is a stationary Gaussian measure with spectral measure μ. We also write as before $\mu(\mathrm{d}x) = f(x)\,\mathrm{d}x + \mu_s(\mathrm{d}x)$. The main result of this section is that the quasi-Markov property introduced in Section 3.4 can easily be read off from the behaviour of the spectral measure μ:

Theorem 3.24 *A generic random dynamical system as above is quasi-Markovian if and only if f is almost everywhere positive and $\int_{-\pi}^{\pi} \frac{1}{f(x)} \, dx$ is finite.*

Proof. Let e_n be the 'unit vectors' defined by $\Pi_m e_n = \delta_{mn}$. Then the condition of $\int_{-\pi}^{\pi} \frac{1}{f(x)} \, dx$ being finite is equivalent to e_n belonging to the reproducing kernel of $\hat{\mathbf{P}}$, a classical result dating back to Kolmogorov (see also [8], p. 83).

To show that the condition is sufficient, denote by $D_w(x)$ the (Gaussian) density of $\bar{\mathcal{P}}(w, \cdot)$ with respect to Lebesgue measure on $\mathbf{R}$. Given any two open sets U and V in $\mathbf{R}$, we can find some x, y, and $r > 0$ such that $\mathcal{B}(x, r) \subset U$ and $\mathcal{B}(y, r) \subset V$. Take then for $\bar{\mathcal{P}}^{U,V}(w, \cdot)$ the push-forward under the map $z \mapsto (z, z+y-x)$ of the measure with density $z \mapsto \min\{D_w(z), D_w(z+y-x)\}$ with respect to Lebesgue measure. Since, by Proposition 3.23, Λ contains all pairs (w, w') which differ by an element of $\hat{\mathcal{H}} \cap \mathcal{W}_-$ and since the condition of the theorem is precisely what is required for e_0 to belong to $\hat{\mathcal{H}}$, this shows the sufficiency of the condition.

To show that the condition is necessary as well, suppose that it does not hold and take for example $U = (-\infty, 0)$ and $V = [0, \infty)$. Since we have the standing assumption that $\sigma^2 > 0$ with σ^2 as in (3.10), one has $\bar{\mathcal{P}}(w, U) > 0$ for $\mathbf{P}$-almost every w and similarly for V. Assume by contradiction that the system is quasi-Markovian, so we can construct a measure on $\mathbf{R}^{\mathbf{Z}} \times \mathbf{R}^{\mathbf{Z}}$ in the following way. Define $\mathcal{W}_+$ as before and define $\mathcal{W}_-$ as the span of Π_n for $n < 0$ so that $\mathcal{W} = \mathcal{W}_- \oplus \mathcal{W}_0 \oplus \mathcal{W}_+$. Let $P_{\pm} : \mathcal{W}_- \oplus \mathcal{W}_0 \to \mathcal{M}_1(\mathcal{W}_+)$ be the conditional probability of $\hat{\mathbf{P}}$ given $\mathcal{W}_- \oplus \mathcal{W}_0$. Let $\bar{\mathcal{P}}^{U,V} : \mathcal{W}_- \to \mathcal{M}_1(\mathcal{W}_0)$ be as in Definition 3.17 and construct a measure M on $\mathcal{W}_-^2 \times \mathcal{W}_+^2$ by

$$M(A_1 \times A_2 \times B_1 \times B_2) = \int_{A_1 \cap A_2} \int_{\bar{\Lambda}} P_{\pm}(w_- \sqcup w_0, B_1) P_{\pm}(w_- \sqcup w_0', B_2)$$
$$\times \bar{\mathcal{P}}^{U,V}(w_-, dw_0 dw_0') \, \mathbf{P}(dw_-).$$

This measure has the following properties:

1. By the properties of $\bar{\mathcal{P}}^{U,V}$ and by the definition of $P_{\pm}$, it is a subcoupling for the projection of $\hat{\mathbf{P}}$ on $\mathcal{W}_- \times \mathcal{W}_+$ with itself, and it is not the trivial measure.

2. Denote by M_1 and M_2 the projections on the two copies of $\mathcal{W}_1 \times \mathcal{W}_+$. Since $P_{\pm}(w_- \sqcup w_0, \cdot) \approx P_{\pm}(w_- \sqcup w_0', \cdot)$ for $\mathbf{P}$-almost every w and for every pair $(w_0, w_0') \in \bar{\Lambda}$, one has $M_1 \approx M_2$.

On the other hand, since e_0 does not belong to the reproducing kernel of $\hat{\mathbf{P}}$ by assumption, there exists a $\hat{\mathbf{P}}$-measurable linear map $\mathfrak{m}\colon \mathcal{W}_- \times \mathcal{W}_+ \to \mathcal{W}_0$ such that the identity $w_0 = \mathfrak{m}(w_-, w_+)$ holds for $\hat{\mathbf{P}}$-almost every triple (w_-, w_0, w_+); see Proposition 3.42. Denote by A the preimage of U under $\mathfrak{m}$ in $\mathcal{W}_- \times \mathcal{W}_+$ and by A^c its complement. Then one has $M_1(A^c) = M_2(A) = 0$, which contradicts property 2 above. $\qquad\square$

Note that, although the condition of this theorem is easy to read off from the spectral measure, it is in general not so straightforward to read off from the behaviour of the correlation function C. In particular, it does *not* translate into a decay condition of the coefficients C_n. Take for example the case

$$
C_n = \begin{cases} 2 & \text{if } n = 0, \\ 1 & \text{if } n = 1, \\ 0 & \text{otherwise.} \end{cases}
$$

This can be realized for example by taking for ξ_n a sequence of i.i.d. normal Gaussian random variables and setting $W_n = \xi_n + \xi_{n+1}$. We can check that one has, for every $N > 0$, the identity

$$
W_0 = \frac{1}{N} \sum_{n=1}^{N} (-1)^n \left(\xi_{n+1} + \xi_{-n}\right) - \frac{1}{N} \sum_{n=1}^{N} (-1)^n (N+1-n)\left(W_n + W_{-n}\right).
$$

Since the first term converges to 0 almost surely by the law of large numbers, it follows that one has the almost sure identity

$$
W_0 = -\lim_{N \to \infty} \frac{1}{N} \sum_{n=1}^{N} (-1)^n (N+1-n)\left(W_n + W_{-n}\right),
$$

which shows that W_0 can be determined from the knowledge of the W_n for $n \neq 0$. In terms of the spectral measure, this can be seen from the fact that $f(x) = 1 + \cos(x)$, so that $1/f$ has a non-integrable singularity at $x = \pi$. This also demonstrates that there are cases in which the reproducing kernel of $\mathbf{P}$ contains all elements with finitely non-zero entries, even though the reproducing kernel of $\hat{\mathbf{P}}$ contains no such elements.

3.5.2 The strong Feller property

It turns out that in the case of discrete stationary Gaussian noise, the quasi-Markov and the strong Feller properties are very closely related.

In this section, we assume that we are again in the framework of Section 3.2.3, but we take $\mathcal{W}_0 = \mathbf{R}^d$ and we assume that the driving noise consists of d independent stationary Gaussian sequences with spectral measures satisfying the condition of Theorem 3.24.

We are going to derive a criterion for the strong Feller property for the Markovian case where the driving noise consists of d independent sequences of i.i.d. Gaussian random variables and we will see that this criterion still works in the quasi-Markovian case.

It will be convenient for the purpose of this section to introduce the Fréchet space $L^\Gamma(\mathbf{R}^d)$ consisting of measurable functions $f\colon \mathbf{R}^d \to \mathbf{R}$ such that the norms $\|f\|_{\gamma,p}^p = \int f^p(w) e^{-\gamma|w|^2}\, dw$ are finite for all $\gamma > 0$ and all $p \geq 1$. For example, since these norms are increasing in p and and decreasing in γ, $L^\Gamma(\mathbf{R}^d)$ can be endowed with the distance

$$d(f,g) = \sum_{p=1}^{\infty} \sum_{n=1}^{\infty} 2^{-n-p}\left(1 \wedge \|f-g\|_{\frac{1}{n},p}\right).$$

With this notation, we will say that a function $g\colon \mathbf{R}^n \times \mathbf{R}^d \to \mathbf{R}^m$ belongs to $\mathcal{C}^{0,\Gamma}(\mathbf{R}^n \times \mathbf{R}^d)$ if, for every $i \in \{1,\dots,m\}$, the map $x \mapsto g_i(x,\cdot)$ is continuous from $\mathbf{R}^n$ to $L^\Gamma(\mathbf{R}^d)$.

Given a function $\Phi\colon \mathbf{R}^n \times \mathbf{R}^d \to \mathbf{R}^n$ with elements of $\mathbf{R}^n$ denoted by x and elements of $\mathbf{R}^d$ denoted by w, we also define the 'Malliavin covariance matrix' of Φ by

$$M_{ij}^{\Phi}(x,w) = \sum_{k=1}^{d} \partial_{w^k} \Phi_i(x,w)\, \partial_{w^k} \Phi_j(x,w).$$

With this notation, we have the following criterion:

Proposition 3.25 *Let $\Phi \in \mathcal{C}^2(\mathbf{R}^n \times \mathbf{R}^d; \mathbf{R}^n)$ be such that the derivatives $D_w\Phi$, $D_w D_x\Phi$, and $D_w^2\Phi$ all belong to $\mathcal{C}^{0,\Gamma}(\mathbf{R}^n \times \mathbf{R}^d)$. Assume furthermore that M_{ij}^{Φ} is invertible for Lebesgue-almost every (x,w) and that $(\det M_{ij}^{\Phi})^{-1}$ belongs to $\mathcal{C}^{0,\Gamma}(\mathbf{R}^n \times \mathbf{R}^d)$. Then the Markov semigroup over $\mathbf{R}^d$ defined by*

$$(\mathcal{P}f)(x) = \int_{\mathbf{R}^d} f(\Phi(x,w))\, \Gamma(dw),$$

where Γ is an arbitrary non-degenerate Gaussian measure on $\mathbf{R}^d$, has the strong Feller property.

Proof. Take a function $f \in \mathcal{C}_0^\infty(\mathbf{R}^n)$ and write (in this proof we use

Einstein's convention of summation over repeated indices):

$$(\partial_i \mathcal{P} f)(x) = \int_{\mathbf{R}^d} \partial_j f(\Phi(x, w)) \partial_{x_i} \Phi_j(x, w) \, \Gamma(dw) . \qquad (3.11)$$

At this point, we note that, since we assumed M^Φ to be invertible, one has for every pair (i, j) the identity

$$\partial_{x_i} \Phi_j(x, w) = \partial_{w^m} \Phi_j(x, w) \Xi_{mi}(x, w), \qquad (3.12)$$

where

$$\Xi_{mi} = \partial_{w^m} \Phi_k(x, w) \big(M^\Phi(x, w) \big)^{-1}_{k\ell} \partial_{x_i} \Phi_\ell(x, w).$$

This allows us to integrate (3.11) by parts, yielding

$$(\partial_i \mathcal{P} f)(x) = - \int_{\mathbf{R}^d} f(\Phi(x, w)) \big(\partial_{w^m} - (Qw)^m \big) \Xi_{mi}(x, w) \, \Gamma(dw),$$

where Q is the inverse of the covariance matrix of Γ. Our assumptions then ensure the existence of a continuous function $K \colon \mathbf{R}^n \to \mathbf{R}$ such that $|(\partial_i \mathcal{P} f)(x)| \le K(x) \sup_y |f(y)|$ which, by a standard approximation argument ([5], Chapter 7), is sufficient for the strong Feller property to hold.   □

Remark 3.26 *We could easily have replaced $\mathbf{R}^n$ by an n-dimensional Riemannian manifold with the obvious changes in the definitions of the various objects involved.*

Remark 3.27 *Just as in the case of the Hörmander condition for the hypoellipticity of a second-order differential operator, the conditions given here are not far from being necessary. Indeed, if $M^\Phi(x, \cdot)$ fails to be invertible on some open set in $\mathbf{R}^d$, then the image of this open set under $\Phi(x, \cdot)$ will be a set of dimension $n' < n$. In other words, the process starting from x will stay in some subset of lower dimension n' with positive probability, so that the transition probabilities will not have a density with respect to the Lebesgue measure.*

Remark 3.28 *Actually, this condition gives quite a bit more than the strong Feller property, since it gives local Lipschitz continuity of the transition probabilities in the total variation distance with local Lipschitz constant $K(x)$.*

We now show that if we construct a skew-product from Φ and take as driving noise d independent copies of a stationary Gaussian process with a covariance structure satisfying the assumption of Theorem 3.24, then the assumptions of Proposition 3.25 are sufficient to guarantee that it also satisfies the strong Feller property in the sense of Definition 3.9. We have indeed that:

Theorem 3.29 *Let* $\mathcal{W}_0 = \mathbf{R}^d$, $\mathcal{W} = \mathcal{W}_0^{\mathbf{Z}^-}$, *let* $\Phi\colon \mathbf{R}^n \times \mathcal{W}_0 \to \mathbf{R}^n$ *satisfy the assumptions of Proposition 3.25, and let* $\mathbf{P} \in \mathcal{M}_1(\mathcal{W})$ *be a Gaussian measure such that there exist measures* $\mu_1, \dots, \mu_d$ *with*

$$\int_{\mathcal{W}} w_n^j w_{n+k}^m \, \mathbf{P}(\mathrm{d}w) = \frac{\delta_{jm}}{2\pi} \int_{-\pi}^{\pi} \mathrm{e}^{\mathrm{i}kx} \mu_j(\mathrm{d}x) \; .$$

Then, if the absolutely continuous part of each of the μ_j *satisfies the condition of Theorem 3.24, the skew-product* $(\mathcal{W}, \mathbf{P}, \mathcal{P}, \mathbf{R}^n, \Phi)$ *has the strong Feller property.*

Proof. Let $\mathfrak{m}$ be as in Lemma 3.21, let $x \in \mathbf{R}^n$, and let $w \in \mathcal{W}$ such that $\mathfrak{m}(w) < \infty$. We want to show that there exists a continuous function $\bar{K}$ depending continuously on x and on $\mathfrak{m}(w)$ such that $\mathcal{S}(x, w; \cdot)$ is locally Lipschitz continuous in the total variation distance (as a function of x) with local Lipschitz constant $\bar{K}(x, \mathfrak{m}(w))$. Since we assumed from the beginning that $\sigma^2 > 0$, where σ is defined as in (3.10), we know that the set of elements in $\mathcal{W}$ with only finitely many non-zero coordinates belongs to the reproducing kernel space of $\mathbf{P}$. Since, by (3.16), the map $\mathfrak{m}$ is bounded from the reproducing kernel space of $\mathbf{P}$ into $\mathbf{R}$, $\mathfrak{m}(w)$ depends continuously on each of the coordinates of w and so the assumptions of Definition 3.9 are verified.

It remains to construct $\bar{K}$. This will be done in a way that is almost identical to the proof of Proposition 3.25. Take a bounded smooth test function $f\colon (\mathbf{R}^n)^{\mathbf{N}} \to \mathbf{R}$ which depends only on its first N coordinates and consider the function $(\mathcal{S}f)(x, w)$ defined by

$$(\mathcal{S}f)(x, w) = \int_{(\mathbf{R}^n)^{\mathbf{N}}} f(y) \mathcal{S}(x, w; \mathrm{d}y) \; .$$

Consider now the splitting $\hat{\mathcal{W}} = \mathcal{W}_- \oplus \mathcal{W}_+$, as well as the measurable linear map $\hat{P}$ and the space $\mathcal{H}_+^c$ introduced for the statement of Proposition 3.23 (note that $\hat{P}$ relates to $\mathfrak{m}$ via $(\hat{P}w)_0 = \mathfrak{m}(w)$). Denote furthermore by $\mathbf{P}_+$ the Gaussian measure on $\mathcal{W}_+$ with reproducing kernel space $\mathcal{H}_+^c$. With these notations at hand, we have the expression

$$(\mathcal{S}f)(x, w) = \int_{\mathcal{W}_+} f(\hat{\Phi}(x, \tilde{w} + \hat{P}w)) \mathbf{P}_+(\mathrm{d}\tilde{w}),$$

where we denoted by $\hat{\Phi}\colon \mathcal{X} \times \mathcal{W}_+ \to \mathcal{X}^{\mathbf{N}}$ the map defined in (3.6). We see that, as in (3.12), one has the identity (again, summation over repeated indices is implied):

$$\partial_{x_i}\hat{\Phi}(x,\tilde{w}) = \partial_{\tilde{w}_0^m}\hat{\Phi}(x,\tilde{w})\,\Xi_{mi}(x,\tilde{w}_0)\,,$$

where the function Ξ is exactly the same as in (3.12). At this point, since the 'coordinate vectors' e_0^m belong to the reproducing kernel of $\hat{\mathbf{P}}$, and therefore also of $\mathbf{P}_+$ by (3.15), we can integrate by parts against the Gaussian measure $\mathbf{P}_+$ ([2], Theorem 5.1.8) to obtain

$$\partial_i \mathcal{S}f(x,w) = -\int_{\mathcal{W}_+} f(\hat{\Phi}(x,\tilde{w}+\hat{P}w))\partial_{\tilde{w}_0^m}\Xi_{mi}(x,\tilde{w}_0+\mathfrak{m}(w))\,\mathbf{P}_+(d\tilde{w})$$

$$+\int_{\mathcal{W}_+} f(\hat{\Phi}(x,\tilde{w}+\hat{P}w))\Xi_{mi}(x,\tilde{w}_0+\mathfrak{m}(w))e_0^m(\tilde{w})\,\mathbf{P}_+(d\tilde{w}).$$

Here, we have abused the notation and interpreted e_0^m as a measurable linear functional on $\mathcal{W}_+$, via the identification (3.14). Since we have $\int |e_0^m(\tilde{w})|^2\mathbf{P}_+(d\tilde{w}) < \infty$ by assumption and since the law of $\tilde{w}_0$ under $\mathbf{P}_+$ is centred Gaussian with variance σ^2, this concludes the proof. $\square$

3.5.3 The off-white noise case

The question of which stationary Gaussian sequences correspond to off-white noise systems was solved by Ibragimov and Solev in the seventies; see [13] and also [25]. It turns out that the correct criterion is:

Theorem 3.30 *The random dynamical system is off-white if and only if the spectral measure μ has a density f with respect to Lebesgue measure and $f(\lambda) = \exp\phi(\lambda)$ for some function ϕ belonging to the fractional Sobolev space $H^{1/2}$.*

Remark 3.31 *By a well-known result by Trudinger [23], later extended in [21], any function $\phi \in H^{1/2}$ satisfies $\int \exp(\phi^2(x))\,dx < \infty$. In particular, this shows that the condition of the previous theorem is therefore much stronger than the condition of Theorem 3.24 which is required for the quasi-Markov property.*

As an example, the (Gaussian) stationary autoregressive process, which has covariance structure $C_n = \alpha^n$, does have the quasi-Markov property since its spectral measure has a density of the form

$$\mu(dx) = \frac{1-\alpha^2}{1+\alpha^2-2\alpha\cos(x)}\,dx,$$

which is smooth and bounded away from the origin. However, if we take a sequence ξ_n of i.i.d. normal random variables and define a process X_n by

$$X_n = \sum_{k=1}^{\infty} k^{-\beta} \xi_{n-k},$$

for some $\beta > 1/2$, then X_n does still have the quasi-Markov property, but it is not an off-white noise.

3.6 Appendix A: Equivalence of the strong and ultra Feller properties

In this section, we show that, even though the ultra Feller property seems at first sight to be stronger than the strong Feller property, the composition of two Markov transition kernels satisfying the strong Feller property always satisfies the ultra Feller property. This fact had already been pointed out in [6] but had been overlooked by a large part of the probability community until Seidler 'rediscovered' it in 2001 [20]. We take the opportunity to give an elementary proof of this fact. Its structure is based on the notes by Seidler, but we take advantage of the simplifying fact that we only work with Polish spaces.

We introduce the following definition:

Definition 3.32 *A Markov transition kernel P over a Polish space $\mathcal{X}$ satisfies the ultra Feller property if the transition probabilities $P(x, \cdot)$ are continuous in the total variation norm.*

Recall first the following well-known fact of real analysis (see for example [28], Example IV.9.3):

Proposition 3.33 *For any measure space $(\Omega, \mathcal{F}, \lambda)$ such that $\mathcal{F}$ is countably generated and any $p \in [1, \infty)$, one has $L^p(\Omega, \lambda)' = L^q(\Omega, \lambda)$ with $q^{-1} + p^{-1} = 1$. In particular, this is true with $p = 1$.*

As a consequence, one has

Corollary 3.34 *Assume that $\mathcal{F}$ is countably generated and let g_n be a bounded sequence in $L^{\infty}(\Omega, \lambda)$. Then there exist a subsequence g_{n_k} and some $g \in L^{\infty}(\Omega, \lambda)$ such that $\int g_{n_k}(x) f(x) \lambda(dx) \to \int g(x) f(x) \lambda(dx)$ for every $f \in L^1(\Omega, \lambda)$.*

Proof. Since $\mathcal{F}$ is countably generated, $L^1(\Omega, \lambda)$ is separable and therefore contains a countable dense subset $\{f_m\}$. Since the g_n are uniformly bounded, a diagonal argument allows us to exhibit an element $g \in L^1(\Omega, \lambda)'$ and a subsequence n_k such that $\int g_{n_k}(x) f_m(x) \lambda(\mathrm{d}x) \to \langle f_m, g \rangle$ for every m. The claim follows from the density of the set $\{f_m\}$ and the previous proposition. $\qquad\square$

Note also that one has

Lemma 3.35 *Let P be a strong Feller Markov kernel on a Polish space $\mathcal{X}$. Then there exists a probability measure λ on $\mathcal{X}$ such that $P(x, \cdot)$ is absolutely continuous with respect to λ for every $x \in \mathcal{X}$.*

Proof. Let $\{x_n\}$ be a countable dense subset of $\mathcal{X}$ and define a probability measure λ by $\lambda(A) = \sum_{n=1}^{\infty} 2^{-n} P(x_n, A)$. Let $x \in \mathcal{X}$ be arbitrary and assume by contradiction that $P(x, \cdot)$ is not absolutely continuous with respect to λ. This implies that there exists a set A with $\lambda(A) = 0$ but $P(x, A) \neq 0$. Set $f = \chi_A$ and consider Pf. One one hand, $(Pf)(x) = P(x, A) > 0$. On the other hand, $(Pf)(x_n) = 0$ for every n. Since P is strong Feller, Pf must be continuous, thus leading to a contradiction. $\qquad\square$

Finally, set $B = \{g \in \mathcal{B}_b(\mathcal{X}) \,|\, \sup_x |g(x)| \leq 1\}$ the unit ball in the space of bounded measurable functions, and note that one has the following alternative formulation of the ultra Feller property:

Lemma 3.36 *A Markov kernel P on a Polish space $\mathcal{X}$ is ultra Feller if and only if the set of functions $\{Pg \,|\, g \in B\}$ is equicontinuous.*

Proof. This is an immediate consequence of the fact that one has the characterisation $\|P(x, \cdot) - P(y, \cdot)\|_{\mathrm{TV}} = \sup_{g \in B} |Pg(x) - Pg(y)|$; see for example [27], Example 1.17. $\qquad\square$

We have now all the ingredients necessary for the proof of the result announced earlier.

Theorem 3.37 *Let $\mathcal{X}$ be a Polish space and let P and Q be two strong Feller Markov kernels on $\mathcal{X}$. Then the Markov kernel PQ is ultra Feller.*

Proof. Applying Lemma 3.35 to Q, we see that there exists a reference measure λ such that $Q(y, dz) = k(x, z)\, \lambda(\mathrm{d}z)$.

Suppose by contradiction that $R = PQ$ is not ultra Feller. Therefore, by Lemma 3.36 there exists an element $x \in \mathcal{X}$, a sequence $g_n \in B$, a sequence x_n converging to x, and a value $\delta > 0$ such that

$$Rg_n(x_n) - Rg_n(x) > \delta \,, \tag{3.13}$$

for every n. Interpreting the g_n s as elements of $L^\infty(\mathcal{X}, \lambda)$, it follows from Corollary 3.34 that, extracting a subsequence if necessary, we can assume that there exists an element $g \in L^\infty(\mathcal{X}, \lambda)$ such that

$$\lim_{n \to \infty} Qg_n(y) = \lim_{n \to \infty} \int k(y, z) g_n(z) \, \lambda(\mathrm{d}z) = \int k(y, z) g(z) \, \lambda(\mathrm{d}z) = Qg(y)$$

for every $y \in \mathcal{X}$. (This is because $k(y, \cdot) \in L^1(\mathcal{X}, \lambda)$.) Let us define the shorthands $f_n = Qg_n$, $f = Qg$, and $h_n = \sup_{m \geq n} |f_m - f|$.

Since $f_n \to f$ pointwise, it follows from Lebesgue's dominated convergence theorem that $Pf_n(x) \to Pf(x)$. The same argument shows that $Ph_n(y) \to 0$ for every $y \in \mathcal{X}$. Since, furthermore, the h_n are positive decreasing functions, one has

$$\lim_{n \to \infty} Ph_n(x_n) \leq \lim_{n \to \infty} Ph_m(x_n) = Ph_m(x),$$

which is valid for every m, thus showing that $\lim_{n \to \infty} Ph_n(x_n) = 0$. This implies that

$$\lim_{n \to \infty} Pf_n(x_n) - Pf(x)$$
$$\leq \lim_{n \to \infty} |Pf_n(x_n) - Pf(x_n)| + \lim_{n \to \infty} |Pf(x_n) - Pf(x)|$$
$$\leq \lim_{n \to \infty} Ph_n(x_n) + 0 = 0 \,,$$

thus creating the required contradiction with (3.13).   □

Example 3.38 *Let us conclude this section with an example of a strong Feller Markov kernels which is not ultra Feller. Take $\mathcal{X} = [0, 1]$ and define P by*

$$P(x, \mathrm{d}y) = \begin{cases} \mathrm{d}y & \text{if } x = 0, \\ c(x)(1 + \sin(y/x)) \, \mathrm{d}y & \text{otherwise.} \end{cases}$$

Here, the function c is chosen in such a way that $P(x, \cdot)$ is a probability measure. It is obvious that, for any $f \in \mathcal{B}_b(\mathcal{X})$, Pf is continuous (even C^∞) outside of $x = 0$. It follows furthermore from the Riemann–Lebesgue lemma that Pf is continuous at $x = 0$. However, the map $x \mapsto P(x, \cdot)$ is discontinuous at 0 in the total variation topology (one has $\lim_{x\to 0} \|P(x,\cdot) - P(0,\cdot)\|_{\mathrm{TV}} = \frac{2}{\pi}$), which shows that P is not ultra Feller.

Remark 3.39 *Since, as seen in the previous example, there are strong Feller Markov kernels that are not ultra Feller, Theorem 3.37 fails in general if one of the two kernels is only Feller (take the identity).*

3.7 Appendix B: Some Gaussian measure theory

This section is devoted to a short summary of the theory of Gaussian measures and in particular on their conditioning. Denote by X some separable Fréchet space and assume that we are given a splitting $X = X_1 \oplus X_2$. This means that the X_i are subspaces of X and every element of X can be written uniquely as $x = x_1 + x_2$ with $x_i \in X_i$, and the projection maps $\Pi_i \colon x \mapsto x_i$ are continuous.

Assume that we are given a Gaussian probability measure $\mathbf{P}$ on X, with covariance operator Q. That is $Q \colon X^* \to X$ is a continuous bilinear map such that $\langle Qf, g \rangle = \int f(x)g(x)\, \mathbf{P}(\mathrm{d}x)$ for every f and g in X^*. (Such a map exists because $\mathbf{P}$ is automatically a Radon measure in our case.) Here, we used the notation $\langle f, x \rangle$ for the pairing between X^* and X. We denote by $\mathcal{H}$ the reproducing kernel Hilbert space of $\mathbf{P}$. The space $\mathcal{H}$ can be constructed as the closure of the image of the canonical map $\iota \colon X^* \to L^2(X, \mathbf{P})$ given by $(\iota h)(w) = h(w)$, so that $\mathcal{H}$ is the space of $\mathbf{P}$-measurable linear functionals on X. If we assume that the support of $\mathbf{P}$ is all of X (replace X by the support of $\mathbf{P}$ otherwise), then this map is an injection, so that we can identify X^* with a subspace of $\mathcal{H}$. Any given $h \in \mathcal{H}$ can then be identified with the (unique) element h_* in X such that $\langle Qg, h \rangle = g(h_*)$ for every $g \in X^*$. With this notation, the scalar product on $\mathcal{H}$ is given by $\langle \iota h, w \rangle = h(w)$, or equivalently by $\langle \iota h, \iota g \rangle = \langle h, Qg \rangle = \langle Qh, g \rangle$. We will from now on use these identifications, so that one has

$$X^* \subset \mathcal{H} \subset X, \tag{3.14}$$

and, with respect to the norm on $\mathcal{H}$, the map Q is an isometry between X^* and its image. For elements x in the image of Q (which can be identified with a dense subset of $\mathcal{H}$), one has $\|x\|^2 = \langle x, Q^{-1}x \rangle$.

Given projections $\Pi_i \colon X \to X_i$ as above, the reproducing kernel Hilbert spaces $\mathcal{H}_i^p$ of the projected measures $\mathbf{P} \circ \Pi_i^{-1}$ are given by $\mathcal{H}_i^p = \Pi_i \mathcal{H} \subset X_i$, and their covariance operators Q_i are given by

$Q_i = \Pi_i Q \Pi_i^* \colon X_i^* \to X_i$. The norm on $\mathcal{H}_i^p$ is given by

$$\|x\|_{i,p}^2 = \inf\{\|y\|^2 \ : \ x = \Pi_i y \ , \ y \in \mathcal{H}\} = \langle x, Q_i^{-1} x \rangle \,,$$

where the last equality is valid for x belonging to the image of Q_i. It is noteworthy that, even though the spaces $\mathcal{H}_i^p$ are not subspaces of $\mathcal{H}$ in general, there is a natural isomorphism between $\mathcal{H}_i^p$ and some closed subspace of $\mathcal{H}$ in the following way. For x in the image of Q_i, define $U_i x = Q \Pi_i^* Q_i^{-1} x \in X$. One has

Lemma 3.40 *For every x in the image of Q_i, one has $U_i x \in \mathcal{H}$. Furthermore, the map U_i extends to an isometry between $\mathcal{H}_i^p$ and $U_i \mathcal{H}_i^p \subset \mathcal{H}$.*

Proof. Since $U_i x$ belongs to the image of Q by construction, one has $\|U_i x\|^2 = \langle U_i x, Q^{-1} U_i x \rangle = \langle Q \Pi_i^* Q_i^{-1} x, \Pi_i^* Q_i^{-1} x \rangle = \langle x, Q_i^{-1} x \rangle = \|x\|_{i,p}^2$. The claim follows because the image of Q_i is dense in $\mathcal{H}_i^p$. $\qquad \square$

We denote by $\hat{\mathcal{H}}_i^p$ the images of $\mathcal{H}_i^p$ under U_i. Via the identification (3.14), it follows that $\hat{\mathcal{H}}_i^p$ is actually nothing but the closure in $\mathcal{H}$ of the image of X_i^* under Π_i^*. Denoting by $\hat{\Pi}_i \colon \mathcal{H} \to \mathcal{H}$ the orthogonal projection (in $\mathcal{H}$) onto $\hat{\mathcal{H}}_i^p$, it is a straightforward calculation to see that one has the identity $\hat{\Pi}_i x = U_i \Pi_i x$. On the other hand, it follows from the definition of Q_i that $\Pi_i U_i x = x$, so that $\Pi_i \colon \hat{\mathcal{H}}_i^p \to \mathcal{H}_i^p$ is the inverse of the isomorphism U_i.

We can also define subspaces $\mathcal{H}_i^c$ of $\mathcal{H}$ by

$$\mathcal{H}_i^c = \overline{\mathcal{H} \cap X_i} = \overline{\mathcal{H} \cap \mathcal{H}_i^p} \,, \tag{3.15}$$

where we used the identification (3.14) and the embedding $X_i \subset X$. The closures are taken with respect to the topology of $\mathcal{H}$. The spaces $\mathcal{H}_i^c$ are again Hilbert spaces (they inherit their structure from $\mathcal{H}$, not from $\mathcal{H}_i^p$!) and they therefore define Gaussian measures $\mathbf{P}_i$ on X_i. Note that for $x \in \mathcal{H}_i^c \cap \mathcal{H}_i^p$, one has $\|x\| \geq \|x\|_{i,p}$, so that the inclusion $\mathcal{H}_i^c \subset \mathcal{H}_i^p$ holds. One has

Lemma 3.41 *One has $\mathcal{H}_1^c = \left(\hat{\mathcal{H}}_2^p \right)^{\perp}$ and vice versa.*

Proof. It is an immediate consequence of the facts that $X = X_1 \oplus X_2$, that $\hat{\mathcal{H}}_1^p$ is the closure of the image of Π_1^* and that, via the identification (3.14) the scalar product in $\mathcal{H}$ is an extension of the duality pairing between X and X^*. $\qquad \square$

We now define a (continuous) operator $P \colon \mathcal{H}_1^p \to \mathcal{H}_2^p$ by $Px = \Pi_2 U_1 x$.

It follows from the previous remarks that P is unitarily equivalent to the orthogonal projection (in $\mathcal{H}$) from $\hat{\mathcal{H}}_1^p$ to $\hat{\mathcal{H}}_2^p$. Furthermore, one has $Px = U_1 x - x$, so that

$$\|Px\|_{\mathcal{H}} \leq \|x\|_{\mathcal{H}_1^p} + \|x\|, \qquad (3.16)$$

which, combined with (3.15), shows that P can be extended to a bounded operator from $\mathcal{H}_1^c$ to $\mathcal{H}_2^c$.

A standard result in Gaussian measure theory states that P can be extended to a $(\mathbf{P} \circ \Pi_1^{-1})$-measurable linear operator $\hat{P} \colon X_1 \to X_2$. With these notations at hand, the main statement of this section is given by:

Proposition 3.42 *The measure* $\mathbf{P}$ *admits the disintegration*

$$\int \phi(x)\,\mathbf{P}(dx) = \int_{X_1} \int_{X_2} \phi(x + \hat{P}x + y)\mathbf{P}_2(dy)\,(\mathbf{P} \circ \Pi_1^{-1})(dx). \quad (3.17)$$

Proof. Denote by ν the measure on the right-hand side. Since ν is the image of the Gaussian measure $\mu = (\mathbf{P} \circ \Pi_1^{-1}) \otimes \mathbf{P}_2$ under the μ-measurable linear operator $A \colon (x, y) \mapsto x + \hat{P}x + y$, it follows from [2, Theorem 3.10] that ν is again a Gaussian measure. The claim then follows if we can show that the reproducing kernel Hilbert space of ν is equal to $\mathcal{H}$. Since the reproducing kernel space $\mathcal{H}(\mu)$ of μ is canonically isomorphic to $\mathcal{H}(\mu) = \mathcal{H}_1^p \oplus \mathcal{H}_2^c \subset \mathcal{H} \oplus \mathcal{H}$, this is equivalent to the fact that the operator $x \mapsto x + Px = x + \Pi_2 U_1 x$ from $\mathcal{H}_1^p$ to $\mathcal{H}$ is an isometry between $\mathcal{H}_1^p$ and $(\mathcal{H}_2^c)^\perp$. On the other hand, we know from Lemma 3.41 that $(\mathcal{H}_2^c)^\perp = \hat{\mathcal{H}}_1^p$ and we know from Lemma 3.40 that U_1 is an isomorphism between $\mathcal{H}_1^p$ and $\hat{\mathcal{H}}_1^p$. Finally, it follows from the definitions that $\Pi_1 U_1 x = Q_1 Q_1^{-1} x = x$ for every $x \in \mathcal{H}_1^p$, so that one has $x + \Pi_2 U_1 x = (\Pi_1 + \Pi_2) U_1 x = U_1 x$, which completes the proof. $\square$

Acknowledgements

The author would like to thank J. Mattingly for countless conversations on the topic of this article, as well as S. Assing and J. Voß for their helpful comments that hopefully lead to greatly improved clarity and the correction of many misprints. The report of a careful referee lead to further clarity. This work was supported by EPSRC fellowship EP/D071593/1.

Bibliography

[1] L. Arnold, *Random Dynamical Systems*, Springer Monographs in Mathematics, Berlin: Springer-Verlag, 1998.

[2] V.I. Bogachev, *Gaussian Measures*, Vol. 62 of Mathematical Surveys and Monographs, American Mathematical Society, 1998.

[3] H. Crauel, 'Invariant measures for random dynamical systems on the circle', *Arch. Math. (Basel)* **78** (2002), no. 2, 145–154.

[4] H. Crauel, *Random Probability Measures on Polish Spaces*, Vol. 11 of Stochastics Monographs, Taylor & Francis, 2002.

[5] G. Da Prato, J. Zabczyk, *Ergodicity for Infinite-dimensional Systems*, Vol. 229 of London Mathematical Society Lecture Note Series, Cambridge University Press, 1996.

[6] C. Dellacherie, P.-A. Meyer, *Probabilités et potentiel. Chapitres IX à XI*, Publications de l'Institut de Mathématiques de l'Université de Strasbourg [Publications of the Mathematical Institute of the University of Strasbourg], XVIII, revised edn, Hermann, Paris. Théorie discrète du potential. [Discrete potential theory], Actualités Scientifiques et Industrielles [Current Scientific and Industrial Topics], 1410. Paris: Hermann, 1983.

[7] H. Dym, H.P. McKean, *Gaussian Processes, Function Theory, and the Inverse Spectral Problem*, New York: Academic Press [Harcourt Brace Jovanovich Publishers]. Probability and Mathematical Statistics, Vol. 31, 1976.

[8] U. Grenander, M. Rosenblatt, *Statistical Analysis of Stationary Time Series*, New York: John Wiley, 1957.

[9] M. Hairer, 'Ergodicity of stochastic differential equations driven by fractional Brownian motion', *Ann. Probab.* **33** (2005), no. 2, 703–758.

[10] M. Hairer, J. C. Mattingly, 'Ergodicity of the 2D Navier-Stokes equations with degenerate stochastic forcing', *Ann. Math. (2)* **164** (2006), no. 3, 993–1032.

[11] M. Hairer, A. Ohashi, Ergodic theory for SDEs with extrinsic memory, *Ann. Probab.* **35** (2007), 1950–1977.

[12] H. Helson, G. Szegö, 'A problem in prediction theory', *Ann. Mat. Pura Appl. (4)* **51** (1960), 107–138.

[13] I.A. Ibragimov, Y.A. Rozanov, *Gaussian random processes*, Vol. 9 of Applications of Mathematics, New York: Springer-Verlag, Translated from the Russian by A. B. Aries, 1978.

[14] Y. Kifer, *Ergodic Theory of Random Transformations*, Vol. 10 of Progress in Probability and Statistics, Bosten, MA: Birkhäuser, 1986.

[15] Y. Le Jan, 'Équilibre statistique pour les produits de difféomorphismes aléatoires indépendants', *Ann. Inst. H. Poincaré Probab. Statist.* **23** (1987), no. 1, 111–120.

[16] G. Maruyama, 'The harmonic analysis of stationary stochastic processes', *Mem. Fac. Sci. Kyūsyū Univ. A.* **4** (1949), 45–106.

[17] S.P. Meyn, R.L. Tweedie, *Markov Chains and Stochastic Stability*, Communications and Control Engineering Series, London: Springer-Verlag, 1993.

[18] R.J. Sacker, G.R. Sell, 'Skew-product flows, finite extensions of minimal transformation groups and almost periodic differential equations', *Bull. Am. Math. Soc.* **79** (1973), 802–805.

[19] R.J. Sacker, G.R. Sell, 'Lifting properties in skew-product flows with applications to differential equations', *Mem. Am. Math. Soc.* **11** (1977), no. 190, iv+67.

[20] J. Seidler, A note on the strong Feller property. Unpublished lecture notes, 2001.

[21] R.S. Strichartz, 'A note on Trudinger's extension of Sobolev's inequalities', *Indiana Univ. Math. J.* **21** (1971/72), 841–842.

[22] G. Szegö, 'Beiträge zur Theorie der Toeplitzschen Formen', *Math. Z.* **6** (1920), no. 3–4, 167–202.

[23] N.S. Trudinger, 'On imbeddings into Orlicz spaces and some applications', *J. Math. Mech.* **17** (1967), 473–483.

[24] B. Tsirelson, From slightly coloured noises to unitless product systems. Preprint, `math.FA/0006165`, 2000.

[25] B. Tsirelson, 'Spectral densities describing off-white noises', *Ann. Inst. H. Poincaré Probab. Statist.* **38** (2002), no. 6, 1059–1069. En l'honneur de J. Bretagnolle, D. Dacunha-Castelle, I. Ibragimov.

[26] M. Valadier, 'Young measures', in *Methods of Nonconvex Analysis* (Varenna, 1989)', Vol. 1446 of Lecture Notes in Math., Springer: Berlin, pp. 152–188, 1990.

[27] C. Villani, *Topics in Optimal Transportation*, Vol. 58 of *Graduate Studies in Mathematics*, Providence, RI: American Mathematical Society, 2003.

[28] K. Yosida, *Functional Analysis*, Classics in Mathematics, Berlin: Springer-Verlag, 1995. Reprint of the sixth (1980) edition.

4

Why study multifractal spectra?

Peter Mörters[†]

Department of Mathematical Sciences
University of Bath
Bath BA2 7AY, England

Abstract

We show by three simple examples how multifractal spectra can enrich our understanding of stochastic processes. The first example concerns the problem of describing the speed of fragmentation in a stick-breaking process, the second concerns the nature of a phase transition in a simple model of statistical mechanics, and the third example discusses the speed of emergence in Kingman's coalescent.

4.1 Introduction

I am often asked why I am interested in Hausdorff dimension. Are there any important problems that can be solved using Hausdorff dimension? Can Hausdorff dimension really add to our understanding of stochastic processes? I believe that the answer is *yes* to both questions, and in this paper I attempt to give some evidence in the case of the second question, by means of three examples. I will focus on the notion of a *multifractal spectrum* or *dimension spectrum*, which in its broadest form refers to the Hausdorff dimension of a parametrized family of sets, seen as a function of the parameter.

The examples are chosen on the one hand for their relative *simplicity*, on the other hand to illustrate the *diversity* of shapes which a multifractal spectrum can take. A common thread in all the examples is the notion of a *tree*, which either features prominently in the initial description or presents a very valuable reformulation of the model. Moreover, all our proofs rely, in some form or other, on one of the most beautiful and powerful ideas of probability theory, the concept of a *martingale*. Nevertheless, I will not give full details of any proofs in this review, but only sketch the basic ideas.

Trends in Stochastic Analysis, ed. J. Blath, P. Mörters and M. Scheutzow.
Published by Cambridge University Press. ©Cambridge University Press 2008.
† Research supported by an EPSRC Advanced Research Fellowship.

In the *first* example of this paper we look at the iterated breaking of a stick of unit length into a (random) finite number of parts. With every point on the stick we associate a fragmentation speed, measuring the rate at which the length of the part containing this point goes to zero. Attempting to plot the fragmentation speed as a function of the point position of the stick we are confronted with an extremely irregular (or fractal) function. A multifractal spectrum turns out to be exactly the right way to organize the information contained in this function in a comprehensible way.

In the *second* example we look at a simple model of a polymer in a random environment. We associate random weights to the vertices of a regular tree and model a polymer chain attracted by large weights and repelled by small weights by a path in the tree. More precisely, for fixed large n, we associate to each path v from the root to the nth generation a probability proportional to $\exp\{\beta H(v)\}$ where $H(v)$ is the sum of the weights along v and $\beta > 0$ is an inverse temperature parameter. This model often has a phase transition, a sudden qualitative change as β increases from zero to infinity, which is noticeable in the limiting behaviour of the normalization factor, or partition function. In the absence of a spatial component in this model, the qualitative difference between the two phases is difficult to grasp. A multifractal spectrum helps to attach a physical meaning to this phase transition.

In the *third* example we look at a famous process arising in the context of mathematical biology. Kingman's coalescent describes the genealogy of a population in terms of a process with values in the set of partitions of $\mathbf{N}$: consider a population consisting at time one of infinitely many individuals, which are represented by the natural numbers. For any $s > 0$ the individuals are then grouped into blocks $B \subset \mathbf{N}$ sharing the same ancestor at time $1 - s$. This model has the interesting feature that at $s = 0$ there is a transition from a partition consisting of infinitely many finite blocks to partitions consisting of finitely many infinite blocks. A multifractal spectrum allows us to better understand how this instant coalescence happens.

The first two examples represent the first steps in ongoing work with current PhD students of mine and I would like to thank them for permission to include this material here and for providing the illustrative pictures: The first example is drawn from joint work with *Adam Kinnison*, the second one from joint work with *Marcel Ortgiese*. Both will publish more substantial accounts of their work when the time is right, and I hope this paper can serve as an advertisement for their work.

The third example was communicated to me by *Julien Berestycki*, and full details are yet to be written. It is an adaptation of results in Berestycki, Berestycki and Schweinsberg [2], which concerns the class of Beta-coalescents. The Kingman coalescent is a limiting case of the Beta-coalescents, which is different in some respects. Its advantage from our point of view is that its treatment can be based on more familiar concepts. Readers interested in the original result and a more sophisticated treatment of Beta-coalescents are recommended to consult [2], and to see also [5] and the contribution of Birkner and Blath in this volume for a survey of related results.

4.2 The speed of fragmentation in stick-breaking

Suppose that N is a nondegenerate random variable with values in the positive integers, and assume that

$$\gamma(\beta) := \log \mathbf{E}\big[N^{1-\beta}\big] < \infty \qquad \text{for every } \beta \in \mathbf{R}.$$

We begin with a stick of length 1, represented by the unit interval. At the first stage we sample N and break this stick into N sticks of length $1/N$. At the nth stage we sample for every stick of the $(n-1)$th stage an independent random variable N and break this stick into N further pieces of equal length. Hence, at any time n, we have a partition of the unit interval into a finite, random number of intervals (or sticks) of random length.

Having done this we can associate with every point $x \in [0,1]$ a decreasing sequence $(\ell_1, \ell_2, \dots)$ where $\ell_n = \ell_n(x)$ is the length of the stick containing x in the nth stage. The fragmentation speed at x is

$$f(x) := -\lim_{n \to \infty} \frac{1}{n} \log \ell_n(x),$$

whenever this limit is defined. It is a natural problem to explore the nature of the random function f for various distributions of N.

We first note that

$$f(x) = \lim_{n \to \infty} \frac{1}{n} \sum_{j=1}^{n} \log N_n(x),$$

where $N_n(x)$ is the number of pieces in which the stick containing x is broken in the nth step. For any fixed x the sequence $N_1(x), N_2(x), \dots$

is i.i.d. and hence, by the law of large numbers, $f(x) = \mathbf{E}[\log N]$ almost surely. By Fubini's theorem we thus get

$$0 = \int_0^1 \mathbf{E}\big|f(x) - \mathbf{E}[\log N]\big|\,\mathrm{d}x = \mathbf{E}\int_0^1 \big|f(x) - \mathbf{E}[\log N]\big|\,\mathrm{d}x,$$

and hence, almost surely,

$$f(x) = \mathbf{E}[\log N] \qquad \text{for Lebesgue-almost all } x.$$

But the analysis does not end with the fact that f is constant almost everywhere. The plot in Figure 4.1 reveals its fractal nature even in the case when N is uniformly distributed. What one might guess from the picture is that f is bounded from above and below, and that values above seem to be a lot more common than values below $\mathbf{E}\log N$.

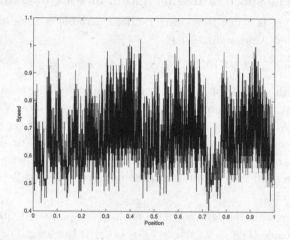

Fig. 4.1. Plot of the speed in the case of the uniform distribution on $\{1, 2, 3\}$ using an approximation based on the fragmentation after 16 steps. The typical value of f is here $\mathbf{E}\log N \approx 0.597$.

The interesting question is therefore: How frequent are the various values of f? Can we measure and compare the size of the sets

$$\mathsf{S}(a) := \Big\{x \in [0,1] \colon -\lim_{n\to\infty} \tfrac{1}{n}\log \ell_n(x) = a\Big\}$$

for all possible values of a? If this can be done, the nontrivial information contained in f would take the form of a function mapping any possible value a to the size of the set $\mathsf{S}(a)$.

The next theorem shows that this is indeed possible when Hausdorff dimension is used as the notion of size. The resulting (deterministic) function is a typical example of a *multifractal spectrum*; see Figure 4.2 for an example plot.

Theorem 4.1 (Kinnison) *For every $a \geq 0$, almost surely*

$$\dim \mathsf{S}(a) = \frac{1}{a} \inf_{\beta}\{a\beta + \gamma(\beta)\},$$

whenever the right-hand side is nonnegative.

Remark Our stick-breaking process is a discrete-time example of a *random fragmentation process*. A thorough multifractal analysis of continuous-time homogeneous fragmentation models has been performed by J. Berestycki [1]. Our result is not contained in his, and our proof uses a different setup, but there is still a great similarity of ideas. For further study in the mathematical theory of random fragmentation I recommend the recent book of Bertoin [3].

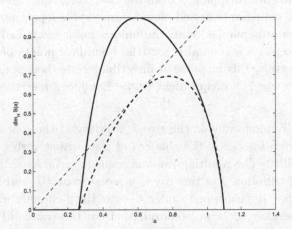

Fig. 4.2. The multifractal spectrum for the stick-breaking process with N uniformly distributed on $\{1, 2, 3\}$ is the bold curve, the broken curves are the diagonal and the function $a \mapsto \inf_{\beta}\{a\beta + \gamma(\beta)\}$ included for comparison.

The idea of the proof of Theorem 4.1

We reinterpret the problem in terms of a tree, which will allow us to introduce the crucial objects in a natural manner. First, represent the closed unit interval $[0,1]$ as the root ρ of the tree, then let $N(\rho)$ be the number of parts into which the unit interval is split, representing the new sticks as the children of the root. Continuing this process, any vertex v in the nth generation of the tree represents a stick arising after n breaking steps. Let $N(v)$ be the number of parts in which this stick is broken, represent the parts as the children of the vertex v, and continue ad infinitum.

Denote the resulting tree by T and, for each vertex $v \in T$, denote by $|v|$ its generation and by $T(v)$ the tree of descendants of v, which is rooted in v. Obviously, T is a Galton–Watson tree with offspring distribution given by the law of N. The *rays* in this tree are sequences of vertices $(v_0, v_1, v_2, \ldots)$ such that $v_0 = \rho$ and v_{i+1} is a child of v_i. The set of rays, called the *boundary* ∂T of the tree, carries a metric structure given by the genealogical distance,

$$d(u,v) = \exp\left\{-\min\{n \in \mathbf{N}: u_n \neq v_n\}\right\} \qquad \text{for } u, v \in \partial T.$$

There is a canonical mapping ϕ from the tree to the unit interval $[0,1]$ such that the sequence of nested closed subintervals represented by the vertices of a ray is mapped onto the unique point contained in every interval. Except on a countable set (the boundary points of the construction intervals) this mapping is invertible. Note, however, that the metric on the tree is not equivalent to the Euclidean metric on the interval.

Now run a random walk on the tree T, starting at the root and moving at each step to each of the children of the current vertex with the same probability. The resulting random sequence $(X_0, X_1, \ldots)$ is a ray. Hence the distribution ν of this ray is a measure on ∂T and it is easy to observe that it is mapped under ϕ onto the Lebesgue measure λ on $[0,1]$. Therefore, ϕ is an isomorphism from the (random) measure space $(\partial T, \mathcal{B}orel, \nu)$ to the (non-random) measure space $([0,1], \mathcal{B}orel, \lambda)$. Moreover, *any* random walk on T, which starts in the root and in each step moves from a vertex to one of its children, generates a measure on ∂T and hence, via the mapping ϕ, also on the interval $[0,1]$.

The key to the proof is the use of a family $\{M_n^{(\beta)}: n \in \mathbf{N}\}$ of nonnegative martingales, defined, for any $\beta \in \mathbf{R}$, by

$$M_n^{(\beta)} = e^{-n\gamma(\beta)} \sum_{|v|=n} \prod_{j=0}^{n-1} N(v_j)^{-\beta}.$$

Let $M^{(\beta)}(T) := \lim M_n^{(\beta)}$. Under our moment conditions this convergence holds in the L^1-sense and the limit is almost surely positive; see for example [17].

We now sketch the proof of the *upper bound* in Theorem 4.1. To describe an efficient covering of $S(a)$, fix some large $m \in \mathbf{N}$ and interpret

$$\mathcal{S} := \left\{ v \in T : |v| = n \geq m, \sum_{j=0}^{n-1} \log N(v_j) \approx an \right\}$$

as a collection of intervals. This collection covers $S(a)$ and its s-value is

$$\sum_{I \in \mathcal{S}} |I|^s \approx \sum_{n=m}^{\infty} \sum_{|v|=n} e^{-ans} \mathbf{1}\left\{ \sum_{j=0}^{n-1} \log N(v_j) \approx an \right\}.$$

Suppose for a moment that $a > \mathbf{E}[\log N]$. Let $\beta < 0$ and, using Chebyshev's inequality, estimate the *expected* s-value of the covering from above by

$$\sum_{n=m}^{\infty} e^{-n(as-a\beta)} \mathbf{E} \sum_{|v|=n} \prod_{j=0}^{n-1} N(v_j)^{-\beta}.$$

By the convergence results for $\{M_n^{(\beta)} : n \in \mathbf{N}\}$ the expectation is of order $e^{n\gamma(\beta)}$. Hence the expected s-value is finite if $s > \frac{1}{a}\left(\gamma(\beta) + a\beta\right)$. Optimizing over $\beta < 0$ gives the required upper bound. For the case $a < \mathbf{E}N$ the analogous argument can be performed choosing $\beta > 0$.

For the sketch of the more delicate *lower bound* in Theorem 4.1 we introduce

$$\mathsf{SpinedTrees} = \left\{ (T, v) : v \in \partial T \right\},$$

the space of trees endowed with a ray acting as a 'spine'. There is a canonical shift $\theta : \mathsf{SpinedTrees} \to \mathsf{SpinedTrees}$ which maps (T, v) to the tree $T(v_1)$ of descendants of the first vertex in the original spine, together with the trace $(v_1, v_2, \ldots)$ of the spine in this tree.

Denote by GW the distribution of our Galton–Watson tree. Given T, we select a spine $(X_0, X_1, \ldots)$ by following a random walk started at the root and, in each step, moving to each of the children w of the current vertex with a probability proportional to $M^{(\beta)}(T(w))$. Let $\mu_T^{(\beta)}$ be the law of this spine in ∂T.

If $\beta \neq 0$, the measure $\mu_T^{(\beta)}$ makes a *size-biased* choice of the trees $T(v_1), T(v_2), T(v_3), \ldots$ along the spine, and hence the measure given by $\mu_T^{(\beta)}(dv)\mathsf{GW}(dT)$ is not shift-invariant on SpinedTrees. However, it is not hard to show (see e.g. [18] for similar arguments) that this size-bias can be compensated entirely by introducing the martingale limit $M^{(\beta)}(T)$ as a density for GW, i.e.

$$\mu_T^{(\beta)}(dv)\ M^{(\beta)}(T)\ \mathsf{GW}(dT)$$

is a shift-invariant and ergodic measure on SpinedTrees. The *ergodic theorem* now allows us to determine the speed of fragmentation as

$$f(\phi(v)) = \lim_{n\to\infty} \frac{1}{n}\sum_{j=0}^{n-1} \log N(X_j) = \iint \log N(\rho)\ M^{(\beta)}(T)\ \mathsf{GW}(dT)$$

$$= \frac{\mathsf{E}[N^{1-\beta}\log N]}{\mathsf{E}[N^{1-\beta}]} \qquad \text{for } \mu_T^{(\beta)}\text{-a.e. } v \text{ and GW-a.e. } T.$$

Every subset of $[0,1]$ which has full measure for $\mu_T^{(\beta)}\circ\phi^{-1}$ has at least the Hausdorff dimension given by a lower bound on the local dimension in each point. The local dimension in $\phi(v)$ equals

$$\dim \mu_T^{(\beta)}\circ\phi^{-1}(\phi(v)) = \lim_{n\to\infty} \frac{\log\left(\prod_{j=1}^{n} \frac{M^{(\beta)}(T(v_j))}{\sum M^{(\beta)}(T(w))}\right)}{-\log\left(\prod_{j=0}^{n-1} N(v_j)\right)},$$

where the sums in the denominators are over all siblings w of the vertex in the argument of the numerator. Denoting

$$a(\beta) = \mathsf{E}[N^{1-\beta}\log N]/\mathsf{E}[N^{1-\beta}]$$

and using the ergodic theorem and a small calculation, this limit equals

$$\frac{-1}{a(\beta)} \iint \log\left(\frac{M^{(\beta)}(T(v_1))}{\sum M^{(\beta)}(T(w))}\right) \mu_T^{(\beta)}(dv)\ M^{(\beta)}(T)\ \mathsf{GW}(dT)$$

$$= \frac{1}{a(\beta)}\left(\log\mathsf{E}[N^{1-\beta}] + \beta\, a(\beta)\right),$$

for $\mu_T^{(\beta)}$-almost every v and GW-almost every T. Hence we get a lower bound for the spectrum as

$$\dim \mathsf{S}(a(\beta)) \geq \tfrac{1}{a(\beta)}\left(\gamma(\beta) + \beta\, a(\beta)\right).$$

As $a(\beta) = -\gamma'(\beta)$, for any $a = a(\beta)$ the parameter β is the minimizer in the variational problem characterizing the spectrum. Hence this lower bound coincides with the upper bound, completing the sketch of the argument.

4.3 A polymer model in a random environment

We look at a very crude model of a polymer in a disordered medium, which was introduced by Derrida and Spohn in [8]. The main interest here is to understand the effect of the disorder on the asymptotic behaviour of the polymer and the occurrence of a phase transition.

To describe the disorder we let V be a random variable such that

$$\phi(t) := \mathbf{E}\left[e^{tV}\right] < \infty \qquad \text{for all } t > 0.$$

Let T be a binary tree with root ρ. We endow T with a disordered medium $\mathcal{V} = (V(v) : v \in T)$ by letting each random weight $V(v)$ be an independent copy of the random variable V. We identify vertices v in the tree with the chain $(v_0, \ldots, v_n)$, starting from the root v_0 and ending at $v_n = v$, such that each v_{i+1} is a child of v_i.

In the 'finite volume' setting, the *polymers* of length n are given by the vertices $v \in T$ in the nth generation. For any inverse temperature $\beta > 0$ the probability of a polymer v of length n is given by

$$P^{(\beta)}(v) = \frac{1}{Z_n(\beta)} \exp\left\{\beta \sum_{j=1}^n V(v_j)\right\},$$

with a normalization factor

$$Z_n(\beta) = \sum_{|v|=n} \exp\left\{\beta \sum_{j=1}^n V(v_j)\right\},$$

which is called the *partition function*.

One expects in this and similar models that the behaviour of the polymer depends on the parameter β in the following manner: If β is small, we are in an *entropy-dominated* regime, where the fluctuations in $\mathcal{V}$ have no big influence and limiting features are largely the same as in the case of a uniformly distributed polymer. For large values of β we may encounter an *energy-dominated* regime where, because of the disorder, the phase space breaks up into pieces separated by free energy barriers. Polymers then follow specific tracks with large probability, an effect often called *localization*.

In such a simple model there are not too many features to distinguish the phases, and a crucial rôle is played by the *free energy*, defined as $\lim(\beta n)^{-1} \log Z_n(\beta)$. We further define, for any $\beta > 0$,

$$h(\beta) = \frac{1}{\beta} \log \mathbf{E} \exp\{\beta V + \log 2\} = \frac{1}{\beta} \log\left(2\phi(\beta)\right).$$

As $\beta \mapsto \beta h(\beta)$ is strictly convex, one can see that h' has at most one positive root. If it exists, we define $\beta_c > 0$ to be this root, and we let $\beta_c = \infty$ otherwise.

Theorem 4.2 (Buffet, Patrick and Pulé [7]) *Almost surely, the free energy is*

$$\lim_{n \to \infty} \frac{1}{\beta n} \log Z_n(\beta) = \begin{cases} h(\beta) & \text{if } \beta \leq \beta_c, \\ h(\beta_c) & \text{if } \beta \geq \beta_c. \end{cases}$$

Remark At the critical temperature $1/\beta_c$ the model has a *phase transition* and, for low temperatures, it is in a frozen state. The two phases are often called the *weak disorder* phase $(\beta < \beta_c)$, and the *strong disorder* phase $(\beta > \beta_c)$.

In the weak disorder phase we have

$$\frac{1}{\beta n} \log Z_n(\beta) \sim \frac{1}{\beta n} \log \sum_{|v|=n} \mathbf{E} \exp\left\{\beta \sum_{j=1}^{n} V(v_j)\right\},$$

and hence, at high temperatures, it may look like *all* polymers $v \in T$ with $|v| = n$ making the same contribution to $Z_n(\beta)$, namely the joint mean

$$\mathbf{E} \exp\left\{\beta \sum_{j=1}^{n} V(v_j)\right\} = \phi(\beta)^n .$$

In fact, this impression is *wrong*, and even at high temperatures only a vanishing proportion of the paths contribute to the free energy. The precise picture is conveyed by a multifractal spectrum.

To describe this spectrum we need a notion of a Hausdorff dimension of a tree. For the purpose of this paper we use the *growth rate*

$$\dim(\tilde{T}) := \lim_{n \to \infty} \frac{1}{n} \log \#\{v \in \tilde{T} : |v| = n\}$$

as a notion of dimension and restrict the discussion to trees $\tilde{T}$ where this notion is well-defined. Let us emphasise that there is the more powerful concept of the *branching number* of a tree, introduced by Lyons, which is the appropriate way to measure the average number of children per vertex in an infinite tree. The logarithm of the branching number coincides with the Hausdorff dimension of the boundary of the tree, which carries a natural metric structure. For sufficiently regular trees this notion of dimension coincides with the growth rate, but in general

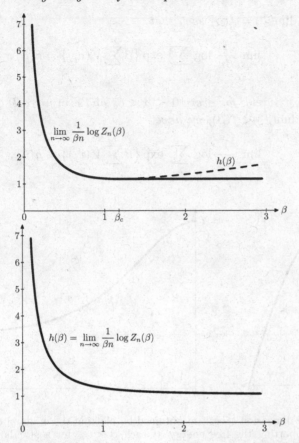

Fig. 4.3. The free energy in the case of a standard normal distribution (top), where $\beta_c < \infty$, and in the case of a binary distribution given by $\mathbb{P}\{V = 1\} = 1 - \mathbb{P}\{V = -1\} = p$ (bottom), where $\beta_c = \infty$.

the growth rate, if it exists, is the larger number. In this paper we may restrict attention to the easier concept.

Theorem 4.3 (Ortgiese) *Define* $f\colon (0, \beta_c) \to [0, \infty)$ *by*

$$f(\beta) = \log 2 + \log \phi(\beta) - \frac{\phi'(\beta)}{\phi(\beta)} \beta\,.$$

(a) For every $0 < \beta < \beta_c$, almost surely, there exists a tree $\tilde{T} \subset T$

with $\dim(\tilde{T}) = f(\beta)$ *such that*

$$\lim_{n \to \infty} \frac{1}{\beta n} \log \sum_{\substack{v \in \tilde{T} \\ |v| = n}} \exp\left\{\beta \sum_{j=1}^{n} V(v_j)\right\} = h(\beta).$$

(b) Almost surely, for every $0 < \beta < \beta_c$ and every tree $\tilde{T} \subset T$ such that $\dim(\tilde{T}) < f(\beta)$, we have

$$\lim_{n \to \infty} \frac{1}{\beta n} \log \sum_{\substack{v \in \tilde{T} \\ |v| = n}} \exp\left\{\beta \sum_{j=1}^{n} V(v_j)\right\} < h(\beta).$$

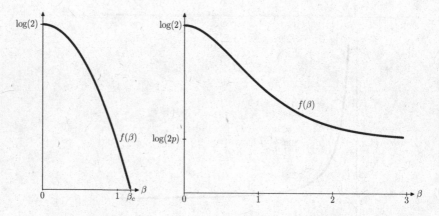

Fig. 4.4. The dimension spectrum obtained by plotting the minimal dimension of a tree supporting the free energy at each $\beta < \beta_c$ for a standard normal distribution (left) and at each $\beta < \infty$ for the binary distribution (right).

Remark If $0 < \beta < \beta_c$ the free energy is supported by a tree of dimension $f(\beta)$. The rays v in this tree have the exceptional behaviour

$$\lim_{n \to \infty} \frac{1}{n} \sum_{j=1}^{n} V(v_j) = a(\beta) := \frac{\phi'(\beta)}{\phi(\beta)} > \mathbf{E}V.$$

If $\beta \uparrow \beta_c < \infty$ the dimension of this tree is going to zero and beyond the critical value there are no more rays with an average weight big enough to sustain a free energy of size $h(\beta)$. Instead, if $\beta > \beta_c$ and n large, a subexponential number of polymers of length n with the maximal weight

$$\frac{1}{n} \sum_{j=1}^{n} V(v_j) \approx a(\beta_c)$$

support $Z_n(\beta)$.

The idea of the proof of Theorem 4.3

The key to the proof of Theorem 4.3 lies in the existence and fractal structure of the *infinite volume Gibbs measure* in the weak disorder regime. This object is of course also of independent interest.

To define the *infinite* volume Gibbs measure as a limit of the *finite* volume Gibbs measures we need to embed, for every $n \in \mathbf{N}$, the nth generation of the binary tree in its boundary ∂T. The easiest way to do this is by extending a vertex $w = (w_0, \ldots, w_n)$ uniquely to a ray $w^+ = (w_0, w_1, \ldots)$ by letting w_{i+1} be the left child of w_i for any $i \geq n$. We can then define the finite volume Gibbs measures on the boundary ∂T as

$$\mu_n^{(\beta)} := \frac{1}{Z_n(\beta)} \sum_{|w|=n} \exp\left\{\beta \sum_{j=1}^n V(w_j)\right\} \delta_{w^+},$$

and, if possible, the infinite volume Gibbs measure as the almost-sure limit in the weak topology of measures

$$\mu^{(\beta)} := \lim_{n \to \infty} \mu_n^{(\beta)}.$$

Martingales play the key role in the proof of existence of this measure in the weak disorder regime. Indeed, it is easy to verify that

$$M_n^{(\beta)}(v) = 2^{-n} \phi(\beta)^{-n} \sum_{\substack{w \in T(v) \\ |w|-|v|=n}} \exp\left\{\beta \sum_{j=|v|+1}^{|v|+n} V(w_j)\right\}$$

defines a martingale $\{M_n^{(\beta)}(v) : n \in \mathbf{N}\}$ for every $\beta > 0$ and $v \in T$. Criteria for uniform integrability of these martingales can be found, for example, in [17] or [7]. They show that precisely if $\beta < \beta_c$ the martingales $\{M_n^{(\beta)}(v) : n \in \mathbf{N}\}$ converge almost surely to a strictly positive limit, which we denote by $M^{(\beta)}(v)$.

Now focus on the weak disorder regime $\beta < \beta_c$ and note that

$$Z_n(\beta) = 2^n \phi(\beta)^n M_n^{(\beta)}(\rho) \sim 2^n \phi(\beta)^n M^{(\beta)}(\rho),$$

and from this we readily get the weak disorder part of Theorem 4.2. For every vertex $v \in T$ we denote by $B(v) \subset \partial T$ the set of all rays passing through the vertex v. The collection $(B(v) : v \in T)$ is exactly the collection of all balls in ∂T. We obtain, with $m := n - |v| \geq 0$,

$$\mu_n^{(\beta)}\big(B(v)\big) = \frac{1}{Z_n(\beta)} \sum_{\substack{w \in T(v) \\ |w|=n}} \exp\Big\{\beta \sum_{j=1}^{n} V(w_j)\Big\}$$

$$= \frac{1}{Z_n(\beta)} 2^m \phi(\beta)^m \exp\Big\{\beta \sum_{j=1}^{|v|} V(v_j)\Big\} M_m^{(\beta)}(v),$$

and combining the last two displays we get

$$\lim_{n \to \infty} \mu_n^{(\beta)}\big(B(v)\big) = 2^{-|v|} \phi(\beta)^{-|v|} \exp\Big\{\beta \sum_{j=1}^{|v|} V(v_j)\Big\} \frac{M^{(\beta)}(v)}{M^{(\beta)}(\rho)}.$$

This suffices to ensure, for every $0 < \beta < \beta_c$, the almost sure existence of the infinite volume Gibbs measure $\mu^{(\beta)}$, which is characterized by

$$\mu^{(\beta)}\big(B(v)\big) = 2^{-|v|} \phi(\beta)^{-|v|} \exp\Big\{\beta \sum_{j=1}^{|v|} V(v_j)\Big\} \frac{M^{(\beta)}(v)}{M^{(\beta)}(\rho)}.$$

The key to Theorem 4.3 is now that, other than in some otherwise similar models such as the *random energy model* discussed in [6], the measures μ_β are fractal measures in the sense that they are supported by a very thin subset of ∂T.

This can be explored using the method of *spined trees* in a way similar to the previous example: let P be the distribution of the environment $\mathcal{V}$ and, given $\mathcal{V}$, we select the spine according to the infinite volume Gibbs measure $\mu_{\mathcal{V}}^{(\beta)}$ (indicating the dependence on the environment by an additional subindex). Slightly abusing the notation of the previous example, we let SpinedTrees be the space of weights attached to the vertices of a binary tree with a marked spine. Recall the definition of the canonical shift θ on SpinedTrees.

Writing $M_{\mathcal{V}}^{(\beta)}$ instead of $M^{(\beta)}(\rho)$, the measure on SpinedTrees given by

$$\mu_{\mathcal{V}}^{(\beta)}(dv) \, M_{\mathcal{V}}^{(\beta)} \, \mathsf{P}(d\mathcal{V}),$$

is shift-invariant and ergodic. From the ergodic theorem we thus obtain

$$\lim_{n \to \infty} \frac{1}{n} \sum_{j=1}^{n} V(v_j) = \int V(\rho) \, M_{\mathcal{V}}^{(\beta)} \, \mathsf{P}(d\mathcal{V}) = \frac{\phi'(\beta)}{\phi(\beta)} = a(\beta);$$

and, recalling the representation of the infinite volume Gibbs measure,

$$\lim_{n\to\infty} \frac{-1}{n} \log \mu^{(\beta)}\big(B(v_n)\big)$$

$$= \log 2 + \log \phi(\beta) - \int \left(\beta V(\rho) + \log \frac{M^{(\beta)}(v_1)}{M^{(\beta)}(\rho)}\right) \mu_{\mathcal{V}}^{(\beta)}(dv) \, M_{\mathcal{V}}^{(\beta)} \, \mathrm{P}(d\mathcal{V})$$

$$= \log 2 + \log \phi(\beta) - \beta \, a(\beta),$$

for $\mu^{(\beta)}$-almost every $v \in \partial T$ and P-almost every medium $\mathcal{V}$.

For the proof of Theorem 4.3 (a) we use Egorov's theorem to select a *compact* set $A \subset \partial T$ such that $\mu^{(\beta)}(A) > 0$ and the two convergences just proved hold uniformly for all rays in A. We then define the tree

$$\tilde{T} = \bigcup_{v\in A} \bigcup_{j=1}^{\infty} \{v_j\} \subset T.$$

Note that, by compactness of A, we have $\partial \tilde{T} = A$. The second convergence readily ensures that

$$\dim\big(\tilde{T}\big) = \log 2 + \log \phi(\beta) - \beta \, a(\beta) = f(\beta),$$

and, using the first convergence, we get

$$\lim_{n\to\infty} \frac{1}{\beta n} \log \sum_{\substack{v\in\tilde{T}\\|v|=n}} \exp\Big\{\beta \sum_{j=1}^{n} V(v_j)\Big\} = \frac{f(\beta)}{\beta} + a(\beta) = h(\beta)$$

and this completes the sketch of the proof of Theorem 4.3 (a).

The consideration of the infinite volume Gibbs measures $\mu^{(\beta)}$ also establishes the lower bound in the 'crude spectrum'

$$\lim_{n\to\infty} \frac{1}{n} \log \#\Big\{v \in T : |v| = n, \sum_{j=1}^{n} V(v_j) \ge n a(\beta)\Big\} = f(\beta),$$

while the corresponding upper bound follows easily from Cramér's theorem. Theorem 4.3 (b) can now be established by studying for any $\eta < f(\beta)$ the 'worst case scenario' of a tree $\tilde{T} \subset T$ with $\dim(\tilde{T}) \leqslant \eta$, which captures in each generation n the maximal possible value of

$$\sum_{j=1}^{n} \exp\{\beta \sum V(v_j)\}.$$

4.4 The speed of emergence in Kingman's coalescent

Kingman's coalescent is probably the most studied object in mathematical genetics and the key problems in this area are certainly more fundamental than the study of multifractal spectra. However, there are also some mathematical aspects on which a multifractal spectrum can shed some light.

We start the investigation from a simple population model. Suppose first that the population consists of n individuals positioned at $\{1, \ldots, n\}$. Each position i carries an independent exponential clock with rate $(n-1)/2$ and, once this clock rings, the individual at position i produces two offspring, one at position i, the other one at a position $j \in \{1, \ldots, n\} \setminus \{i\}$ chosen uniformly at random. At the same time the individual that used to be at position j dies.

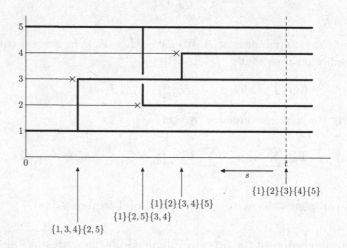

Fig. 4.5. An illustration of Kingman's coalescent restricted to five individuals. The population evolves from left to right, a cross at position j indicates that the particle in the position j dies and is replaced by the offspring of the particle in the position connected to the cross by a vertical line. The coalescent is obtained by looking from right to left.

For any fixed time $t > 0$ this population model gives rise to a natural Markov process of *ancestral partitions* $(\Pi_s^{(n)}: 0 \leq s \leq t)$ called the n-*coalescent*. It takes values in the space of partitions of $\{1, \ldots, n\}$ and we declare $i, j \in \{1, \ldots, n\}$ to be in the same partition set of $\Pi_s^{(n)}$ if the individuals in position i and j at time t have the same ancestor at time $t - s$. Note that this process starts at time $s = 0$ with the trivial partition consisting entirely of singletons.

Kingman [14] has shown that there exists a unique Markov process $(\Pi_s : s \geq 0)$ with values in the space of partitions of $\mathbf{N}$ such that, for all $n \in \mathbf{N}$ and $t > 0$, the process running for t time units obtained by restricting partitions to $\{1, \ldots, n\}$ is an n-coalescent. This process is called *Kingman's coalescent*.

One of the key features of Kingman's coalescent is that it *comes down from infinity*, which means that for every time $s > 0$ the number of partition sets, or *blocks*, in Π_s is almost surely finite. Define the *frequency* of a block $B \subset \mathbf{N}$ as

$$\lim_{n \to \infty} \frac{1}{n} \#(B \cap \{1, \ldots, n\}).$$

At all times $s > 0$ all blocks of Π_s have positive frequency, almost surely. In order to understand the instant transition from a state of dust at time $s = 0$, when all blocks are singletons, to a state at time $s > 0$ when the partition consists of finitely many blocks of positive frequency, we would like to follow the sequences of nested blocks as $s \downarrow 0$, and study the possible rates of decrease of the block frequencies in the form of a multifractal spectrum.

We need to rigorously define a metric space representing the sequences of nested blocks. This construction is due to Evans [9]. We first define a (random) metric on $\mathbf{N}$ by letting

$$\delta(i, j) = \inf\left\{s > 0 : i, j \in B \text{ for some } B \in \Pi_s\right\}.$$

The required metric space $(\mathcal{S}, \delta)$ is the completion of $(\mathbf{N}, \delta)$. Indeed, the set $\mathcal{S}$ is simply the boundary of a rooted, binary tree and all the interesting random structure enters in the metric δ. It is shown in [9] that dim $\mathcal{S} = 1$.

Given an element $x \in \mathcal{S}$ we need to make sense of the speed of coalescence at x. For this purpose we define a probability measure η on $(\mathcal{S}, \mathcal{B}orel)$ by letting

$$\eta(B(x, s)) = \lim_{n \to \infty} \frac{1}{n} \#\{i \in \{1, \ldots, n\} : \delta(i, x) \leq s\}.$$

This uniquely defines a probability measure η on $\mathcal{S}$. We define the *lower* and *upper speed of emergence* of $x \in \mathcal{S}$ as

$$\underline{\mathrm{speed}}(x) = \liminf_{s \downarrow 0} \frac{\log \eta(B(x, s))}{\log s}, \qquad \overline{\mathrm{speed}}(x) = \limsup_{s \downarrow 0} \frac{\log \eta(B(x, s))}{\log s}.$$

From the results of [9] we infer that, almost surely,

$$\underline{\text{speed}}(x) = \overline{\text{speed}}(x) = 1,$$

for η-almost every x. The multifractal spectrum shows the presence of points of exceptional upper speed of emergence.

Theorem 4.4 (Berestycki _et al._) _Almost surely, if $1 \leq a \leq 2$, then_

$$\dim \left\{ x \in \mathcal{S} : \overline{\text{speed}}(x) \geq a \right\} = \frac{2}{a} - 1,$$

and there are no points $x \in \mathcal{S}$ with $\overline{\text{speed}}(x) > 2$ or $\overline{\text{speed}}(x) < 1$.

Remark

(a) This is a very different kind of spectrum compared with those in the previous sections, as variations in the speed only happen at exceptional times $s > 0$ and _lower_ speeds exceeding one can not occur. Spectra of this kind have been associated with the 'breakdown of the multifractal formalism' in various examples; see e.g. [20] or [15].

(b) It is conjectured that the lower speed of emergence equals 1 for all $x \in \mathcal{S}$, which would be in contrast with the case of Beta-coalescents with parameter $1 < \alpha < 2$. In that case exceptionally small lower speeds can occur; see [2, Theorem 5.1].

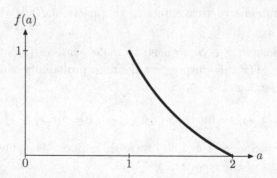

Fig. 4.6. The dimension spectrum for the upper speed of emergence in Kingman's coalescent.

The idea of the proof of Theorem 4.4

The key to the proof is to establish a link between the metric space S equipped with the measure η and a tree of excursions embedded in a Brownian motion equipped with a local time measure. The multifractal structure of the local time can then be analysed, for example using a percolation technique.

For an (unfortunately very superficial) sketch of this link, we start with a Brownian motion $\{B(t) \colon 0 \le t \le \tau\}$ stopped at

$$\tau = \inf\{t > 0 \colon L^0(t) = 1\},$$

where $\{L^s(t) \colon t \ge 0\}$ is the local time process at level s. Write $Z_t = L^t(\tau)$ and recall from the Ray-Knight theorem that $\{Z_t \colon t \ge 0\}$ is a Feller diffusion. Then define an increasing process $\{R(t) \colon 0 \le t < T\}$ by

$$R(t) = \int_0^t \frac{1}{Z_s}\, \mathrm{d}s.$$

Note that this process explodes at time $T = \max_{0 \le t \le \tau} B(t)$. Therefore its inverse $\{R^{-1}(t) \colon t \ge 0\}$ maps the positive halfline onto $[0, T)$. For any $0 < s < 1$ we let $E(s)$ be the set of excursions of the Brownian motion above level $R^{-1}(s)$ that reach level $R^{-1}(1)$. Note that $E(s)$ is a finite set, but the number of elements in $E(s)$ is increasing to infinity as $s \uparrow 1$.

We denote by Ξ the set of functions $(\zeta(s) \colon 0 \le s < 1)$ such that

- $\zeta(s) \in E(s)$ for any $0 \le s < 1$,
- $\zeta(s_2)$ is contained in $\zeta(s_1)$ for any $0 < s_1 < s_2 < 1$.

Ξ is a complete metric space when endowed with the metric

$$d(\zeta_1, \zeta_2) = \inf\left\{1 - s \,:\, \zeta_1(s) = \zeta_2(s)\right\},$$

and there exists a metric isomorphism $\Phi \colon \Xi \to S$, such that

$$\eta\big(B(\Phi(\zeta), s)\big) = Z^{-1}_{R^{-1}(1)}\, \ell\big(\zeta(1 - s)\big) \qquad \text{for all } \zeta \in \Xi \text{ and } 0 < s < 1,$$

where $\ell(e)$ denotes the local time of the excursion $e \in E(1 - s)$ at level $R^{-1}(1)$.

We now rescale the paths $(\zeta(s) \colon 0 < s < 1)$ by letting

$$\zeta'(s) = \zeta\big(R(sR^{-1}(1))\big) \qquad \text{for } 0 < s < 1,$$

so that $\zeta'(s)$ is an excursion above level $sR^{-1}(1)$ which reaches level $R^{-1}(1)$. Note that the mapping $\psi \colon \Xi \to \Xi'$ which maps every path

ζ to its rescaling ζ' is a bijection, but does not preserve the metric d. However, because

$$R^{-1}(1) - R^{-1}(1-t) \sim qt \qquad \text{as } t \downarrow 0,$$

for some (random) constant q, small distances are only linearly affected by ψ and hence $\dim(\psi(A)) = \dim A$ for any $A \subset \Xi$ and

$$\overline{\text{speed}}(\Phi(\zeta)) = \limsup_{s \downarrow 0} \frac{\log \ell(\zeta'(1-s))}{\log s}.$$

Starting from this representation, the idea of the proof is to test for which deletion parameters the set of paths with high upper speed of emergence has a positive probability of surviving a (suitably defined) percolation process. This technique is the continuous time analogue of the percolation technique based on [16] and used for a similar problem in a discrete setup in [19].

To be precise, fix the deletion parameter $0 < \lambda < 1$. With any path $(\zeta' : 0 \leq s \leq 1)$ we associate a Poisson process, or clock, with intensity measure

$$\lambda \frac{ds}{1-s} \qquad \text{on } (0,1),$$

which kills the path at the first strike of the clock. Any paths $\zeta'_1, \zeta'_2 \in \Xi'$,

- share the same clock for $0 < s \leq d(\zeta'_1, \zeta'_2)$, but
- have independent clocks for $d(\zeta'_1, \zeta'_2) < s < 1$.

We say that a set $A \subset \Xi'$ of paths survives percolation, if there exists a path in A which has lifetime 1. Then any analytic set has positive probability of surviving the percolation if its Hausdorff dimension is $> \lambda$, but zero probability if it is $< \lambda$.

To find, for $a \geq 1$, the critical deletion parameter of the set

$$\left\{ \zeta' \in \Xi' : \limsup_{s \downarrow 0} \frac{\log \ell(\zeta'(1-s))}{\log s} \geq a \right\}$$

one proves the following key estimate: for an excursion $e \in E(0)$,

$$\lim_{x \downarrow 0} \frac{1}{\log x} \log \mathbb{P}(\{\ell(e) \leq x\} \cap \{\{\zeta' : \zeta'(0) = e\} \text{ survives}\}) = 1 + \lambda.$$

Roughly speaking, this holds because the typical local time of an excursion above level $1 - x$ which reaches height 1 is of order x. Therefore the optimal strategy to obtain the event $\{\ell(e) < x\}$ is to ensure that only one of the excursions above level $1 - x$, which are embedded in e, reaches height 1. The probability of this event is of order x and, given this, the probability that the set of paths starting with $\zeta'(0) = e$ survives percolation is, up to a constant multiple, equal to the probability that the path $(\zeta'(s): 0 < s < 1 - x)$ survives, which is

$$\exp\left\{ -\lambda \int_0^{1-x} \frac{ds}{1-s} \right\} = x^\lambda.$$

With the key estimate at hand we roughly argue as follows: typically the number of excursions above level $1 - s$ that reach level 1 is of order $1/s$. Out of the paths $(\zeta'(t): 0 \le t \le 1 - s)$ ending in these excursions typically a proportion of order s^λ survives up to this time. Hence, for a fixed small $s > 0$, we can expect of order $s^{\lambda-1}$ conditionally independent trials to realize the event

$$\left\{ \ell(\zeta'(1-s)) \le s^a \right\} \cap \left\{ B(\zeta', s) \text{ survives} \right\}.$$

By scaling the key estimate, this event has probability of order

$$\left(\frac{s^a}{s} \right)^{1+\lambda} = s^{(a-1)(1+\lambda)}.$$

Hence, the expected number of paths $(\zeta'(t): 0 \le t \le 1-s)$ which survive the percolation procedure and satisfy $\ell(\zeta'(1-s)) \le s^a$ is

$$s^{\lambda-1} s^{(a-1)(1+\lambda)},$$

indicating that the threshold for the existence of paths of length 1 occurs when $\lambda - 1 + (a - 1)(1 + \lambda) = 0$ or, equivalently, when $\lambda = 2/a - 1$. This argument readily gives the upper bound in Theorem 4.4, while the lower bound follows with only marginally more effort, exploiting the self-similarity of the Brownian structure by means of Baire's theorem.

4.5 Conclusion

We have reviewed different forms of multifractal spectra using some simple but interesting examples. At least for me, the spectra (and their derivation) have been helpful in forming an intuition for the studied processes. Many of the martingale and tree ideas used in our proofs can be considered as part of the folklore and have been rediscovered many times in different guises. While it is impossible to give a full list of the relevant publications here, [4, 10, 11, 12, 13, 16, 18, 21] represent a good selection of the pioneering papers, from which these ideas have been formed.

Bibliography

[1] Berestycki, J. Multifractal spectra of fragmentation processes. *J. Stat. Phys.* **113** (2002) 100–111.

[2] Berestycki, J.; Berestycki, N.; Schweinsberg, J. Beta-coalescents and continuous stable trees. *Ann. Probab.* **35** (2007) 1835–1887.

[3] Bertoin, J. *Random Fragmentation and Coagulation Processes.* Cambridge University Press (2006).

[4] Biggins, J.D. Martingale convergence in the branching random walk. *J. Appl. Probab.* **14** (1977) 25–37.

[5] Birkner, M.; Blath, J.; Capaldo, M.; Etheridge, A.; Möhle, M.; Schweinsberg, J.; Wakolbinger, A. Alpha-stable branching and beta-coalescents. *Electron. J. Probab.* **10** (2005) 303–325.

[6] Bovier, A. *Statistical Mechanics of Disordered Systems.* Cambridge University Press (2006).

[7] Buffet, E.; Patrick, A.; Pulé, J.V. Directed polymers on trees: a martingale approach. *J. Phys. A* **26** (1993) 1823–1834.

[8] Derrida, B.; Spohn, H. Polymers on disordered trees, spin glasses, and traveling waves. *J. Stat. Phys.* **51** (1988) 817–840.

[9] Evans, S. N. Kingman's coalescent as a random metric space. In: *Stochastic Models* (Ottawa, 1998), CMS Conf. Proc., 26, Amer. Math. Soc., Providence, (2000) 105–114.

[10] Hawkes, J. Trees generated by a simple branching process. *J. London Math. Soc.* **24** (1981) 373–384.

[11] Joffe, A.; Le Cam, L.; Neveu, J. Sur la loi des grands nombres pour des variables aléatoires de Bernoulli attachées à un arbre dyadique. *C. R. Acad. Sci. Paris* **277** (1973) A963–A964.

[12] Kahane, J.-P.; Peyrière, J. Sur certaines martingales de Benoit Mandelbrot. *Adv. Math.* **22** (1976) 131–145.

[13] Kingman, J.F.C. The first birth problem for an age-dependent branching process. *Ann. Probab.* **3** (1975) 790–801.

[14] Kingman, J.F.C. The coalescent. *Stoch. Proc. Appl.* **13** (1982) 235–248.

[15] Klenke, A.; Mörters, P. The multifractal spectrum of Brownian intersection local time. *Ann. Probab.* **33** (2005) 1255–1301.

[16] Lyons, R. Random walks and percolation on trees. *Ann. Probab.* **18** (1990) 931–958.

[17] Lyons, R. A simple path to Biggins' martingale convergence for branching random walks. In: *Classical and Modern Branching Processes*, K. Athreya and P. Jagers (eds.), Springer, New York, (1997) 217–222.

[18] Lyons, R.; Pemantle, R.; Peres, Y. Ergodic theory on Galton-Watson trees: speed of random walk and dimension of harmonic measure. *Ergodic Theory Dynam. Systems* **15** (1995) 593–619.

[19] Mörters, P.; Shieh, N.R. On the multifractal spectrum of the branching measure on a Galton–Watson tree. *J. Appl. Probab.* **41** (2004) 1223–1229.

[20] Perkins, E.; Taylor, S.J. The multifractal structure of super-Brownian motion. *Ann. Inst. Henri Poincaré* **34** (1998) 97–138.

[21] Waymire, E.; Williams, S.C. A general decomposition theory for random cascades. *Bull. Am. Math. Soc.* **31** (1994) 216–222.

II. Construction, simulation, discretization of stochastic processes

5

Construction of surface measures for Brownian motion

Nadia Sidorova

Department of Mathematics
University College London
Gower Street, London WC1E 6BT, UK

Olaf Wittich

Technische Universiteit Eindhoven
Fakulteit Wiskunde en Informatica
Den Dolech 2 5612 AZ Eindhoven, NL

Abstract

Given the intrinsic Wiener measure on the path space of a Riemannian manifold, we describe several ways to construct the induced surface measures on path spaces of Riemannian submanifolds. It turns out that they are related to the intrinsic Wiener measure of the submanifolds by a Radon-Nikodym density of Feynman-Kac type associated to a potential which depends on in- and extrinsic geometric properties of the submanifold. We demonstrate that this potential also describes the effective dynamic of a quantum particle of bounded energy confined to small tubes around the submanifold by an infinite hard-wall potential. Finally, we briefly discuss some possible extensions of these results using conditioning with respect to tubes of variable shapes, other constraining potentials or conditioning to submanifolds with singularities, and some of the complications that are to expect.

Trends in Stochastic Analysis, ed. J. Blath, P. Mörters and M. Scheutzow.
Published by Cambridge University Press. ©Cambridge University Press 2008.

5.1 Introduction

As a general setup for the notion of surface measure, we consider a measured metric space (Ω, d, μ) consisting of a metric space (Ω, d) and a Borel probability measure μ on Ω. A subset $A \subset \Omega$ is called *Minkowski-regular* if, as ε tends to zero, the sequence

$$\mu_\varepsilon := \frac{1}{\mu(A(\varepsilon))}\mu|_{A(\varepsilon)}$$

of probability measures supported by the ε-neighbourhoods

$$A(\varepsilon) := \{\omega \in \Omega \,:\, d(\omega, A) < \varepsilon\}$$

converges weakly to a probability measure μ_0 supported by A. The measure μ_0 is called the *induced Minkowski-* or *surface measure*.

In this paper, we investigate Minkowski-regularity and surface measures in the case where the measured metric space is the space of continuous paths in a Riemannian manifold equipped with Wiener measure, the law of Brownian motion on this manifold. The subsets we are interested in are path spaces of regularly embedded closed submanifolds. If we consider spaces of paths defined up to a finite time horizon T, the measures μ_ε are immediately identified with the laws of Brownian motion on the ambient manifold conditioned to the event that it stays within the ε-neighbourhood of the submanifold up to time T. This suggests looking at the problem from a probabilistic (or measure-theoretic) point of view, trying to construct the limit measure by considering the processes as limits of solutions of stochastic differential equations or by successively pinning the ambient Brownian motion to the submanifold. We will consider these ideas in the second part of the paper.

In the first part of the paper, we consider a different, more analytical approach. It is based on the observation that the law of conditioned Brownian motion is intimately connected to the law of Brownian motion *absorbed* at the boundary of the tube. This is explained in Section 5.3. But the main point with this approach is that it reveals a connection between two a priori different concepts, the construction of surface measures for Brownian motion and the construction of effective Hamiltonians which describe the dynamics of a quantum particle confined to a small tubular neighbourhood.

Consider a free particle in the Euclidean space $\mathbb{R}^n$ whose motion is driven by the Hamiltonian $-\Delta/2$ on $L^2(\mathbb{R}^n, dV_{\mathbb{R}^n})$, where $dV_{\mathbb{R}^n}$ is the Lebesgue measure on $\mathbb{R}^n$. We are interested in the limit behaviour of this particle when it is forced to stay increasingly close to some Riemannian manifold L isometrically embedded into $\mathbb{R}^n$.

The motion of a free particle on L is believed to be driven by the Hamiltonian $-\Delta_L/2$, where Δ_L is the Laplace–Beltrami operator on $L^2(L, \mathrm{d}V_L)$ and $\mathrm{d}V_L$ denotes the induced Lebesgue measure on L. Since the Laplace–Beltrami operator is determined by the metric on the manifold L, the free motion on L is completely intrinsic and does not depend on the embedding of L into $\mathbb{R}^n$. The free motion on L corresponds to an *idealized concept of constraining* when the particle is forced to lie in the manifold exactly.

However, constraining in reality consists of introducing stronger and stronger forces pushing the particle to the manifold. For each $\varepsilon > 0$, one would consider the motion driven by a Hamiltonian $H_\varepsilon = -\Delta/2 + U_\varepsilon$, where U_ε is a family of non-negative potentials such that $U_\varepsilon|_L = 0$ and $U_\varepsilon(x) \to \infty$ for $x \notin L$, and study the limiting dynamics as $\varepsilon \to 0$. In this paper, we consider the hard-wall potential defined as

$$U_\varepsilon(x) = \begin{cases} 0 & x \in L(\varepsilon), \\ \infty & \text{otherwise,} \end{cases}$$

where $L(\varepsilon)$ denotes the tubular neighbourhood of L in $\mathbb{R}^n$ of radius $\varepsilon > 0$. The soft (quadratic) potentials have been studied in [5]. Considering the hard-wall potential is equivalent to imposing *Dirichlet boundary conditions* on the boundary $\partial L(\varepsilon)$ of the tube.

Surprisingly, an effective description of the dynamics generated by the Dirichlet Laplacian on tubes of small radius will be different from the dynamics of the free (that is, *ideally constrained*) particle on L associated to Δ_L. In contrast, and in a sense that has to be made precise, the new dynamics will be determined by a Hamiltonian $-\Delta_L/2 + W$, where $W \in C^\infty(L)$ is a smooth effective potential which depends on the intrinsic geometry of the manifold and of the embedding.

The paper is organised as follows: In Section 5.2 we study the semigroups corresponding to absorbed Brownian motion. To make the essential steps more transparent, we consider the example of a curve in $\mathbb{R}^2$. In this way, lengthy differential geometric calculations are kept short. Since absorbed Brownian motion is submarkovian, which implies that the probability not to be absorbed decays exponentially as time goes on, one has to renormalize the generators to avoid degeneration. Then one can study their strong convergence with the help of epiconvergence of the corresponding quadratic forms. For the Schrödinger group, the limit result turns out to be slightly weaker. Strong convergence can only be achieved after projecting to some particular subspace of the full Hilbert space which corresponds to eigenstates for which the energy remains finite as ε tends to zero.

In Section 5.3 we discuss the relation between conditioned and absorbed Brownian motion. In particular, we show that the conditioned Brownian motion is a Brownian motion with a time-dependent drift, and we compute its transition probabilities in terms of the Dirichlet Laplacian in the tube $L(\varepsilon)$.

In Section 5.4 we state the results for a more general situation, where we isometrically embed L into another Riemannian manifold M instead of $\mathbb{R}^n$. One can observe the same effect here: the limit dynamics of the conditioned Brownian motions is no longer intrinsic and it is described by an effective potential W, which is given in terms of both intrinsic (such as the scalar curvature) and extrinsic (such as mean and sectional curvatures) characteristics of L.

In Section 5.5 we explain the main idea of the probabilistic approach to the surface measures. We also discuss the original approach to Wiener surface measures, which was suggested in [14]. We show that it leads to the same surface measure as the conditioned Brownian motions and can be treated using the same technique.

So far, all subsets considered are Minkowski-regular and the surface measures are equivalent to the intrinsic measures on the subsets. Then, in Section 5.6, we finally discuss two related open problems, one about conditioning to tubes of variable diameter, which is related to considering *soft constraints*, and one about conditioning to singular submanifolds, where we expect a completely different behaviour. The path spaces of these subsets are not even Minkowski-regular in general, if one restricts oneself to continuous paths.

5.2 Example: confinement to a curved planar wire

One possible application of surface measures is the description of quantum particles such as electrons which are confined to move within small spatial structures such as thin layers. As an example, we consider an electron that is confined to a planar wire. The wire is described as the tubular neighbourhood of a real line which is isometrically embedded into $\mathbb{R}^2$. The electron is forced to stay within the wire by introducing Dirichlet boundary conditions for the free particle Hamiltonian on the tube. If the diameter $\varepsilon > 0$ of the tube is small enough, the asymptotic dynamic letting ε tend to zero provides an effective description of the actual behaviour of the electron. Somewhat surprisingly, it turns out that the motion of the electron is influenced by the geometry of the embedding and thus differs considerably from the behaviour of a free electron on the real line.

The asymptotic result presented at the end of this section is valid in much greater generality. However, considering tubular neighbourhoods around submanifolds of Riemannian manifolds results in a huge bookkeeping problem, keeping track of Fermi coordinates, second fundamental form, metric and so forth. We decided to discuss the problem at hand basically for two reasons. On the one hand, the geometric situation is sufficiently involved to show most of the complications that are present in the general situation. On the other hand, the analysis of the problem leading to the necessity of rescaling and renormalization, and the general line of the proof using epiconvergence and weak compactness in the boundary Sobolev space becomes more visible since it is no longer hidden behind lengthy differential geometric calculations. Besides, the underlying physical intuition is also helpful to grasp the structure of the problem.

It is, however, important to remark that we mainly want to investigate the notion of surface measure. Thus we will consider the case of semigroups associated to Brownian motion on small tubular neighbourhoods taking the interpretation of the Dirichlet problem as the confinement of a quantum particle merely as a motivation. But although electrons may certainly not be interpreted as Brownian particles, the surface measure contains valuable information about the ground state of the quantum particle (see e.g. [13], p. 57) and the stationary Gibbs measure induced by it. Thus, in a sense that we will make precise in Theorem 5.1 at the end of this section, the solution of the heat equation with smooth initial condition $f \in C^\infty(\mathbb{R}^2)$ and Dirichlet boundary conditions on a small tube around a curved planar wire is approximately given by

$$u(t) \approx u_0 \left\{ e^{-\frac{t}{2}(\Delta_{\mathbb{R}} - \frac{1}{4}\kappa^2)} f|_L \right\} \circ \pi$$

where u_0 is some given fixed function on the tube (*background*), κ is the *curvature* of the embedded wire, $\Delta_{\mathbb{R}}$ its intrinsic Laplacian, and $f|_L$ the restriction of the initial condition to the wire. For the Schrödinger equation, a similar result holds only when we project onto a certain subspace of the full Hilbert space.

5.2.1 Geometry: an embedded curved wire

Let $\phi : \mathbb{R} \to \mathbb{R}^2$ be a smooth curve parametrized by arc-length. That means we consider an isometric embedding of the line into the plane. We denote the embedded line by $L := \{\phi(s) : s \in \mathbb{R}\}$. Let

$$e_1(s) := \dot{\phi}(s), \quad e_2(s) := \ddot{\phi}(s)/\|\ddot{\phi}(s)\|$$

be the 2-frame for the curve. We assume that the embedding is *regular* in the sense that the curvature

$$\kappa(s) := \langle \ddot{\phi}, e_2 \rangle(s)$$

is uniformly bounded, i.e. there is some $K > 0$ with $\|\ddot{\phi}\| < K$ for all $s \in \mathbb{R}$.

To confine the electron to the vicinity of the curve, we introduce an infinite hard-wall potential where the walls are given by the boundary components of the *tubular ε-neighbourhood*

$$L(\varepsilon) := \{x \in \mathbb{R}^2 : d(x, L) < \varepsilon\},$$

$\varepsilon > 0$, of the embedded line L. That means the confining potential is given by

$$U_C(x) := \begin{cases} 0 & x \in L(\varepsilon), \\ \infty & \text{otherwise.} \end{cases}$$

We now assume that $\varepsilon < K^{-1}$ is smaller than the *radius of curvature*. That implies that the ε-neighbourhood is diffeomorphic to the product of the real line and a small interval, i.e. $L(\varepsilon) = \mathbb{R} \times (-\varepsilon, \varepsilon)$. A diffeomorphism $\Phi : \mathbb{R} \times (-\varepsilon, \varepsilon) \to L(\varepsilon)$ is explicitly given by the so called *Fermi coordinates*

$$\Phi(s, w) := \phi(s) + we_2(s)$$

for the tube. In these local coordinates, the metric is given by

$$g(s, w) := \begin{pmatrix} \langle \Phi_s, \Phi_s \rangle & \langle \Phi_s, \Phi_w \rangle \\ \langle \Phi_w, \Phi_s \rangle & \langle \Phi_w, \Phi_w \rangle \end{pmatrix} = \begin{pmatrix} (1 - w\kappa(s))^2 & 0 \\ 0 & 1 \end{pmatrix}$$

where Φ_s, Φ_w denote the partial derivatives with respect to s, w. The associated Riemannian volume is, in the same local coordinates, given by

$$dV(s, w) = \sqrt{\det g}(s, w) \, ds \, dw = (1 - w\kappa(s)) \, ds \, dw$$

and, using this measure, we can define the space $L^2(L(\varepsilon), g)$ and the *boundary Sobolev space* $H_0^1(L(\varepsilon), g)$ (cf. [16]).

The Laplacian with Dirichlet boundary conditions on the tube is the unique self-adjoint operator on $L^2(L(\varepsilon), g)$ which is associated to the quadratic form

$$q_\varepsilon(f) := \int_{L(\varepsilon)} dV \, g(df, df)$$

with domain $\mathcal{D}(q_\varepsilon) = \mathrm{H}_0^1(L(\varepsilon), g)$. As ε tends to zero, this expression will simply tend to zero. This is already obvious from the fact that $L \subset \mathbb{R}^2$ is a zero set. Therefore, to understand the details of the dynamic behaviour of an electron or a Brownian particle in a small tube around L, we have to apply some normalizing transformation of the actual situation which is reminiscent of looking at the particles on the tube using some special kind of microscope.

5.2.2 Rescaling

From now on, we will always assume for simplicity that the curvature radius K^{-1} is greater than 1 so that $L(1) = \mathbb{R} \times (-1, 1)$ via Φ. Now we want to make precise what we mean by a normalizing transformation. To do so, we first use the fact that $\Phi : \mathbb{R} \times (-1, 1) \to L(1)$ yields a global map of the 1-tube and that we can therefore use these local coordinates to define other Riemannian structures on and maps between the tubes. First of all, we define another metric on $L(1)$ which, in general, is different from the metric induced by the embedding.

Definition 5.1 *The* reference metric *on $L(1)$ is the metric which is given in local Fermi coordinates by*

$$g(s, w) := \begin{pmatrix} 1 & 0 \\ 0 & 1 \end{pmatrix}.$$

The associated Riemannian volume is $dV_0 = ds\, dw$.

The second component that is necessary to construct the desired microscope is the *rescaling map*.

Definition 5.2 *Let $0 < \varepsilon < 1$. The* rescaling map *$\sigma_\varepsilon : L(\varepsilon) \to L(1)$ is the diffeomorphism given in local Fermi coordinates by*

$$\sigma_\varepsilon(s, w) := (s, w/\varepsilon).$$

The induced map $\sigma_\varepsilon^* : C(L(1)) \to C(L(\varepsilon))$ on the respective spaces of continuous functions is given by $\sigma_\varepsilon^*(f) := f \circ \sigma_\varepsilon$.

With the help of the rescaling map, we will now transform the family $Q_{\varepsilon,\varepsilon>0}$ of quadratic forms on different domains into a family of quadratic forms on a fixed domain. Let

$$\rho(s,w) := \frac{dV}{dV_0} = 1 - w\kappa(s)$$

be the *Radon–Nikodym density* of the two Riemannian volume measures. The map σ_ε, $\varepsilon > 0$, can be extended to $L^2(L(\varepsilon), g)$ and

$$\Sigma_{\varepsilon} : L^2(L(1), g) \to L^2(L(\varepsilon), g_0), \tag{5.1}$$

given by $\Sigma_\varepsilon(f) := \{\varepsilon\rho\}^{-1/2}\sigma_\varepsilon^*(f)$, is actually a *unitary* map. This is the microscope under which we want to consider the dynamics on the small tubes. A crucial observation is that since $\rho > 1 - K > 0$ is smooth, Σ_ε also yields a homeomorphism

$$\Sigma_\varepsilon : H_0^1(L(1), g) \to H_0^1(L(\varepsilon), g_0).$$

That means we can use the maps $\Sigma_{\varepsilon,\varepsilon>0}$ to transform the family $q_{\varepsilon,\varepsilon>0}$ to a family of quadratic forms on a fixed Hilbert space.

Definition 5.3 *The* rescaled family *associated to the family* $q_{\varepsilon,\varepsilon>0}$ *of quadratic forms is the family* $Q_{\varepsilon,\varepsilon>0}$ *given by* $Q_\varepsilon := q_\varepsilon \circ \Sigma_\varepsilon$ *with common domain* $\mathcal{D}(Q_\varepsilon) = H_0^1(L(1), g_0)$.

To summarize: The rescaling map works like a microscope that just enlarges the direction perpendicular to the embedded wire. We use it to transform the perturbation problem for the forms q_ε into a corresponding perturbation problem for a family of forms on a fixed Hilbert space.

It will turn out that we still obtain a singular perturbation problem. The reason is that the eigenvalues of the associated operators tend to infinity as ε tends to zero. That means that there is no reasonable limit dynamic. To avoid this kind of degeneration, one has to suitably *renormalize*.

5.2.3 Renormalization

First of all, let us see what the rescaled family looks like. A short calculation yields for the exterior derivative of the transformed function

$$d\Sigma_\varepsilon(f) = (\varepsilon\rho)^{-1/2}\left(d\sigma_\varepsilon^*(f) - \frac{1}{2}\sigma_\varepsilon^*(f)\,d\log\rho\right).$$

That implies

$$
\begin{aligned}
&Q_\varepsilon \circ \Sigma_\varepsilon(f)\\
&= \int_{L(\varepsilon)} dV\, g(d\Sigma_\varepsilon(f), d\Sigma_\varepsilon(f))\\
&= \frac{1}{\varepsilon} \int_{L(\varepsilon)} dV_0 \left\{ g(d\sigma_\varepsilon^*(f), d\sigma_\varepsilon^*(f)) + \frac{1}{4}\sigma_\varepsilon^*(f)^2\, g(d\log\rho, d\log\rho) \right\}\\
&\quad - \frac{1}{\varepsilon} \int_{L(\varepsilon)} dV\, \rho^{-1} g(\sigma_\varepsilon^*(f)\, d\log\rho, d\sigma_\varepsilon^*(f)).
\end{aligned}
$$

The first two terms already yield a quadratic form that one would expect for an elliptic operator with a zero-order term that can be interpreted as a potential. After partial integration, the third term will fit into that picture. To do so, we use

$$
\sigma_\varepsilon^*(f)^2 = \sigma_\varepsilon^*(f^2), \quad \sigma_\varepsilon^*(f)\, d\sigma_\varepsilon^*(f) = \tfrac{1}{2} d\sigma_\varepsilon^*(f^2), \quad \rho^{-1}\, d\log\rho = d\rho^{-1},
$$

and the fact that $\sigma_\varepsilon^*(f) \in \mathrm{H_0}^1(L(\varepsilon), g)$ has generalized boundary value zero. That implies for the third term that the boundary contribution to Green's formula vanishes and we obtain

$$
\begin{aligned}
&\int_{L(\varepsilon)} dV\, \rho^{-1} g(\sigma_\varepsilon^*(f)\, d\log\rho, d\sigma_\varepsilon^*(f))\\
&\qquad = -\frac{1}{2} \int_{L(\varepsilon)} dV\, g(d\rho^{-1}, d\sigma_\varepsilon^*(f^2)) = \frac{1}{2} \int_{L(\varepsilon)} dV\, \Delta\rho^{-1}\, \sigma_\varepsilon^*(f^2)\\
&\qquad = \frac{1}{2} \int_{L(\varepsilon)} dV_0\, (\|d\log\rho\|^2 - \Delta\log\rho)\, \sigma_\varepsilon^*(f^2).
\end{aligned}
$$

In the last step we used $\rho\,\Delta\rho^{-1} = \|d\log\rho\|^2 - \Delta\log\rho$, where Δ denotes the (non-negative) Laplace–Beltrami operator and $\|-\|$ the norm on the cotangent bundle, both associated to the metric g. Applying the transformation formula for integrals to the expression for Q_ε established so far yields as a first major step the following description of the rescaled family.

Proposition 5.4 *Let $W : L(1) \to \mathbb{R}$ be the smooth potential*

$$
W(x) := \frac{1}{2}\Delta\log\rho - \frac{1}{4}\|d\log\rho\|^2 \tag{5.2}
$$

and $W_\varepsilon := W \circ \sigma_\varepsilon$ the rescaled potential. Furthermore, let g_ε denote the rescaled metric on $L(1)$, i.e. $g_\varepsilon(df, dh) := g(d\sigma_\varepsilon^*(f), d\sigma_\varepsilon^*(h))$. Then, the *rescaled family is given by*

$$Q_\varepsilon(f) = \int_{L(1)} dV_0 \left(g_\varepsilon(df, df) + W_\varepsilon f^2 \right).$$

Thus the transformed perturbation problem on $L(1)$ yields a family of quadratic forms that consists of the form associated to the Dirichlet Laplacian on $L(1)$ which is associated to the rescaled metric g_ε together with a potential that also depends on the geometry of the configuration. To see this, it might be helpful to consider the rescaled family in our special situation in local coordinates. First of all, the rescaled metric is given by (from here on $\kappa \equiv \kappa(s)$)

$$
\begin{aligned}
g_\varepsilon(s, w) &= \begin{pmatrix} (1 - \varepsilon w\kappa)^2 & 0 \\ 0 & \varepsilon^{-2} \end{pmatrix} \\
&= \begin{pmatrix} 1 & 0 \\ 0 & \varepsilon^{-2} \end{pmatrix} - \varepsilon \begin{pmatrix} 2w\kappa & 0 \\ 0 & 0 \end{pmatrix} + \varepsilon^2 \begin{pmatrix} w^2\kappa^2 & 0 \\ 0 & 0 \end{pmatrix}.
\end{aligned}
$$

Denoting by $g_{0,\varepsilon}(df, dh) := g_0(d\sigma_\varepsilon^*(f), d\sigma_\varepsilon^*(h))$ the rescaled reference metric, we can write

$$g_\varepsilon(s, w) - g_{0,\varepsilon}(s, w) = \varepsilon w \begin{pmatrix} 2\kappa & 0 \\ 0 & 0 \end{pmatrix} + \varepsilon^2 w^2 \begin{pmatrix} \kappa^2 & 0 \\ 0 & 0 \end{pmatrix}. \tag{5.3}$$

Hence, the rescaled metric degenerates as ε tends to zero. The degeneration is the same as for the rescaled reference metric and the difference between the two families tends to zero.

This observation and the fact that the Laplacian is naturally associated to the metric make us believe that such a perturbation ansatz is also successful for the operators. Thus we will later consider the Laplacian to the rescaled metric as a perturbation of the Laplacian to the reference metric and deduce its asymptotic behaviour from that.

The rescaled potential is thus given by

$$
\begin{aligned}
W_\varepsilon(s, w) &= -\frac{1}{2}\varepsilon(1 - \varepsilon w\kappa)w\ddot{\kappa} + \frac{3}{4}\varepsilon^2 w^2 \dot{\kappa}^2 - \frac{1}{4}\frac{\kappa^2}{(1 - \varepsilon w\kappa)^2} \\
&= -\frac{1}{4}\kappa^2 + R_\varepsilon(s, w),
\end{aligned}
$$

where

$$R_\varepsilon(s, w) = \frac{\varepsilon w}{2}\left[\frac{\kappa^3}{(1 - \varepsilon w\kappa)^2} - \ddot{\kappa} \right] + \frac{\varepsilon^2 w^2}{4}\left[2\kappa\ddot{\kappa} + 3\dot{\kappa}^2 - \frac{\kappa^4}{(1 - \varepsilon w\kappa)^2} \right].$$

For compact submanifolds, i.e. an embedding of the one-sphere $S^1 \subset \mathbb{R}^2$ it would be clear that the remainder is actually $O(\varepsilon)$ with respect to the supremum norm. But for the embedding of the real line, we have to assume that it is *geometrically finite* meaning that the absolute values of first and second derivatives of the curvature are bounded, i.e.

$$|\dot{\kappa}|, |\ddot{\kappa}| < C. \tag{5.4}$$

In that case, $R_\varepsilon(s, w) = O(\varepsilon)$ with respect to the supremum norm on the tube such that the potential W_ε tends to the potential

$$W_0(s, w) = -\frac{1}{4}\kappa^2(s) \tag{5.5}$$

which is constant on the fibres $F_s := \{(s, w) : w \in (-1, 1)\}$ of the tubular neighbourhood.

By equation (5.3), the difference of the rescaled metric and the rescaled reference metric tends to zero as ε tends to zero. On the other hand, the potential also converges uniformly to W_0 as ε tends to zero. It is therefore quite a natural idea to first solve the asymptotic problem for the reference family and then to treat the induced family by perturbation-theoretical methods. Let us thus consider the family of quadratic forms

$$Q_{0,\varepsilon}(f) := \int_{L(1)} dV_0 \, g_{0,\varepsilon}(df, df),$$

$\varepsilon > 0$, with domain $\mathcal{D}(Q_{0,\varepsilon}) = \mathrm{H}_0^1(L(1), g_0)$. In local Fermi coordinates, we may use the fact that the tubular neighbourhood is *trivial* in the sense that it is globally diffeomorphic to the product $\mathbb{R} \times (-1, 1)$, which implies that

$$\mathrm{H}_0^1(L(1), g_0) = \mathrm{H}^1(\mathbb{R}) \otimes \mathrm{H}_0^1(-1, 1),$$

and with respect to that decomposition we may write

$$Q_{0,\varepsilon}(f) = \int_{\mathbb{R}} ds \int_{-1}^1 ds \, |\partial_s \otimes 1(f)|^2 + \frac{1}{\varepsilon^2} \int_{\mathbb{R}} ds \int_{-1}^1 dw \, |1 \otimes \partial_w(f)|^2.$$

From this expression, it is obvious that $Q_{0,\varepsilon}(f)$ will tend to infinity as ε tends to zero as long as

$$\int_{\mathbb{R}} ds \int_{-1}^1 dw \, |1 \otimes \partial_w(f)|^2 > 0.$$

To investigate when this type of degeneration happens and how it can be avoided, we consider the family of quadratic forms

$$Q_{0,\varepsilon}^+(f) := \frac{1}{\varepsilon^2} \int_{\mathbb{R}} ds \int_{-1}^1 dw \, |\partial_w(f)|^2,$$

$\varepsilon > 0$, with natural domain $\mathcal{D}(Q_{0,\varepsilon}^+) = L^2(\mathbb{R}) \otimes \mathrm{H}_0^1(-1,1)$. The main observation is now that this quadratic form is *decomposable* with respect to the *direct integral decomposition* [8] of $\mathcal{D}(Q_{0,\varepsilon}^+)$ given by

$$\int_{s\in\mathbb{R}}^{\oplus} ds\, \mathrm{H}_0^1(F_s) := \left\{ (f_s)_{s\in\mathbb{R}} : f_s \in \mathrm{H}_0^1(F_s), \int_{\mathbb{R}} ds\, \|f_s\|_{\mathrm{H}_0^1(F_s)}^2 < \infty \right\}$$

where, in addition, $s \mapsto f_s$ is supposed to be measurable. If we use

$$\|f\|_{\mathrm{H}_0^1(F_s)}^2 := \int_{-1}^{1} dw\, |\partial_w f|^2(w,s)$$

as Sobolev-norm on the space $\mathrm{H}_0^1(F_s)$, the decomposed quadratic form is hence given by

$$Q_{0,\varepsilon}^+(f) := \frac{1}{\varepsilon^2} \int_{\mathbb{R}} ds\, \|f_s\|_{\mathrm{H}_0^1(F_s)}^2. \tag{5.6}$$

Since the Sobolev-norm is strictly positive definite, the first conclusion that we can draw from this representation of the quadratic form is that for all $f \neq 0$ we have

$$\lim_{\varepsilon\to 0} Q_{0,\varepsilon}^+(f) = \infty,$$

i.e. the family $Q_{0,\varepsilon}$ of quadratic forms degenerates in the most dramatic way. Thus, the perturbation problem obtained by considering the rescaled families will still not yield a sensible answer.

To overcome this difficulty, we have to *renormalize* the rescaled families. To see how this can be done, we need the following two basic observations:

(i) The operator associated to the quadratic form $q_s(f) := \|f_s\|_{\mathrm{H}_0^1(F_s)}^2$ with domain $\mathcal{D}(q_s) = \mathrm{H}_0^1(F_s)$ is the *Dirichlet Laplacian* on F_s, i.e. the Laplace–Beltrami operator $-\Delta_s$ on F_s with respect to the metric induced by g_0 with domain $\mathcal{D}(\Delta_s) = \mathrm{H}_0^1 \cap \mathrm{H}^2(F_s)$,

(ii) In Fermi coordinates, we see that the induced metric on F_s is simply the flat metric such that all fibres F_s are isometric to the interval $(-1,1) \subset \mathbb{R}$ equipped with the flat metric. In particular, the associated Dirichlet Laplacians are all *isospectral*.

These two statements together imply by standard results [8] about decomposable operators

Proposition 5.5 *The operator associated to the quadratic form $Q_{0,\varepsilon}^+$ is the decomposable operator*

$$D := -\int_L^\oplus ds\, \Delta_s$$

with domain $\mathcal{D}(D) := \int_L^\oplus ds\, \mathrm{H}_0^1 \cap \mathrm{H}^2(F_s)$. On $L^2(L(1), g_0)$, the operator is unbounded and if we denote the spectral decomposition of the Dirichlet Laplacian $-\Delta_{(-1,1)}$ on $(-1,1) \subset \mathbb{R}$ by

$$-\Delta_{(-1,1)} f = \sum_{k \geq 0} \lambda_k u_k \langle u_k, f \rangle_{L^2(-1,1)},$$

then the spectral decomposition of D is given by

$$Df(s,w) = \sum_{k \geq 0} \lambda_k \langle u_k, f_s \rangle_{L^2(F_s)} u_k(w) \tag{5.7}$$

where we identify a function $f \in \mathcal{D}(D)$ with its decomposition $(f_s)_{s \in \mathbb{R}}$.

By this result, we can now introduce a *renormalization* to the effect that the family of forms $Q_{0,\varepsilon}^+$ degenerates everywhere except on a nontrivial subspace. From the spectral decomposition of the associated Laplace operators, we obtain

$$
\begin{aligned}
Q_{0,\varepsilon}^+(f) - \frac{\lambda_0}{\varepsilon^2} \langle f, f \rangle_0 &= \frac{1}{\varepsilon^2} \int_\mathbb{R} ds\, \left[\|f_s\|_{\mathrm{H}_0^1(F_s)}^2 - \lambda_0 \int_{-1}^1 dw\, |f_s|^2 \right] \\
&= \frac{1}{\varepsilon^2} \int_\mathbb{R} ds\, \left[\|f_s\|_{\mathrm{H}_0^1(F_s)}^2 - \lambda_0 \|f_s\|_{L^2(F_s)}^2 \right] \\
&= \sum_{k \geq 1} \frac{\lambda_k - \lambda_0}{\varepsilon^2} \int_\mathbb{R} ds\, |\langle u_k, f_s \rangle_{L^2(F_s)}|^2
\end{aligned}
$$

where $\langle -, - \rangle_0$ denotes the scalar product on $L^2(L(1), g_0)$. That implies that the quadratic forms $Q_{0,\varepsilon}^+$, $\varepsilon > 0$ have a non-trivial common *null space*

$$\mathcal{N} := \{ f \in \mathcal{D}(Q_{0,\varepsilon}^+) : E_0 f = f \} \tag{5.8}$$

where the $L^2(L(1), g_0)$-orthogonal projection E_0 onto the null space is given by

$$E_0 f(s,w) := \langle u_0, f_s \rangle_{L^2(F_s)} u_0(w).$$

As ε tends to zero, we have thus

$$\lim_{\varepsilon \to 0} Q_{0,\varepsilon}^+(f) - \frac{\lambda_0}{\varepsilon^2} \langle f, f \rangle_0 = \begin{cases} 0 & f \in \mathcal{N}, \\ \infty & \text{otherwise.} \end{cases}$$

pointwise. This is the reason to expect that — after renormalization — we will obtain some non-trivial limit dynamic on the subspace $\mathcal{N}$. To obtain a proper asymptotic problem we therefore introduce the following two modified families of quadratic forms.

Definition 5.6 *The* renormalized rescaled *families of quadratic forms with parameter $\varepsilon > 0$ are given by*

$$Q_{0,\varepsilon}^R(f) := \int_{L(1)} dV_0 \, g_{0,\varepsilon}(df, df) - \frac{\lambda_0}{\varepsilon^2} \langle f, f \rangle_0$$

for the reference metric and by

$$Q_\varepsilon^R(f) := \int_{L(1)} dV_0 \left(g_\varepsilon(df, df) + W_\varepsilon f^2 \right) - \frac{\lambda_0}{\varepsilon^2} \langle f, f \rangle_0$$

for the induced metric. $\mathrm{H}_0^1(L(1), g_0)$ is the common domain of both families.

Starting from the idea of looking at electrons on small tubes using a special kind of microscope, we thus arrived at a problem that we finally can solve: we have to calculate a renormalized limit dynamic that concentrates on the subspace $\mathcal{N} \subset L^2(L(1), g_0)$.

5.2.4 Epiconvergence

In the preceding subsection, we computed that the modified families $Q_{0,\varepsilon}^+$ converge pointwise to a non-trivial limit as ε tends to zero. However, pointwise convergence of the quadratic forms is useless if one strives to say something about convergence of the associated operators. The proper notion for this is *epiconvergence* (see [1], [3]).

To explain this notion, let H be a Hilbert space and q_n, $q_\infty : H \to \mathbb{R}$, $n \geq 1$, be non-negative but not necessarily densely defined quadratic forms with domains $\mathcal{D}(q_n)$, $\mathcal{D}(q_\infty)$ respectively. Denote by H_n, $H_\infty \subset H$ the closures of the domains with respect to the norm in H.

Definition 5.7 *The sequence q_n of non-negative closed quadratic forms on H* epiconverges *to the closed quadratic form q with respect to the weak topology on H iff*

   (i) *For all $u \in \mathcal{D}(q_\infty)$ there is a weakly convergent sequence $u_n \to u$ such that $\lim_n q_n(u_n) = q_\infty(u)$.*
   (ii) *For all $u \in \mathcal{D}(q_\infty)$ and for all weakly convergent sequences $u_n \to u$ we have $\liminf_n q_n(u_n) \geq q_\infty(u)$.*

In our case, the Hilbert space will be $H := \mathrm{H}_0^1(L(1), g_0)$. We consider the problem for the reference family first. Note that $f \in E_0 \cap \mathcal{D}(Q_{0,\varepsilon}^R)$ implies that $f(s, w) = u_0(w)\, h(s)$ with $h \in \mathrm{H}^1(\mathbb{R})$. We want to prove that $Q_{0,\varepsilon}^R$ converges to

$$Q_0^R(f) := \begin{cases} \int_{\mathbb{R}} ds |\partial_s h|^2 & f \in \mathcal{N}, \\ \infty & \text{otherwise.} \end{cases}$$

in the sense of epiconvergence with respect to the weak topology of $\mathrm{H}^1(L(1), g_0)$. The first requirement on epiconvergence follows from the fact that the forms converge pointwise, i.e. we may use the sequence $f_n \equiv f$ and obtain

$$
\begin{aligned}
\lim_{\varepsilon \to 0} Q_{0,\varepsilon}^R(f) &= \lim_{\varepsilon \to 0} Q_{0,\varepsilon}^+(f) - \frac{\lambda_0}{\varepsilon^2} \langle f, f \rangle_0 + \int_{\mathbb{R}} ds \int_{-1}^{1} dw\, |\partial_s \otimes 1(f)|^2 \\
&= \begin{cases} \int_{\mathbb{R}} ds \int_{-1}^{1} dw\, |\partial_s \otimes 1(f)|^2 & f \in \mathcal{N} \\ \infty & \text{otherwise} \end{cases} \\
&= \begin{cases} \int_{-1}^{1} dw\, |u_0|^2 \int_{\mathbb{R}} ds |\partial_s h|^2 & f \in \mathcal{N} \\ \infty & \text{otherwise} \end{cases} \\
&= \begin{cases} \int_{\mathbb{R}} ds |\partial_s h|^2 & f \in \mathcal{N}, \\ \infty & \text{otherwise} \end{cases} \\
&= Q_0^R(f)
\end{aligned}
$$

since the eigenfunction $\int_{-1}^{1} dw\, |u_0|^2 = 1$ is normalized. For the second requirement, we have to prove another result first.

Lemma 5.8 *Let $\alpha > \lambda_0$. Then we have for all $f \in \mathrm{H}_0^1(L(1), g_0)$ and all $\varepsilon > 0$*

$$Q_{0,\varepsilon}^R(f) + \alpha \langle f, f \rangle_0 \geq \|f\|_{\mathrm{H}_0^1(L(1), g_0)}^2.$$

Proof. We have for $\varepsilon^2 < 1 - \lambda_0/\lambda_1 < 1 - \lambda_0/\lambda_k$ for all k

$$
\begin{aligned}
&Q_{0,\varepsilon}^R(f) \\
&= Q_{0,\varepsilon}^+(f) - \frac{\lambda_0}{\varepsilon^2} \langle f, f \rangle_0 + \int_{\mathbb{R}} ds \int_{-1}^{1} dw\, |\partial_s \otimes 1(f)|^2 \\
&= \sum_{k \geq 1} \frac{\lambda_k - \lambda_0}{\varepsilon^2} \int_{\mathbb{R}} ds\, |\langle u_k, f_s \rangle_{L^2(F_s)}|^2 + \int_{\mathbb{R}} ds \int_{-1}^{1} dw\, |\partial_s \otimes 1(f)|^2
\end{aligned}
$$

and hence

$$Q_{0,\varepsilon}^R(f) + \alpha\langle f,f\rangle_0$$

$$\geq \sum_{k\geq 1}\lambda_k \int_{\mathbb{R}} ds\,|\langle u_k,f_s\rangle_{L^2(F_s)}|^2 + \int_{\mathbb{R}} ds\int_{-1}^{1} dw\,|\partial_s\otimes 1(f)|^2$$

$$\quad +\lambda_0\langle f,f\rangle_0$$

$$\geq \sum_{k\geq 0}\lambda_k \int_{\mathbb{R}} ds\,|\langle u_k,f_s\rangle_{L^2(F_s)}|^2 + \int_{\mathbb{R}} ds\int_{-1}^{1} dw\,|\partial_s\otimes 1(f)|^2$$

$$= \int_{\mathbb{R}} ds\int_{-1}^{1} dw\,|1\otimes\partial_w(f)|^2 + \int_{\mathbb{R}} ds\int_{-1}^{1} dw\,|\partial_s\otimes 1(f)|^2$$

$$= \|f\|_{\mathrm{H}_0^1(L(1),g_0)}^2.$$

$\square$

This inequality means that the sequence $Q_{0,\varepsilon}^R$, $\varepsilon > 0$, is *equicoercive* (see [1], [3]). To establish the second property of epiconvergence, the lim inf-inequality, we consider a slightly different estimate, namely for $\varepsilon < 1 - \lambda_0/\lambda_1 < 1 - \lambda_0/\lambda_k$,

$$Q_{0,\varepsilon}^R(f)$$

$$= Q_{0,\varepsilon}^+(f) - \frac{\lambda_0}{\varepsilon^2}\langle f,f\rangle_0 + \int_{\mathbb{R}} ds\int_{-1}^{1} dw\,|\partial_s\otimes 1(f)|^2$$

$$= \sum_{k\geq 1}\frac{\lambda_k - \lambda_0}{\varepsilon^2}\int_{\mathbb{R}} ds\,|\langle u_k,f_s\rangle_{L^2(F_s)}|^2 + \int_{\mathbb{R}} ds\int_{-1}^{1} dw\,|\partial_s\otimes 1(f)|^2$$

$$\geq \sum_{k\geq 1}\frac{\lambda_k}{\varepsilon}\int_{\mathbb{R}} ds\,|\langle u_k,f_s\rangle_{L^2(F_s)}|^2 + \int_{\mathbb{R}} ds\int_{-1}^{1} dw\,|\partial_s\otimes 1(f)|^2$$

$$= \frac{1}{\varepsilon}\int_{\mathbb{R}} ds\int_{-1}^{1} dw\,|1\otimes\partial_w(E_0^\perp f)|^2 + \int_{\mathbb{R}} ds\int_{-1}^{1} dw\,|\partial_s\otimes 1(f)|^2$$

where we denote by $E_0^\perp$ the $L^2(L(1),g_0)$-orthogonal projection onto the orthogonal complement of $\mathcal{N}$. Now note that this sum consists of a linear combination of two quadratic forms

$$Q_{0,\varepsilon}^R(f) \geq \frac{1}{\varepsilon}Q_1(f) + Q_2(f)$$

which are all *continuous* and *non-negative* on $\mathrm{H}_0^1(L(1),g_0)$. That implies by *polarization*

$$Q_i(f_n) - 2B_i(f_n,f) + Q_i(f) = Q_i(f_n - f) \geq 0$$

and hence $Q_i(f_n) \geq 2B_i(f_n, f) - Q_i(f)$ which by continuity of the bilinear forms B_i associated to Q_i and by weak convergence of f_n to f finally implies

$$\liminf_{\varepsilon \to 0} Q_i(f_n) \geq Q_i(f)$$

for $i = 1, 2$. Now, to finally prove the second assertion, note that we have established so far that

$$\liminf_{\varepsilon \to 0} Q_{0,\varepsilon}^R(f_n) \geq Q_2(f) + \liminf_{\varepsilon \to 0} \frac{1}{\varepsilon} Q_1(f_\varepsilon).$$

For $f \in \mathcal{N}$ there is nothing else to prove since by $Q_1(f_\varepsilon)/\varepsilon \geq 0$ we then have

$$\liminf_{\varepsilon \to 0} Q_{0,\varepsilon}^R(f_n) \geq Q_2(f) = Q_0^R(f).$$

For $f \notin \mathcal{N}$ we have $\liminf Q_1(f_n) = a > 0$ and hence for some suitable $a > \delta > 0$

$$\liminf_{\varepsilon \to 0} Q_{0,\varepsilon}^R(f_n) \geq Q_2(f) + \liminf_{\varepsilon \to 0} \frac{1}{\varepsilon}(a - \delta) = \infty.$$

That establishes *epiconvergence* of the family $Q_{0,\varepsilon}^R$, $\varepsilon > 0$.

The reason to consider epiconvergence of quadratic forms is that it implies strong resolvent convergence of the self-adjoint operators associated to these forms. This will now be explained.

5.2.5 Strong convergence of the generators

So far we proved that the quadratic forms epiconverge in the weak topology of $H_0^1(L(1), g_0)$. But what is this good for — we actually want to prove that the associated operators converge in the strong resolvent sense in $L^2(L(1), g_0)$? It turns.out that this follows by a rather general chain of arguments that we will present below. We again use the setup from the beginning of the last subsection, referring to the implications to our special situation in between. The reasoning consists essentially of four steps:

(i) Epiconvergence in the weak topology implies convergence of the minimizers of the quadratic forms q_n in the following sense: denote by u_n^* a minimizer of q_n. Then we have the following statement [3]:

Lemma 5.9 *Let the sequence of quadratic forms $q_{n,n\geq1}$ epiconverge to q_∞ in the weak topology on H. If a sequence $u_{n,n\geq1}^*$ of minimizers of the q_n converges in the weak sense to some $u^* \in H$, then u^* is a minimizer of q_∞.*

(ii) The question of whether the sequence $u^*_{n,n\geq 1}$ of minimizers really converges is in general very hard to answer. However, if as in our situation (see Lemma 5.8) the sequence of quadratic forms is *equicoercive*, meaning that

$$q_n(u) \geq A\|u\|^2_H + B,$$

with some $A > 0$, $B \in \mathbb{R}$, we have that, for all $t \in \mathbb{R}$, the set

$$K_t := \bigcap_{n\geq 1}\{u \in H : q_n(u) \leq t\} \subset H$$

is relatively bounded in H. If we now assume that the family does not degenerate in the sense that we exclude the case $\lim_{n\to\infty} q_n(u) = \infty$ for all $u \in H$ then there is some $t_0 \in \mathbb{R}$ such that K_t is non-empty for all $t > t_0$. But Hilbert spaces are always *reflexive* so that we can apply the *Banach–Alaoglu theorem* also to the weak topology on H. Hence the sets $K_t \subset\subset H$ are relatively compact in H with respect to the weak topology. That means that every sequence of points in K_t — and $u^*_{n,n\geq 1} \subset K_t$ is such a sequence if $t > t_0$ — contains a weakly convergent subsequence. Thus, if the minimizer u^*_∞ of the limiting quadratic form q_∞ is *unique*, Lemma 5.9 implies that all these weakly convergent subsequences must converge to u^*_∞. That means nothing but:

Lemma 5.10 *Assume that the family of quadratic forms $q_{n,n\geq 1}$ epiconverges to the quadratic form q_∞ and that there is one $t \in \mathbb{R}$ with $K_t \neq \emptyset$. If the minimizer u^*_∞ of q_∞ is unique then every sequence $u^*_{n,n\geq 1}$ of minimizers of the $q_{n,n\geq 1}$ converges weakly to u^*_∞.*

In our case, the minimizer of the limiting quadratic form is unique since Q^R_0 is strictly convex on the subset $\mathcal{N}$ where it is not infinite.

(iii) Weak convergence of the sequence of minimizers also implies weak convergence of the resolvents of the operators that are associated to the quadratic forms. To see this, let $v \in H$ and consider the functions

$$Q_{n,v}(u) := \frac{1}{2}q_n(u) - \langle v, u\rangle_X.$$

By the *Cauchy–Schwarz inequality*, equicoercivity of $q_{n,n\geq 1}$ also implies equicoercivity of $Q_{n,v}$, $n \geq 1$ for all $v \in H$. On the other hand, the limiting function $Q_{\infty,v}$ is still strictly convex on $\mathcal{N}$ and therefore has a unique minimizer. Since $u \mapsto \langle v, u\rangle_X$ is just a bounded linear map, weak epiconvergence of the forms $q_{n,n\geq 1}$ to q_∞ implies epiconvergence of the functions $Q_{n,v}$ to $Q_{\infty,v}$. Thus we can draw the same conclusions as before for all sequences of minimizers $u_{n,v}^*$, $n \geq 1$ of the functions $Q_{n,v}$. But by *Friedrich's construction* [6] we can associate to every quadratic form q_n a closed unbounded operator A_n on H_n which is densely defined on some domain $\mathcal{D}(A_n) \subset H_n$. The equation that determines the minimizer (meaning the equation for the zeroes of the *subdifferential operator* (see [1], [2]) as in the finite-dimensional case) is thus given by $u_{n,v}^* \in \mathcal{D}(A_n)$ with $A_n u_{n,v}^* = v$ or $u_{n,v}^* = A_n^{-1} v$, which implies that for all $v \in H$ the inverse operators associated to the quadratic forms converge pointwise in the weak topology of H. By Lemma 5.8, even the sequence $Q_{0,\varepsilon}^R(f) + \alpha \langle f, f \rangle_0$ is equicoercive for $\alpha > \lambda_0$. Thus if we denote the operator associated to $Q_{0,\varepsilon}^R$ by $\Delta_0(\varepsilon)$, the minimizer of

$$\frac{1}{2} \left\{ Q_{0,\varepsilon}^R(f) + \alpha \langle f, f \rangle_0 \right\} - \langle v, f \rangle_0$$

is given by the resolvent $f_{\varepsilon,v,\alpha}^* = (\Delta_0(\varepsilon) + \alpha)^{-1} v$ at $\alpha > \lambda_0$. Now we compute the operator associated to the epi-limit $Q_0^R(f) + \alpha \langle f, f \rangle_0$. Since the limiting quadratic form is infinite outside $\mathcal{N}$, we may restrict ourselves to the calculation of the minimizer on this subspace. Denoting again the $L^2(L(1), g_0)$-orthogonal projection on $\mathcal{N}$ by E_0 we have

$$
\begin{aligned}
&\frac{1}{2} \left\{ Q_{0,\varepsilon}^R(f) + \alpha \langle f, f \rangle_0 \right\} - \langle v, f \rangle_0 \\
=\ &\frac{1}{2} \left\{ Q_{0,\varepsilon}^R(E_0 f) + \alpha \langle E_0 f, E_0 f \rangle_0 \right\} - \langle v, E_0 f \rangle_0 \\
=\ &\frac{1}{2} \int_{\mathbb{R}} ds \left\{ |\partial_s h|^2 + \alpha |h|^2 \right\} - \langle \langle v, u_0 \rangle_{L^2(F_s)}, h \rangle_{L^2(\mathbb{R})}
\end{aligned}
$$

since every function $f \in \mathcal{N}$ can be written $f(s,w) = u_0(w) h(s)$ and thus

$$
\begin{aligned}
\langle v, E_0 f \rangle_0 &= \langle E_0 v, f \rangle_0 = \langle u_0 \langle u_0, v \rangle_{F_s}, u_0 h \rangle_0 \\
&= \langle \langle v, u_0 \rangle_{L^2(F_s)}, h \rangle_{L^2(\mathbb{R})}.
\end{aligned}
$$

The operator associated to the quadratic form Q_0^R is the Laplacian $\Delta_{\mathbb{R}}$ on $\mathbb{R}$. Thus we obtain, differentiating with respect to h and letting the differential be equal to zero:

$$\langle (\Delta_{\mathbb{R}} + \alpha) h, - \rangle_{L^2(\mathbb{R})} = \langle \langle v, u_0 \rangle_{L^2(F_s)}, - \rangle_{L^2(\mathbb{R})}.$$

Multiplying both sides by u_0 and solving for $u_0 h$ yields finally the more convenient form

$$f_{0,v,\alpha}^* = E_0 (\Delta_{\mathbb{R}} + \alpha)^{-1} E_0 v = (\Delta_{\mathbb{R}} + \alpha)^{-1} E_0 v.$$

Thus, the above general considerations about convergence imply

$$\text{w} - \lim_{\varepsilon \to 0}(\Delta_0(\varepsilon) + \alpha)^{-1}v = E_0\,(\Delta_{\mathbb{R}} + \alpha)^{-1}\,E_0 v$$

weakly in $\text{H}_0^1(L(1), g_0)$.

(iv) The last observation is now that $\text{H}^1(L(1), g_0) \subset\subset L^2(L(1), g_0)$, i.e. the inclusion is compact by the *Sobolev embedding theorem* [16]. Thus, weak convergence in the boundary Sobolev space implies strong convergence in $L^2(L(1), g_0)$. That means finally

Proposition 5.11 *In $L^2(L(1), g_0)$, we have strong resolvent convergence of the renormalised Laplacian associated to the rescaled reference metric on $L(1)$ to an operator*

$$E_0\,\Delta_{\mathbb{R}}\,E_0 = \Delta_{\mathbb{R}}\,E_0$$

given by the projection E_0 onto $\mathcal{N}$ followed by the Laplacian on $\mathbb{R}$. Here, the Laplacian $\Delta_{\mathbb{R}}$ acts on $\mathcal{N}$ by $\Delta_{\mathbb{R}}f = \Delta_{\mathbb{R}}u_0 h = u_0 \Delta_{\mathbb{R}} h$.

Since the operators $\Delta_0(\varepsilon)$ and $E_0\,\Delta_{\mathbb{R}}\,E_0$ are self-adjoint, strong resolvent convergence of the operators implies strong convergence of the associated semigroups on compact subintervals of $(0, \infty)$ (see [6], [2]). Therefore, we finally obtain

Proposition 5.12 *For all $v \in L^2(L(1), g_0)$ and all compact subintervals $I \subset (0, \infty)$, we have*

$$\lim_{\varepsilon \to 0}\sup_{t \in I} \|e^{-\frac{t}{2}\Delta_0(\varepsilon)}v - E_0\,e^{-\frac{t}{2}\Delta_{\mathbb{R}}}\,E_0 v\|_0 = 0.$$

Thus, the semigroups generated by the rescaled and renormalized Dirichlet Laplacians on the reference tube converge to a limit semigroup obtained by *homogenization* along the fibres. To see this, recall that $E_0 v(s, w) = h(s)u_0(w)$ and note that

$$\left\{E_0\,e^{-\frac{t}{2}\Delta_{\mathbb{R}}}\,E_0 v\right\}(s, w) = u_0(w)\left\{e^{-\frac{t}{2}\Delta_{\mathbb{R}}}h\right\}(s).$$

The time-dependent part of the dynamic is thus provided by a dynamic on the submanifold alone, which modulates the function u_0 on each fibre.

5.2.6 The result for the induced metric

As said above, the idea is to consider the rescaled and renormalized quadratic forms associated to the induced metric g as perturbations of the forms associated to the reference metric. Recall that the assumption (5.4) of *geometric finiteness*, i.e. that $|\kappa|$, $|\dot\kappa|$ and $|\ddot\kappa|$ are uniformly bounded, implies that

$$\sup_{s\in\mathbb{R},|w|<1} |W_\varepsilon(s,w) - W_0(s,w)| \leq M_1\varepsilon.$$

However, by the explicit calculation (5.3) above, we have that

$$g_\varepsilon(\mathrm{d}f,\mathrm{d}f) - g_{0,\varepsilon}(\mathrm{d}f,\mathrm{d}f) = (\varepsilon^2 w^2\kappa^2 - 2w\varepsilon\kappa)\,|\partial_s f|^2 \leq M_2\,\varepsilon\,|\partial_s f|^2.$$

Hence

$$Q_\varepsilon^R(f) - Q_{0,\varepsilon}^R(f) - \int_{L(1)} \mathrm{d}V_0\, W_0 f^2$$

$$= \int_{L(1)} \mathrm{d}V_0\,\{g_\varepsilon - g_{0,\varepsilon}\}(\mathrm{d}f,\mathrm{d}f) + \int_{L(1)} \mathrm{d}V_0\,(W_\varepsilon - W_0)\,f^2$$

$$\leq M_2\,\varepsilon \int_{L(1)} \mathrm{d}V_0\,|\partial_s f|^2 + M_1\,\varepsilon \int_{L(1)} \mathrm{d}V_0\,f^2$$

$$\leq \varepsilon M\,(Q_{0,\varepsilon}^R(f) + \alpha\langle f,f\rangle_0)$$

where $M := \max\{M_1, M_2\}$. This inequality provides us with a *Kato-type estimate* [6] from perturbation theory for the difference of $Q_\varepsilon^R(f)$ and $Q_{0,\varepsilon}^R(f) + \int_{L(1)} \mathrm{d}V_0\, W_0 f^2$. This estimate enables us to show that both families converge to the same epi-limit. Using essentially the same steps as in Section 5.2.4, we obtain:

Proposition 5.13 *The rescaled and renormalized family of quadratic forms $Q_\varepsilon^R(f) + \alpha\langle f,f\rangle_0$, $\varepsilon > 0$ epiconverges to*

$$Q_0^R(f) + \int_{L(1)} \mathrm{d}V_0\, W_0 f^2 + \alpha\langle f,f\rangle_0$$

$$= \begin{cases} \int_{\mathbb{R}} \mathrm{d}s\,\{|\partial_s h|^2 + (W_0 + \alpha)\,h^2\} & f = u_0 h \in \mathcal{N}, \\ \infty & \text{otherwise} \end{cases}$$

as ε tends to zero.

The operator associated to the limit form is the self-adjoint unbounded operator $(\Delta_L + W_0 + \alpha)\,E_0 = E_0\,(\Delta_L + W_0 + \alpha)\,E_0$ on $L^2(\mathbb{R})$ where $\Delta_L + W_0 + \alpha$ again acts on $\mathcal{N}$ by $(\Delta_L + W_0 + \alpha)u_0 h = u_0(\Delta_L + W_0 + \alpha)h$. Recall that $W_0(s,w) = W_0(s)$ depends only on the s-variable. Now denote the operator associated to the quadratic form Q_ε^R by $\Delta(\varepsilon)$. Deducing strong resolvent convergence from epiconvergence by the same mechanism as in Section 5.2.5 and inserting the expression (5.5) for W_0, we end up with the final homogenization result for the dynamics associated to the induced metric

Theorem 5.1 *For all* $f \in L^2(L(1), g_0)$ *and all compact subintervals* $I \subset (0, \infty)$, *we have*

$$\lim_{\varepsilon \to 0} \sup_{t \in I} \| e^{-\frac{t}{2}\Delta(\varepsilon)} f - E_0 e^{-\frac{t}{2}(\Delta_\mathbb{R} - \frac{1}{4}\kappa^2)} E_0 f \|_0 = 0.$$

As in the case of the reference measure, we may $(E_0 f = u_0 h)$ write

$$\left\{ E_0 e^{-\frac{t}{2}(\Delta_\mathbb{R} - \frac{1}{4}\kappa^2)} E_0 f \right\} (s, w) = u_0(w) \left\{ e^{-\frac{t}{2}(\Delta_\mathbb{R} - \frac{1}{4}\kappa^2)} h \right\} (s).$$

The limit dynamic along the submanifold is thus generated by a Schrödinger operator with a potential that reflects geometric properties of the embedding. For the unitary Schrödinger group, the statement analogous to Theorem 5.1 would be

$$\lim_{\varepsilon \to 0} \sup_{t \in I} \| E_0 (e^{\frac{it}{2}\Delta(\varepsilon)} f - e^{\frac{it}{2}(\Delta_\mathbb{R} - \frac{1}{4}\kappa^2)} E_0 f) \|_0 = 0.$$

E_0 is the projection onto the subspace $\mathcal{N}$ which contains all states in the domains of the quadratic forms associated to the Hamiltonians $\Delta(\varepsilon)$ for which the energy stays finite as ε tends to zero. This projection is necessary since the norm of eigenstates corresponding to eigenvalues tending to infinity in the Schrödinger case does not tend to zero exponentially fast for $\varepsilon \to 0$ as in the case of the heat equation.

5.3 Conditioned Brownian motion

So far, we considered the heat equation on the tube with Dirichlet boundary conditions. This corresponds to Brownian motion on the tube with absorbing boundary conditions, i.e. the Brownian particle only exists until it reaches the boundary for the first time. In this section we will recall some facts that show how intimately the absorbed and the *conditioned* Brownian motions are connected. Let $L \subset M$ be a Riemannian submanifold of the Riemannian manifold M and $\varepsilon > 0$. By the ε-*conditioned Brownian motion with finite time horizon* $T > 0$, we denote the process associated to the measure μ_ε which is the Wiener measure on M conditioned to the event that the paths do not leave the ε-tube $L(\varepsilon)$ up to time T, i.e.

$$\mu_\varepsilon(d\omega) := \mathbb{W}_M(d\omega \, | \, \omega(s) \in L(\varepsilon), \forall_{s \leq T}).$$

By the Markov property of $\mathbb{W}_M$, we obtain a time-dependent version of a well-known formula for μ_ε (see [4]). Let $0 < t < T$ and $q \in L(\varepsilon)$ be a *starting point*. In what follows, we will write $\omega \in \Omega_{u,v}(\varepsilon)$ for the event $\{\omega(s) \in L(\varepsilon), \forall_{u < s \leq v}\}$. Then

$$\mu_\varepsilon^q(d\omega) = \frac{\mathbb{W}_M^{\omega(t)}(d\omega, \omega \in \Omega_{s,T}(\varepsilon)) \mathbb{W}_M^q(\omega \in \Omega_{0,t}(\varepsilon))}{\mathbb{W}_M^q(\omega \in \Omega_{0,T}(\varepsilon))}.$$

Now let τ_ε be the *first exit time* of a Brownian particle from the tubular neighbourhood $L(\varepsilon)$. Then $\{\omega \in \Omega_{0,t}(\varepsilon)\}$, $\{\omega(t) \in L(\varepsilon)\}$ and $\{\tau_\varepsilon > t\}$ simply denote the same event. Hence if ν_ε denotes the measure associated to the absorbed Brownian motion, it can be described via its restrictions to the sigma algebras $\mathcal{F}_t$ by

$$\nu_\varepsilon(d\omega \cap \mathcal{F}_t) = \mathbb{W}_M(d\omega \cap \mathcal{F}_t, \tau_\varepsilon < t).$$

The key observation is that in fact

$$\mu_\varepsilon^q(d\omega) = \frac{\nu_\varepsilon^q(d\omega \cap \mathcal{F}_t, \omega(t) \in L(\varepsilon))}{\nu_\varepsilon^q(\omega(T) \in L(\varepsilon))} \nu_\varepsilon^{\omega(t)}(d\omega \cap \mathcal{F}_{T-t}, \tau_\varepsilon > T - t) \quad (5.9)$$

and hence all terms on the right-hand side can be expressed in terms of the absorbed Brownian motion. That implies that if, for instance, we are interested in the distribution of the position of the conditioned particle at time $t < T$ given that the particle starts at time $s < t$ in q or, equivalently, in the associated flow $P_{s,t}^\varepsilon f$ for some bounded measurable f, we obtain

$$
\begin{aligned}
P_{s,t}^\varepsilon f(q) &= \int_{\Omega(\varepsilon)} \mu_\varepsilon(d\omega \mid \omega(s) = q) f(\omega(t)) \\
&= \int_{L(\varepsilon)} \frac{\nu_\varepsilon^q(\omega(t - s) \in dx)}{\nu_\varepsilon^q(\omega(T - s) \in L(\varepsilon))} \nu_\varepsilon^{\omega(t)}(\tau_\varepsilon > T - t) f(\omega(t)) \\
&= \frac{\int_{L(\varepsilon)} \int_{L(\varepsilon)} p_{t-s}^\varepsilon(q, dx) f(x) p_{T-t}^\varepsilon(x, dy)}{\int_{L(\varepsilon)} p_{T-s}^\varepsilon(q, dz)}
\end{aligned}
$$

where $p_t^\varepsilon(x, dy)$ denotes the transition kernel of Brownian motion absorbed at the boundary of the tube. Since the generator of this process is the Dirichlet Laplacian $-\frac{1}{2}\Delta_\varepsilon$ on the tube, we may also write

$$P_{s,t}^\varepsilon f = \frac{e^{-\frac{t-s}{2}\Delta_\varepsilon}\left(f \, e^{-\frac{T-t}{2}\Delta_\varepsilon} 1\right)}{e^{-\frac{T-s}{2}\Delta_\varepsilon} 1}. \quad (5.10)$$

This relation shows that, in order to understand the limit of the conditioned Brownian motions as ε tends to zero, one has to understand the properties of the absorbed Brownian motion for small tube diameters. This establishes the connection to what was considered before, namely the renormalized limit of Dirichlet operators.

But there is another property of the path measure that is rather immediate from (5.10). The associated process is a *time-inhomogeneous Markov process*. To see this, we formally compute a time-dependent generator, namely

$$G_\varepsilon(s)f = \frac{\mathrm{d}}{\mathrm{d}h}\left[P^\varepsilon_{s,s+h}f\right]_{h=0} = \frac{\frac{\mathrm{d}}{\mathrm{d}h}\left[e^{-\frac{h}{2}\Delta_\varepsilon}\left(f\,e^{-\frac{T-s-h}{2}\Delta_\varepsilon}1\right)\right]_{h=0}}{e^{-\frac{T-s}{2}\Delta_\varepsilon}1}.$$

In what follows, we will write $\pi^\varepsilon_u(q) = e^{-\frac{u}{2}\Delta_\varepsilon}1(q)$, which denotes the *probability that a particle starting from point q is not absorbed up to time u*. Assume now that $f \in C^2(M)$ and note that, by elliptic regularity, for $t > 0$ the absorption probability π^ε_{T-s-h}, which solves the heat equation with Dirichlet boundary conditions for initial value $\pi^\varepsilon_0 = 1$ is smooth in x and zero on the boundary $\partial L(\varepsilon)$. Hence

$$f\,\pi^\varepsilon_{T-s-h} \in \{g \in C^2(L(\varepsilon)) : h|_{\partial L(\varepsilon)} = 0\} \subset \mathcal{D}(\Delta_\varepsilon)$$

even though $f \notin \mathcal{D}(\Delta_\varepsilon)$ and we actually have

$$\Delta_\varepsilon(f\,\pi^\varepsilon_{T-s-h}) = \pi^\varepsilon_{T-s-h}\,\Delta f + 2g(\mathrm{d}f, \mathrm{d}\pi^\varepsilon_{T-s-h}) + f\,\Delta_\varepsilon\pi^\varepsilon_{T-s-h}$$

where the expression Δf is understood as simply applying the Laplacian to the function and no conditions are imposed on the behaviour of f at the boundary. Thus, after dividing by π^ε_u which is positive on $L(\varepsilon)$, the generator is formally given by

$$G_\varepsilon(s)f = -\frac{1}{2}\Delta f + g(\mathrm{d}f, \mathrm{d}\log\pi^\varepsilon_{T-s})$$

and describes a Brownian motion with time-dependent drift given by the *logarithmic derivative of the probability to stay within the tube for a given time*. Of course, this relation also holds for Brownian motion conditioned to an arbitrary set with sufficiently smooth boundary. The vector field is singular at the boundary pushing the particle back into the tube.

5.4 The case of Riemannian submanifolds

In this section, we consider the sequence of measures $\mu_{\varepsilon, \varepsilon > 0}$ on the path space of M. As shown above, they are supported by the path spaces of the respective tubes $L(\varepsilon)$ and correspond to time-inhomogeneous Markovian processes given by the Brownian motion conditioned to the tubes. As ε tends to zero, it seems natural to ask whether these measures converge in a suitable sense to a measure supported by the path space of L. In the case that $L \subset M$ is a *closed* (i.e. compact without boundary) Riemannian submanifold, we can answer this question to the affirmative and provide an explicit description of the limit measure. Namely, we have the following statements:

(i) The sequence μ_ε converges *weakly* on the path space of M to a measure μ_0 which is supported by the path space of the submanifold L.
(ii) μ_0 is equivalent to the Wiener measure $\mathbb{W}_L$ on the submanifold.
(iii) The Radon–Nikodym density depends on geometric properties of the submanifold and of the embedding and is of Feynman–Kac Gibbs type given by

$$\frac{\mathrm{d}\mu_0}{\mathrm{d}\mathbb{W}_L}(\omega) = \frac{1}{Z} \mathrm{e}^{-\int_0^T \mathrm{d}s\, W(\omega(s))} \tag{5.11}$$

where the *effective potential* $W \in C^\infty(L)$ is given by

$$W = \frac{1}{4}\mathrm{Scal}_L - \frac{1}{8}\|\tau\|^2 - \frac{1}{12}\left(\mathrm{Scal}_M + \overline{\mathrm{Ric}}_{M|L} + \overline{\mathrm{R}}_{M|L}\right),$$

Scal denotes the *scalar curvature*, τ the *tension vector field* and $\overline{\mathrm{Ric}}_{M|L}$, $\overline{\mathrm{R}}_{M|L}$ denote the traces of the Riemannian curvature tensor and of the Ricci tensor of M only with respect to the subbundle $TL \subset TM$. To be precise, let $j : TL \subset TM$ be the embedding and $j_x : T_xL \subset T_xM$ the induced map on the fibres. Then

$$\overline{\mathrm{Ric}}_{M|L}(x) := \mathrm{tr}(\mathrm{Ric}_x \circ j_x \otimes j_x) = \sum_{r=1}^{l} \mathrm{Ric}_x(e_r, e_r),$$

$$\overline{\mathrm{R}}_{M|L} := \mathrm{tr}(\mathrm{R}_x \circ j_x \otimes j_x \otimes j_x \otimes j_x) = \sum_{r,s=1}^{l} \mathrm{R}_x(e_r, e_s, e_r, e_s)$$

where $e_1, \ldots, e_l$ denotes an orthonormal base of T_xL. Z is a normalization constant.

Thus, the limit of the conditioned Brownian motions, which we can in fact consider as a version of the regular conditional probability given the sigma algebra generated by the distance on the path space, is an intrinsic Brownian motion on the submanifold subject to a certain potential that depends on the geometry of the submanifold and of the embedding.

The proof of these facts consists of a combination of the representation (5.10) with the homogenisation result Theorem 5.1 for convergence in finite-dimensional distributions together with a tightness result based on a moment estimate. For details, we refer to the forthcoming paper [12]. To understand the significance of the different terms in this density, we consider several examples.

Example 5.1 Submanifolds $L \subset \mathbb{R}^n$ in euclidean space. In this case, the density simplifies to

$$\frac{\mathrm{d}\mu_0}{\mathrm{d}W_L}(\omega) = \frac{1}{Z}\mathrm{e}^{\int_0^T \mathrm{d}s \left[\frac{1}{8}\|\tau\|^2 - \frac{1}{4}\mathrm{Scal}_L\right](\omega(s))}$$

which is an expression depending on the norm of the tension vector field, which depends on the embedding and the scalar curvature, which is an intrinsic geometrical property of the submanifold. The other terms vanish owing to the fact that euclidean space is flat.

Example 5.2 The unit sphere $S^{n-1} \subset \mathbb{R}^n$. For the unit sphere, the norm of the tension vector field which is proportional to the mean curvature of the sphere hypersurface is constant. The scalar curvature of a round sphere is constant, too. Thus, the integral along the path that shows up in the Radon–Nikodym density is given by a constant times T and does not depend on the given path any more. Therefore, by normalization, the total density is in fact equal to 1. Hence, Brownian motion on the sphere can be constructed as the weak limit of the conditioned processes on the tubular neighbourhoods.

Example 5.3 Totally geodesic submanifolds. First of all, owing to our compactness assumption on the submanifolds, we have to make sure that there are relevant cases where our result can be applied. For example, *large spheres $S^l \subset S^m$, $l < m$,* in spheres are closed and totally geodesic. Owing to the validity of the *Gauss embedding equations*, there are several equivalent ways to simplify the effective potential. We choose the expression

$$W = \frac{1}{4}\left(\sigma + \frac{1}{3}\mathrm{Scal}^\perp\right)$$

where $\mathrm{Scal}^\perp$ is the scalar curvature of the fibre $\pi^{-1}(p)$ at p and

$$\sigma_p := \sum_{k=1,\ldots,l;\alpha=1,\ldots,m-l} K(e_k \wedge n_\alpha)$$

is the sum of the *sectional curvatures* $K(-)$ of all two-planes $e_k \wedge n_\alpha \subset T_p M$ which are spanned by one element of the orthonormal base $e_1, \ldots, e_l$ of $T_p L$ and one element of the orthonormal base $n_1, \ldots, n_{m-l}$ of $N_p L$. Note that if $L \subset M$ is a totally geodesic submanifold and M is *locally symmetric*, a property that implies that the curvature tensor R of M is parallel, then the potential W is constant implying, as in the case of the sphere in euclidean space, that $\mu \equiv W_L$. This holds, for instance, for the large spheres mentioned above.

Example 5.4 Plane curves. Let $\varphi : S^1 \to \mathbb{R}^2$ be an isometric embedding, i.e. the curve is parametrized by arc length. The intrinsic curvature of a one-dimensional object is zero, hence the effective potential is given by

$$W = -\frac{1}{8} \|\tau\|^2$$

and since φ is parametrized by arc length, the tension vector field is actually given by $\tau = \ddot{\varphi}$. In total, that yields a density for the conditioned motion given by

$$\frac{d\mu_0}{dW_{S^1}}(\omega) = \frac{1}{Z} e^{\frac{1}{8} \int_0^T ds \, \|\ddot{\varphi}\|^2(\omega(s))}.$$

For an ellipse $\Psi(s) = (a \cos s, b \sin s)$, $s \in [0, 2\pi)$ (note that this is not a parametrization by arc length), we will have, for instance,

$$\|\tau\|^2 = \frac{a^2 b^2}{(a^2 \sin^2 s + b^2 \cos^2 s)},$$

where $2a, 2b > 0$ are the lengths of the major axes. The *sojourn probability* of the Brownian particle is thus largest at the intersection of the ellipse and the longer major axis, where we have the largest curvature, and lowest at the intersection of the ellipse and the smaller major axis.

5.5 Two limits

So far we presented the quadratic form approach to the surface measure because it provides us with the fastest approach to the calculation of the effective potential, directly emphasizing the role played by the reference metric and the logarithmic density $\log \rho$. However, the first approach to surface measures was different.

In [14], the following scheme was introduced. Let $\mathcal{P} = \{0 = t_0 < \cdots < t_k = T\}$ be a partition of the time interval and let x_0 be a fixed point in a smooth compact l-dimensional Riemannian manifold L isometrically embedded into the euclidean space $\mathbb{R}^m$. Then the measure $\mathbb{W}_{\mathcal{P}}$ on $C_{x_0}([0,T], \mathbb{R}^m)$ is defined as the law $\mathbb{W}$ of a Brownian motion in $\mathbb{R}^m$ which is conditioned to be in the manifold L at all times t_i. More precisely, given a cylinder set

$$B := B_{A_1,\ldots,A_m}^{s_1,\ldots,s_m} = \{\omega \in C_{x_0}([0,T], \mathbb{R}^m) \colon \omega_{s_i} \in A_i, 1 \le i \le m\},$$

with all s_j being different from all t_i, its measure $\mathbb{W}_{\mathcal{P}}$ is defined by

$$\mathbb{W}_{\mathcal{P}}(B) = c_{\mathcal{P}} \int_{\underline{B}} p(\Delta u_0, \Delta x_0)..p(\Delta u_{m+k-1}, \Delta x_{m+k-1}) \xi_1 \otimes .. \otimes \xi_{m+k}(dx)$$

where $\mathcal{U} = \{0 = u_0 < \cdots < u_{m+n} = T\}$ is the union of the partitions $\mathcal{P}$ and $\mathcal{S} = \{0 < s_1 \le \cdots < s_m\}$, $x = (x_1, \ldots, x_{m+k})$, $\Delta u_i = u_{i+1} - u_i$, $\Delta x_i = x_{i+1} - x_i$, $\underline{B} = B_1 \times \cdots \times B_{m+k}$ and

$$B_i = \begin{cases} A_j & \text{if } u_i = s_j, \\ L & \text{if } u_i = t_j \text{ for some } j, \end{cases} \qquad \lambda^i = \begin{cases} V_{\mathbb{R}^m} & \text{if } u_i = s_j \text{ for some } j, \\ V_L & \text{if } u_i = t_j \text{ for some } j. \end{cases}$$

$V_{\mathbb{R}^n}$ and V_L are the Riemannian volumes on $\mathbb{R}^m$ and L, respectively, $p(t,x,y)$ is the density of the n-dimensional normal distribution $N(\|y - x\|, t)$, and the normalization constant $c_{\mathcal{P}}$ is chosen so that $\mathbb{W}_{\mathcal{P}}$ becomes a probability measure. The following theorem has been proved in a more general setting in [14].

Theorem 5.2 *As $|\mathcal{P}| \to 0$, the measures $\mathbb{W}_{\mathcal{P}}$ converge weakly to a probability measure $\mathbb{S}_{bb}$ on $C_a([0,T], L)$, which is absolutely continuous with respect to the Wiener measure $\mathbb{W}_L$ on that space, and its Radon–Nikodym density is given by*

$$\rho(\omega) = \frac{1}{Z} e^{\int_0^T ds \left[\frac{\|\tau\|^2}{8} - \frac{\text{Scal}_L}{4} \right] (\omega(s))}, \tag{5.12}$$

where Scal_L is a scalar curvature of L, τ is the tension vector field of the embedding $L \hookrightarrow \mathbb{R}^m$, and Z is the normalizing constant.

Note that this is exactly the same expression as in Example 5.1 above. Thus, we obtain the same limiting surface measure by a completely different ansatz. In what follows, we will discuss the difference between the two approaches and why they yield the same result. The full result from Section 5.4 can also be obtained from this method. This is explained in [15].

The subscript 'bb' in $\mathbb{S}_{bb}$ refers to Brownian bridges, which are a cornerstone of the construction. The idea of the proof is to first show that, as the *mesh* $|\mathcal{P}|$ of the partition tends to zero, the corresponding marginal measures on cylinder functions tend to the marginals of the limit measure which is supported by the path space of the submanifold. The main tool is a careful analysis of the short-time asymptotic of the semigroup associated to the conditioned kernel using heat kernel estimates based on the *Minakshisundaram–Pleijel expansion* [7].

For some fixed partition, the full conditioned measure is given by the marginals constructed as explained above together with *interpolating Brownian bridges* in euclidean space which fix the path measure between those time-points, where the particle is pinned to the submanifold. Along these Brownian bridges, the particle may still leave the submanifold. Finally, a Large Deviation result for Brownian bridges implies tightness for every sequence $\mathbb{W}_{\mathcal{P}_n}$ of conditioned measures for which the meshes $|\mathcal{P}_n|$ tend to zero. Thus, the sequence of measures converges in the weak sense and the limit measure is determined by the limit of the marginals.

Let us now go back to the surface measure $\mathbb{S}_{\mathrm{hc}}$ corresponding to a particle moving under hard constraints. Recall that it is defined as the weak limit

$$\mathbb{S}_{\mathrm{hc}} = \lim_{\varepsilon \to 0} \mathbb{W}_\varepsilon, \qquad (5.13)$$

where $\mathbb{W}_\varepsilon = \mathbb{W}\left(\,\cdot\,|\,\omega_t \in L(\varepsilon) \text{ for all } t \in [0, T]\right)$ is the law of a flat Brownian motion conditioned to stay in the ε-neighbourhood of the manifold for the whole time. The following theorem has been proven in [11].

Theorem 5.3 *The limit in* (5.13) *exists and so the surface measure* $\mathbb{S}_{\mathrm{hc}}$ *is well-defined. It is absolutely continuous with respect to the Wiener measure* $\mathbb{W}_L$ *on that space, and its Radon–Nikodym density is given by* (5.12).

Theorems 5.2 and 5.3 hence imply immediately that $\mathbb{S}_{\mathrm{hc}} = \mathbb{S}_{\mathrm{bb}}$.

Before explaining the main idea of the proof of Theorem 5.3, let us discuss the reason why the two measures $\mathbb{S}_{\mathrm{hc}}$ and $\mathbb{S}_{\mathrm{bb}}$ turn out to be equal: note that the probability for a Brownian motion to be in L at a certain fixed time is zero, and the measures $\mathbb{W}_{\mathcal{P}}$ have been defined using iterated integral of the heat kernel in order to avoid conditioning of the flat Wiener measure to a set of measure zero. Alternatively, one can first force a particle to be in the ε-neighbourhood $L(\varepsilon)$ of L at all times

$t_i \in \mathcal{P}$ and then let ε go to zero. More precisely, define

$$\mathbb{W}_{\mathcal{P},\varepsilon} = \mathbb{W}\left(\,\cdot\mid \omega_{t_i} \in L(\varepsilon) \text{ for all } t_i \in \mathcal{P}\right).$$

Then, in the weak sense, $\mathbb{W}_{\mathcal{P}} = \lim_{\varepsilon \to 0} \mathbb{W}_{\mathcal{P},\varepsilon}$, and Theorem 5.2 can be reformulated as

$$\lim_{|\mathcal{P}| \to 0} \lim_{\varepsilon \to 0} \mathbb{W}_{\mathcal{P},\varepsilon} = \mathbb{S}_{\mathrm{bb}}. \tag{5.14}$$

On the other hand, by the continuity of paths, $\mathbb{W}_\varepsilon = \lim_{|\mathcal{P}| \to 0} \mathbb{W}_{\mathcal{P},\varepsilon}$, and hence Theorem 5.3 is equivalent to

$$\lim_{\varepsilon \to 0} \lim_{|\mathcal{P}| \to 0} \mathbb{W}_{\mathcal{P},\varepsilon} = \mathbb{S}_{\mathrm{hc}}. \tag{5.15}$$

Thus, (5.14) and (5.15) illustrate the fact that $\mathbb{S}_{\mathrm{bb}}$ and $\mathbb{S}_{\mathrm{hc}}$ are obtained by interchanging the two limits. This suggests a more general definition of a surface measure

$$\mathbb{S} = \lim_{\substack{|\mathcal{P}| \to 0 \\ \varepsilon \to 0}} \mathbb{W}_{\mathcal{P},\varepsilon}.$$

It has been proven in [10] that this general limit exists and then, of course, is also absolutely continuous with respect to the Wiener measure $\mathbb{W}_L$ with the Radon–Nikodym density given by (5.12), since it must coincide with both particular limits $\mathbb{S}_{\mathrm{bb}}$ and $\mathbb{S}_{\mathrm{hc}}$. The proof of this statement follows the same lines of the proof of Theorem 5.3 in [11] using additionally the continuity of paths of a Brownian motion.

The intuition behind the proof is the decomposition of the generator $\Delta/2$ of a flat Brownian motion into three components: an operator close to the half of the Laplace–Beltrami operator $\Delta_L/2$, the half of the Laplace operator along the fibres of the tubular neighbourhood, and a differential operator of the first order. More precisely, let us assume without loss of generality that the radius of curvature of L is greater than 1 and so the orthogonal projection π is well-defined on $L(1)$. For each $x \in L(1)$, denote by $L_x \subset L(1)$ an l-dimensional Riemannian manifold containing x and parallel to L, that is, for any $y \in L_x$ the tangent space $T_y L_x$ is parallel to $T_{\pi(y)}L$. Further, denote by $\tau(x)$ the tension vector of the embedding $L_x \hookrightarrow \mathbb{R}^m$ at the point x. Then, for any $f \in C^2(\mathbb{R}^m)$ and $x \in L(1)$,

$$(\Delta f)(x) = \left(\Delta_{L_x} f + \Delta_{N_x L} f\right)(x) + \langle \tau, \nabla f \rangle(a), \tag{5.16}$$

where $N_x L$ denotes the orthogonal space to L at $\pi(x)$. Note that for this decomposition the parallel manifolds L_x need only be defined locally in a neighbourhood of x. Now the projections of the process with generator $(\Delta_{L_x} + \Delta_{N_x L})/2$ yield almost a Brownian motion on the manifold, and precisely a Brownian motion in the orthogonal direction, and those two components are almost independent. This makes it plausible that conditioning this process to $L(\varepsilon)$ would lead, in the limit as $\varepsilon \to 0$, to a Brownian motion on the manifold. Finally, the first order term $\langle \tau, \nabla f \rangle$ can be dealt with by a Girsanov transformation, and it would lead to a non-trivial density.

However, there are certain difficulties in realizing this program, in particular the fact that the parallel manifolds L_x do not always exist. Namely, they exist if and only if the normal bundle NL is flat, which is always the case for embeddings into $\mathbb{R}^2$ and $\mathbb{R}^3$ but rather exceptional for the higher-dimensional spaces. Hence, the operators Δ_{L_x} and the vector field τ are not well-defined. In fact, this is also the basic observation that yield the construction of the unitary rescaling map in Section 5.2.2. By this transformation, the vector field is automatically removed. We will now present an alternative ansatz to overcome this difficulty, together with an alternative and purely probabilistic proof.

It turns out that there is a vector field v on $L(1)$ such that $(\Delta - l\langle v, \nabla \rangle)/2$ is the generator of a stochastic process which converges to a Brownian motion on the manifold, even though it can no longer be written in the form $(\Delta_{L_x} + \Delta_{N_x L})/2$ as in the decomposition (5.16) above. It is defined by

$$v(x) = \nabla \phi(x) \quad \text{with} \quad \phi(x) = \log \frac{\mathrm{d}V_{\mathbb{R}^n}}{\mathrm{d}V_0} = \log \rho,$$

where V_0 is the Riemannian volume associated to the reference metric which can also be thought of as the product measure on $L(1)$ defined by

$$V_0(A) = \int_{\pi(A)} V_{\mathbb{R}^m - l}(A_x)\mathrm{d}V_L(x), \qquad A \subset L(1)\text{Borel},$$

where $A_x = \pi^{-1}(x) \cap A$. In the particular case when the normal bundle NL is flat and the parallel manifolds exist, v coincides with the vector field τ. Moreover, even without that assumption, $v(x)$ coincides with τ on the manifold L where the tension field is defined.

In contrast to the analytical approach above, where one uses mainly perturbation theory for the variational representation of the generators, the approach here is purely probabilistic and one mainly works with stochastic differential equations and convergence of their solutions. However, in both approaches, the Radon–Nikodym density of the Riemannian volumes associated to the induced and reference metrics plays a crucial role. That is, of course, no surprise since the effective potential is computed from this density.

Let us start by writing down the process $(Y_t)_{t \leq T}$ associated to the generator $(\Delta - \langle v, \nabla \rangle)/2$ as a solution of the equation

$$\left. \begin{aligned} \mathrm{d}Y_t &= \mathrm{d}B_t - \frac{1}{2}v(Y_t)\mathrm{d}t, \\ Y_0 &= a. \end{aligned} \right\} \tag{5.17}$$

In order to be able to condition (Y_t) to $L(\varepsilon)$ it would be convenient to first decompose it into its component along the manifold and the orthogonal component. The first one can be naturally defined by $X_t = \pi(Y_t)$, where from now on we assume without loss of generality that (Y_t) denotes the solution of (5.17) stopped at the time when it leaves $L(1)$. The second component has to describe the difference $Y_t - X_t$, which, at every time t, is an element of the $(m - l)$-dimensional orthogonal space $N_{X_t}L$. If there were a smooth globally defined family of orthonormal bases $(e_i(x)_{1 \leq i \leq m})_{x \in L}$ then the orthogonal component Z_t of Y_t could be defined by the coordinates of $Y_t - X_t$ with respect to $(e_i(X_t)_{l+1 \leq i \leq m})$. However, such a global family of bases in general does not exist and a way out is to fix an orthonormal basis $(e_i(a)_{1 \leq i \leq m})$ at the starting point a and move it along the semimartingale (X_t) using the notion of stochastic parallel translation. Then the initial basis will be transformed to a basis at X_t by an orthogonal matrix U_t, which, as a matrix-valued process, is a solution of the Stratonovich equation of stochastic parallel transport

$$\left\{ \begin{aligned} \mathrm{d}U_t &= \Gamma_{X_t}(\delta X_t), \\ U_0 &= I, \end{aligned} \right.$$

where Γ is the Levi-Civita connection on L. Now (Z_t) can be defined as a $\mathbb{R}^{m-l}$-valued process defined by the last $m - l$ coordinates of the vector $Y_t - X_t$ with respect to the moving basis $(e_i(X_t)_{l+1 \leq i \leq m})$, or, equivalently, by the last $n - l$ coordinates of the vector $U_t^{-1}(Y_t - X_t)$ with respect to the fixed basis $(e_i(a)_{l+1 \leq i \leq n})$. The $M \times \mathbb{R}^{m-l}$-valued process (X_t, Z_t) fully characterizes (Y_t) and is called its Fermi decomposition.

The next step is to replace the Brownian motion (B_t) by another Brownian motion $(\hat{B}_t)$, which is better adjusted to the moving frames. We define it by

$$\hat{B}_t = \int_0^t U_s^{-1}\mathrm{d}B_s.$$

With respect to this new Brownian motion, the triple (X_t, Z_t, U_t) satisfies the system of stochastic differential equations

$$\begin{cases} dX_t = \sigma(X_t, Z_t, U_t)d\hat{B}'_t + c(X_t, Z_t, U_t)dt \\ dZ_t = d\hat{B}''_t \\ dU_t = \Gamma_{X_t}(\delta X_t), \end{cases}$$

where $\hat{B}'_t$ are the first l and $\hat{B}''_t$ the last $m - l$ coordinates of $\hat{B}_t$, and the coefficients σ and c can be computed explicitly. The main feature of this system is that the processes (X_t) and (Z_t) are driven by independent Brownian motions and hence conditioning (Z_t) does not affect the Brownian motion driving the process (X_t). Hence, in order to prove that the law of the process (Y_t) conditioned to be in $L(\varepsilon)$ at all times $t_i \in \mathcal{P}$ converges to the Wiener measure on paths in L, it suffices to show that the solution $(X_t^{\mathcal{P}, \varepsilon})$ of

$$\begin{cases} dX_t^{\mathcal{P}, \varepsilon} = \sigma(X_t^{\mathcal{P}, \varepsilon}, Z_t^{\mathcal{P}, \varepsilon}, U_t^{\mathcal{P}, \varepsilon})d\hat{B}'_t + c(X_t^{\mathcal{P}, \varepsilon}, Z_t^{\mathcal{P}, \varepsilon}, U_t^{\mathcal{P}, \varepsilon})dt \\ dU_t^{\mathcal{P}, \varepsilon} = \Gamma_{X_t^{\mathcal{P}, \varepsilon}}(\delta X_t^{\mathcal{P}, \varepsilon}), \end{cases}$$

converges to a Brownian motion on L, where $(Z_t^{\mathcal{P}, \varepsilon})$ denotes an $(m - l)$-dimensional Brownian motion conditioned to be in the ε-disc around zero at all times $t_i \in \mathcal{P}$. This can be done using moment estimates, the continuity of paths of $(Z_t^{\mathcal{P}, \varepsilon})$, and the explicit form of the coefficients of the equation.

Once it is proven that the surface measure corresponding to the process (Y_t) is the Wiener measure $\mathbb{W}_L$, the rest can be done using the Girsanov transformation. Since the law $\mathcal{L}(Y)$ and the Wiener measure are equivalent with

$$\frac{d\mathbb{W}}{d\mathcal{L}(Y)}(\omega) = e^{\frac{1}{2}\int_0^T \langle \nabla\phi(\omega_t), d\omega_t \rangle + \frac{1}{8}\int_0^T \|\nabla\phi(\omega_t)\|^2 dt}$$

$$= e^{\frac{\phi(\omega_1) - \phi(\omega_0)}{2} + \int_0^T \left(-\frac{\Delta\phi(\omega_t)}{4} + \frac{\|\nabla\phi(\omega_t)\|^2}{8}\right)dt},$$

it suffices to show that $\phi|_L = 0$, $\Delta\phi|_L = \mathrm{Scal}_L$, and $\nabla\phi_L = \tau$, which is a routine computation. This leads to the non-trivial density (5.12) in the surface measure.

5.6 Two open problems

So far, the surface measures that we considered turned out to be regular in the sense that they are equivalent to the Wiener measure of the

respective submanifold. Further investigations indicate that this situation is in fact exceptional and that new interesting phenomena appear.
Therefore, we want to conclude the paper with two open problems where
the limit measures most likely show exceptional, or, better, non-regular
behaviour.

5.6.1 Tubes with fibres of variable shape

In the previous sections, we considered surface measures constructed by
conditioning a Brownian motion to neighbourhoods of constant diame-
ter, meaning that all fibres $L_p(\varepsilon) := \pi^{-1}(p) \cap L(\varepsilon)$, $p \in L$, of the tube
were balls in $\pi^{-1}(p)$ centred around p with radius ε. It seems that the
situation changes drastically if the radius of each $L_p(\varepsilon)$ depends on p, or,
if the fibres $L_p(\varepsilon)$ are even of variable shape. For example, one can take
a smooth potential $V : \mathbb{R}^n \to [0, \infty)$ such that $V|_L = 0$ and define the
ε-neighbourhoods by $L^V(\varepsilon) = \{x \in \mathbb{R}^n : V(x) \leq \varepsilon\}$, which would correspond to a natural conditioning of a Brownian motion B to the event
$\{\sup_{t \leq T} V(B_t) \leq \varepsilon\}$. If, for instance, the potential grows quadratically
with respect to the distance from the submanifold, the fibres $L_p^V(\varepsilon)$ are
ellipses whose principal axes, in general, depend on the base point. This
is connected to considering *soft constraints* (not the hard-wall potential)
forcing the particle to remain on the submanifold. Another natural example is to take a smooth function $\alpha : L \to (0, \infty)$ and consider tubular
neighbourhoods such that their radius over a point $x \in L$ is given by
$\varepsilon\alpha(x)$.

First of all, it is not even clear if the corresponding surface measures
exist. However, the leading term of the energy of the conditioned Brownian particle in the second example is believed to be

$$E_\varepsilon(\omega) = -\frac{\lambda}{\varepsilon^2} \int_0^T \frac{dt}{\alpha(\omega_t)^2},$$

where $\lambda < 0$ is the largest eigenvalue of $\Delta_{B_p(1)}/2$ in the ball of radius
1 in the orthogonal space. Hence, as ε becomes smaller, the particle
should try to minimize the energy E_ε and is expected to spend more
and more time in the regions where the neighbourhood is wide. Thus,
the limit measure is expected to be singular with respect to the Wiener
measure and to be concentrated on the paths staying in (probably local)
maxima of α. In particular, one needs to pass to the Skorokhod space in
order to study Brownian motion not starting in the maxima of α, since

we expect a Brownian particle to jump instantaneously to one of the minima as ε tends to zero.

5.6.2 Non-smooth manifolds

Another challenging question is to study conditioning to non-smooth manifolds. Since the projection to the manifold is no longer well-defined, the techniques discussed in the previous sections break down. Moreover, similarly to the previously considered case of non-uniform neighbour-hoods, in most natural cases the surface measures are believed not to be supported by the space of continuous functions alone. The situation is not even clear for one-dimensional manifolds such as, for example, polygons in $\mathbb{R}^2$. For L-shaped domains, it has been proved (see [9]) that the conditioned Brownian motions converge in finite-dimensional distributions to the Dirac measure on the path staying in the corner. In particular, if the surface measure exists it is in this case the Dirac measure. The main reason for degeneration of the surface measure is the fact that the integrated curvature of the manifold, which is the main ingredient of the Radon–Nikodym density, is infinite for such manifolds. There are also other reasons for non-smoothness than singular curvature. For example, for a cross of two orthogonal lines, the particle will try to escape to infinity, and there will be no limit at all. The intuitive reason for this difference in behaviour is that the amount of space in the ε-neighbourhood around the singularity compared to the amount of manifold is large for an L-shaped domain and small for the cross.

Bibliography

[1] Attouch, H. (1984). *Variational Convergence for Functions and Operators.* London: Pitman.

[2] Brezis, H. (1973). *Operateurs Maximaux Monotones.* Amsterdam: North Holland.

[3] Dal Maso, G. (1992). *An Introduction to Γ-Convergence.* Birkhäuser, Basel.

[4] Doob, J. L. (1984). *Classical potential Theory and its Probabilistic Counter-part.* New York: Springer.

[5] Froese, R., Herbst, I. (2001). Realizing holonomic constraints in classical and quantum mechanics. *Commun. Math. Phys.*, **220**, 489–535.

[6] Kato, T. (1980). *Perturbation Theory for Linear Operators.* 2nd ed. Berlin: Springer.

[7] Minakshisundaram, S., Pleijel, A. (1949). Some properties of the eigenfunc-tions of the Laplace operator on Riemannian manifolds. *Can. J. Math.*, **1**, 242–256.

[8] Reed, M.,Simon, B. (1978). *Methods of Modern Mathematical Physics IV: Analysis of Operators*. San Diego: Academic Press.

[9] Seifried, F. (2006). Surface limits of planar Brownian motion with drift, diploma thesis, TU Kaiserslautern.

[10] Sidorova, N. (2004). The Smolyanov surface measure on trajectories in a Riemannian manifold. *Infin. Dimens. Anal. Quantum Probab. Relat. Top.*, **7**, 461–471.

[11] Sidorova, N., Smolyanov, O. G., v. Weizsäcker, H., Wittich, O. (2004). The surface limit of Brownian motion in tubular neighborhoods of an embedded Riemannian manifold. *J. Funct. Anal.*, **206**, 391–413.

[12] Sidorova, N., Smolyanov, O. G., v. Weizsäcker, H., Wittich, O. (in preparation). Brownian motion conditioned to tubular neighborhoods.

[13] Simon, B. (1979). *Functional Integration and Quantum Physics*. New York: Academic Press.

[14] Smolyanov, O. G., v. Weizsäcker, H., Wittich, O. (2000), Brownian motion on a manifold as limit of stepwise conditioned standard Brownian motions. *Stochastic Processes, Physics and Geometry: New Interplays, II (Leipzig, 1999)*, 589–602, CMS Conf. Proc., 29, Amer. Math. Soc., Providence, RI.

[15] Smolyanov, O. G., v. Weizsäcker, H., Wittich, O. (2007). Chernoff's theorem and discrete time approximations of Brownian motion on manifolds. *Pot. Anal.* **26**, 1–29.

[16] Taylor, M. E. (1999). *Partial Differential Equations I: Basic Theory*. 2nd corrected printing, New York: Springer.

6

Sampling conditioned diffusions

Martin Hairer, Andrew Stuart and Jochen Voß
Warwick Mathematics Institute
University of Warwick
Coventry CV4 7AL, UK

Abstract

For many practical problems it is useful to be able to sample conditioned diffusions on a computer (e.g. in filtering/smoothing to sample from the conditioned distribution of the unknown signal given the known observations). We present a recently developed, SPDE-based method to tackle this problem. The method is an infinite-dimensional generalization of the Langevin sampling technique.

6.1 Introduction

In many situations, understanding the behaviour of a stochastic system is greatly aided by understanding its behaviour conditioned on certain events. This allows us, for example, to study rare events by conditioning on the event happening or to analyse the behaviour of a composite system when only some of its components can be observed. Since properties of conditional distributions are often difficult to obtain analytically, it is desirable to be able to study these distributions numerically. This allows us to develop meaningful conjectures about the distribution in question or, in a more applied context, to derive quantitative information about it. In this text we present a general technique to generate samples from conditional distributions on infinite-dimensional spaces. We give several examples to illustrate how this technique can be applied.

Sampling, i.e. finding a mechanism which produces random values distributed according to a prescribed target distribution, is generally a difficult problem. There exist many 'tricks' to sample from specific distributions, ranging from very specialized methods, like the Box–Müller method for generating one-dimensional standard Gaussian distributed values, to generic methods, like rejection sampling, which can be applied to whole classes of distributions. In situations where none of the direct

Trends in Stochastic Analysis, ed. J. Blath, P. Mörters and M. Scheutzow.
Published by Cambridge University Press. ©Cambridge University Press 2008.

methods apply in a useful way, Markov Chain Monte Carlo (MCMC) methods are commonly applied. These techniques work by constructing a Markov chain (or, more generally, a Markov process) which has the target distribution as its stationary distribution. Assuming that the process converges to stationarity fast enough, the states of the Markov chain at 'large' times can be used as approximate samples from the target distribution. While MCMC methods are only approximate methods, they can be used in many situations where no other methods are available. This is particularly true in high-dimensional problems and thus it is natural to employ MCMC methods for infinite-dimensional sampling problems. Indeed, the main tool described in this text is an MCMC method for distributions on infinite-dimensional spaces.

The stochastic systems of interest here are diffusion processes described by stochastic differential equations. The trajectories of these processes can be considered to be random functions and thus the probability distributions we consider typically live on function spaces like $L^2(I, \mathbb{R}^d)$ or $C(I, \mathbb{R}^d)$ where $I \subseteq \mathbb{R}$ is some interval. Thus, in order to construct an MCMC method for these distributions, we have to find Markov processes which have prescribed distributions on these function spaces as their invariant measures. In the context of our framework these Markov processes are given as solutions of stochastic partial differential equations (SPDEs), where the interval I is the 'space' direction of the SPDE.

Throughout this text we give several concrete examples of conditioned diffusions and how to sample from them. A simple case is to condition the process on its value at a fixed time, so that the resulting paths are bridges. Sampling bridges could, for example, be interesting when studying transitions between meta-stable states of some physical system: while these transitions will eventually happen, the times between transitions might be so big that they 'never' occur during an unconditioned numerical simulation. By conditioning on a transition actually happening, one can numerically study the transition mechanism.

A second application presented here will be 'smoothing', i.e. reconstructing a signal from a noisy observation. Since all information which is available about such a signal is contained in the conditional distribution of the signal given the observation, one can solve smoothing problems by understanding this conditioned distribution.

The text is structured as follows: we start by presenting some well-known sampling techniques in Section 6.2, namely Metropolis sampling and the Langevin method. In Section 6.3 we introduce an infinite-dimensional generalization of Langevin sampling. Section 6.4 explains how this technique can be used to study conditioned diffusions in general, and Section 6.5 considers the special case of smoothing problems. Finally, in Section 6.6, we show how the infinite-dimensional Langevin method can be combined with Metropolis sampling to obtain numerically efficient methods. The conclusion in Section 6.7 contains some pointers to extensions of the method and open problems.

6.2 Sampling techniques

Sampling is the process of constructing random values, distributed according to a prescribed target distribution. Since our aim is to derive a numerically useful method, we are specifically interested in constructions which can be implemented on a computer. Generating random values in a computer program is usually done in two steps: first one uses a pseudo-random number generator to generate 'random' values for some simple distribution (usually the uniform distribution on the unit interval) and then, in a second step, these values are transformed to obtain the desired target distribution. In this text we will only consider the second step, i.e. we will assume the availability of a source of uniform or Gaussian distributed random numbers and describe methods to transform given random values in order to obtain values with the correct distribution.

We give an overview of some established sampling techniques which we will use later in the text. Since our aim is to sample distributions on infinite-dimensional spaces, we restrict the presentation to techniques which can be applied in this context.

6.2.1 The Metropolis–Hastings algorithm

A commonly used sampling technique is based on the *Metropolis–Hastings algorithm*. The idea behind this method is to modify a given Markov chain, using a rejection mechanism, in order to obtain a Markov chain with a given stationary distribution. This new Markov chain can then be used as the basis of an MCMC algorithm.

Theorem 6.1 *Let P be the transition kernel of a Markov chain taking values in some measurable space $(\mathcal{X}, \mathcal{F}, \mu)$. Let μ be a probability measure on $\mathcal{X}$. Assume that $\mu(\mathrm{d}y)P(y, \mathrm{d}x)$ is absolutely continuous* *w.r.t.* $\mu(\mathrm{d}x)P(x, \mathrm{d}y)$ *on* $\mathcal{X} \times \mathcal{X}$. *Inductively construct a process* $(X_n)_{n \in \mathbb{N}}$ *as follows: for* $n \in \mathbb{N}$ *let* $Y_n \sim P(X_{n-1}, \cdot)$ *and* U_n *be uniformly distributed on* $[0, 1]$, *where* Y_n, *given* X_{n-1}, *is conditionally independent of* $X_1, \ldots, X_{n-2}$ *and* U_n *is independent of everything else, and let*

$$X_n = \begin{cases} Y_n & \text{if } U_n \le \alpha(X_{n-1}, Y_n), \\ X_{n-1} & \text{otherwise,} \end{cases}$$

where α *is the (truncated) Radon–Nikodym derivative*

$$\alpha(x, y) = 1 \wedge \frac{\mu(\mathrm{d}y)P(y, \mathrm{d}x)}{\mu(\mathrm{d}x)P(x, \mathrm{d}y)}.$$

Then $(X_n)_{n \in \mathbb{N}}$ *is a Markov chain with stationary distribution* μ.

The value $\alpha(X_{n-1}, Y_n)$ is called the *acceptance probability* at step n, the value Y_n is called a *proposal*.

This theorem allows us to change the distribution of any Markov chain which visits a large enough part of the state space, by rejecting some of the steps, in order to obtain a given stationary distribution. Then, assuming the resulting Markov chain is ergodic, one can compute expectations w.r.t. the stationary distribution μ, by taking ergodic averages:

$$\mathbb{E}_\mu(f) = \lim_{N \to \infty} \frac{1}{N} \sum_{n=1}^{N} f(X_n).$$

The usefulness of this method depends strongly on the magnitude of the acceptance rates: if $\alpha(X_{n-1}, Y_n)$ is often very small, convergence of the ergodic average will be very slow. For practical use, the transition kernel P has to be chosen in a way such that the acceptance probabilities are reasonably large.

A special case of the Metropolis–Hastings algorithm is when the transition kernel P does not depend on X_{n-1}. This corresponds to the case when the proposals are generated from an i.i.d. sequence. Because the acceptance probability at step n depends on the value X_{n-1}, the resulting Markov chain is no longer i.i.d. This method is called the *independence sampler*.

The independence sampler can for example be used to sample bridges of diffusion processes: if the target distribution μ is absolutely continuous w.r.t. Brownian bridges, one can use independent Brownian bridges as proposals. The independence sampler then gives a Markov chain with the bridge-distribution μ as its stationary distribution. See [5] for a discussion of this method.

6.2.2 Langevin sampling

Another method to obtain samples from a distribution on $\mathbb{R}^d$ with a density w.r.t. Lebesgue measure, called *Langevin sampling*, is given in the next theorem.

Theorem 6.2 *Let $\varphi \in C^2(\mathbb{R}^d, \mathbb{R})$ be a strictly positive probability density w.r.t. the Lebesgue measure λ. Then the SDE*

$$\mathrm{d}X_t = \nabla \log \varphi(X_t) \, \mathrm{d}t + \sqrt{2} \, \mathrm{d}W_t,$$

where W is a standard Brownian motion, has $\varphi \, \mathrm{d}\lambda$ as its stationary distribution.

The SDE in the theorem is called the *Langevin equation*. One observation which often turns out to be very useful in practice is the fact that, similar to the situation for the Metropolis–Hastings algorithm, the density φ needs to be known only up to a multiplicative constant: changing the constant does not change the resulting Langevin equation.

While this method is known to work well in high dimensions, at first it seems difficult to extend this technique to more general spaces, since the theorem uses a densities w.r.t. Lebesgue measure; the latter does not exist in infinite dimensions. But it transpires that there is a variant of the idea which can be generalized.

Theorem 6.3 *Let $L \in \mathbb{R}^{d \times d}$ be a symmetric matrix such that the SDE*

$$\mathrm{d}Z_t = LZ_t \, \mathrm{d}t + \sqrt{2} \, \mathrm{d}W_t$$

has a stationary distribution ν. Let $\varphi \in C^2(\mathbb{R}^d, \mathbb{R})$ be a strictly positive probability density w.r.t. ν. Then the SDE

$$\mathrm{d}X_t = \left(LX_t + \nabla \log \varphi(X_t)\right) \mathrm{d}t + \sqrt{2} \, \mathrm{d}W_t,$$

where W is standard Brownian motion, has $\varphi \, \mathrm{d}\nu$ as its stationary distribution.

A generalization of this theorem to infinite-dimensional spaces, presented in the next section, forms the basis of our sampling framework. Later, in Section 6.6, we will see how a discretized version of the Langevin equation can be used to generate proposals for the Metropolis–Hastings algorithm, thus combining the two methods presented in this section.

6.3 Langevin equations on path space

In this section we introduce the infinite-dimensional analogue of the Langevin equation from Section 6.2.2. The abstract setting is as follows: the SDEs in Theorem 6.3 are replaced by stochastic evolution equations taking values in a real Banach space E, continuously embedded into a real separable Hilbert space $\mathcal{H}$. In our applications the space $\mathcal{H}$ will mostly be the space $L^2([0,1],\mathbb{R}^d)$ and E will be some subspace of $C([0,1],\mathbb{R}^d)$.

6.3.1 Linear equations

In this section we derive a Hilbert space valued, linear SDE to sample from Gaussian distributions on $\mathcal{H}$. The results of this section can all be stated and proved in $\mathcal{H}$ without reference to the embedded Banach space E. A more detailed analysis can be found in [9].

Recall that a random variable X taking values in a separable Hilbert space $\mathcal{H}$ is said to be *Gaussian* if the law of $\langle y, X\rangle$ is Gaussian for every $y \in \mathcal{H}$. It is called *centred* if $\mathbb{E}\langle y, X\rangle = 0$ for every $y \in \mathcal{H}$. Gaussian random variables are determined by their mean $m = \mathbb{E}X \in \mathcal{H}$ and their covariance operator $\mathcal{C}\colon \mathcal{H} \to \mathcal{H}$ defined by

$$\langle y, \mathcal{C}x\rangle = \mathbb{E}\big(\langle y, X - m\rangle\langle X - m, x\rangle\big).$$

For details see e.g. [3]. We denote the Gaussian measure with mean m and covariance operator $\mathcal{C}$ by $\mathcal{N}(m,\mathcal{C})$.

We consider the $\mathcal{H}$-valued SDE

$$\mathrm{d}z_t = \mathcal{L}z_t\,\mathrm{d}t + \sqrt{2}\,\mathrm{d}w_t, \tag{6.1}$$

where w is a cylindrical Wiener process on $\mathcal{H}$ and $\mathcal{L} = -\mathcal{C}^{-1}$. A process z is a mild solution of (6.1), if it satisfies

$$z_t = \mathrm{e}^{\mathcal{L}t}z_0 + \sqrt{2}\int_0^t \mathrm{e}^{\mathcal{L}(t-s)}\,\mathrm{d}w_s.$$

Since this equation is linear, solutions are Gaussian processes and its invariant measure is a Gaussian measure on $\mathcal{H}$:

Theorem 6.4 *Let $\mu = \mathcal{N}(0,\mathcal{C})$ be a centred Gaussian measure on a separable Hilbert space $\mathcal{H}$. Then the corresponding evolution equation (6.1) with $\mathcal{L} = -\mathcal{C}^{-1}$ has continuous $\mathcal{H}$-valued mild solutions. Furthermore, it has μ as the unique invariant measure and there exists a constant K*

such that for every initial condition $x_0 \in \mathcal{H}$ one has

$$\left\| \mathcal{L}(z_t) - \mu \right\|_{\mathrm{TV}} \leq K \left(1 + \|x_0\|_{\mathcal{H}}\right) \exp\left(-\|\mathcal{C}\|_{\mathcal{H} \to \mathcal{H}}^{-1} t\right),$$

where $\| \cdot \|_{\mathrm{TV}}$ denotes the total variation distance between measures.

By the theorem, equation (6.1) can be used to sample from centred Gaussian measures and by considering the process $(z_t + m)_{t \geq 0}$ we have a sampling equation for arbitrary Gaussian measures $\mathcal{N}(m, \mathcal{C})$ on $\mathcal{H}$. To implement this method one has to identify the operator $\mathcal{L}$. The following example shows how this can be done in the cases which are the focus of our interest here.

Example 6.1. Consider the $\mathbb{R}^d$-valued, linear SDE

$$dZ_u = AZ_u \, du + B \, dW_u, \quad Z_0 = z^- \tag{6.2}$$

on the time interval $[0, 1]$, where $A, B \in \mathbb{R}^{d \times d}$ are matrices and $x^- \in \mathbb{R}^d$ is the starting point. The solution is a Gaussian process with mean $m(u) = \mathbb{E}(Z_u) = e^{uA} x^-$ and covariance function

$$C(u, v) = \mathrm{Cov}(X_u, X_v) = e^{uA} \left(\int_0^{u \wedge v} e^{-rA} BB^* e^{-rA^*} \, dr \right) e^{vA^*}$$

(see e.g. [10], Section 5.6) for reference). It is easy to check that the corresponding covariance operator $\mathcal{C}$ is given by

$$(\mathcal{C}x)(u) = \int_0^1 C(u, v) x(v) \, dv$$

for all $u \in [0, 1]$, $x \in L^2([0, 1], \mathbb{R}^d)$ and, assuming BB^* is invertible, the negative of its inverse $\mathcal{L} = -\mathcal{C}^{-1}$ is the restriction of the distributional differential operator

$$L = (\partial_u + A^*)(BB^*)^{-1}(\partial_u - A) \tag{6.3}$$

to the domain

$$\mathcal{D}(\mathcal{L}) = \left\{ f \in H^2([0, 1], \mathbb{R}^d) \mid f(0) = 0, \partial_u f(1) = Af(1) \right\}.$$

Thus, the stationary distribution of

$$dz_t = \mathcal{L}(z_t - m) \, dt + \sqrt{2} \, dw_t \tag{6.4}$$

is $\mathcal{N}(m, \mathcal{C})$.

Since $\mathcal{L}$ is a differential operator, we can write (6.4) as an SPDE. Using the fact that $Lm = 0$ on $(0, 1)$, this formally leads to the equation

$$\partial_t z(t, u) = Lz(t, u) + \sqrt{2}\,\partial_t w(t, u) \qquad \forall (t, u) \in (0, \infty) \times (0, 1)$$
$$z(t, 0) = z^-, \quad \partial_u z(t, 1) = Az(t, 1) \qquad \forall t \in (0, \infty)$$

where $\partial_t w$ is space-time white noise. By Theorem 6.4, the stationary distribution of this SPDE coincides with the distribution of the process Z.

6.3.2 Semilinear equations

In this subsection we will derive the infinite-dimensional analogue of Theorem 6.3. Here, the process $(z_t)_{t \geq 0}$ from (6.1) will correspond the $(Z_t)_{t \geq 0}$ in Theorem 6.3. The equation for $(X_t)_{t \geq 0}$ will be replaced by a semilinear equation of the form

$$dx_t = \mathcal{L}x_t\,dt + F(x_t)\,dt + \sqrt{2}\,dw_t, \tag{6.5}$$

where $\mathcal{L}$ is a linear operator on $\mathcal{H}$, the drift F maps E into E^*, w is a cylindrical Wiener process on $\mathcal{H}$, and the process x takes values in E. As in the previous subsection, we consider mild solutions of this equation.

For our application of sampling conditioned diffusions, presented in the next section, we will have a distribution-valued drift function F which is only defined on the Banach space of continuous functions. Thus we need the setting described above and cannot use the Hilbert space based theory as found e.g. in [6]. Proofs of the results presented here can be found in [8].

We start the presentation by giving the assumptions which we will require for our results. There are two assumptions on the linear operator $\mathcal{L}$:

(A1) The operator $\mathcal{L}$ is a self-adjoint, strictly dissipative operator on $\mathcal{H}$ which generates an analytic semigroup $S(t)$. The semigroup $S(t)$ can be restricted to a C_0-semigroup of contraction operators on E.

(A2) Let $\mathcal{H}^\alpha$ be the domain of $(-\mathcal{L})^\alpha$, equipped with the inner product $\langle x, y \rangle_\alpha = \langle (-\mathcal{L})^\alpha x, (-\mathcal{L})^\alpha y \rangle$. Then there exists an $\alpha \in (0, 1/2)$ such that $\mathcal{H}^\alpha \subset E$ densely, $(-\mathcal{L})^{-2\alpha}$ is nuclear in $\mathcal{H}$, and the Gaussian measure $\mathcal{N}\big(0, (-\mathcal{L})^{-2\alpha}\big)$ is concentrated on E.

We write $\mathcal{H}^{-\alpha}$ for the dual of $\mathcal{H}^\alpha$ and identify $\mathcal{H}^*$ with $\mathcal{H}$ in the usual way to get the following chain of inclusions:

$$\mathcal{H}^{1/2} \hookrightarrow \mathcal{H}^{\alpha} \hookrightarrow E \hookrightarrow \mathcal{H} \hookrightarrow E^* \hookrightarrow \mathcal{H}^{-\alpha} \hookrightarrow \mathcal{H}^{-1/2}.$$

To formulate our conditions on the drift F we will also use the subdifferential of the norm $\| \cdot \|_E$, defined as

$$\partial \|x\|_E = \left\{ x^* \in E^* \mid x^*(x) = \|x\|_E \text{ and } x^*(y) \leq \|y\|_E \ \forall y \in E \right\}$$

for every $x \in E$. We require the following conditions.

(A3) The nonlinearity $F \colon E \to E^*$ is Fréchet differentiable with

$$\|F(x)\|_{E^*} \leq C(1 + \|x\|_E)^N, \quad \text{and} \quad \|DF(x)\|_{E \to E^*} \leq C(1 + \|x\|_E)^N.$$

for every $x \in E$.

(A4) There exists a sequence of Fréchet differentiable functions $F_n \colon E \to E$ such that

$$\lim_{n \to \infty} \|F_n(x) - F(x)\|_{-\alpha} = 0$$

for all $x \in E$. For every $C > 0$ there exists a $K > 0$ such that for all $x \in E$ with $\|x\|_E \leq C$ and all $n \in \mathbb{N}$ we have $\|F_n(x)\|_{-\alpha} \leq K$. Furthermore, there is a $\gamma > 0$ such that

$$\langle x^*, F_n(x + y) \rangle \leq -\gamma \|x\|_E$$

holds for every $x^* \in \partial \|x\|_E$ and every $x, y \in E$ with $\|x\|_E \geq C(1 + \|y\|_E)^N$.

Our results currently require another, quite technical condition on the drift F which is given here as (A5). While this condition looks quite artificial, it is easy to verify that it holds for all applications discussed in this text.

(A5) For every $R > 0$, there exists a Fréchet differentiable function $F_R \colon E \to E^*$ such that

$$F_R(x) = \begin{cases} F(x) & \text{for } \|x\|_E \leq R, \\ 0 & \text{for } \|x\|_E \geq 2R, \end{cases} \tag{6.6}$$

and such that there exist constants C and N with

$$\|F_R(x)\|_{E^*} + \|DF_R(x)\|_{E \to E^*} \leq C(1 + R)^N,$$

for every $x \in E$.

Definition 6.5 *An E-valued and $(\mathcal{F}_t)$-adapted process x is called a* mild solution *of equation* (6.5), *if almost surely*

$$x_t = S(t)x_0 + \int_0^t S(t-s)F(x_s)\,\mathrm{d}s + z_t \qquad \forall t \geq 0$$

holds where z is the solution of the linear equation (6.1).

The drift F in (6.5) takes only values in E^* while the operator $\mathcal{L}$ will have a smoothing effect. There is a balance between these two effects and it is not *a priori* clear in which space the resulting process takes its values. The following theorem asserts that our assumptions are strong enough so that the solution is continuous with values in E.

Theorem 6.6 *Let $\mathcal{L}$ and F satisfy assumptions (A1)–(A4). Then for every initial value $x_0 \in E$ the equation (6.5) has a global, E-valued, unique mild solution.*

From Theorem 6.4 we know that the linear equation (6.1) has stationary distribution $\nu = \mathcal{N}(0, -\mathcal{L}^{-1})$. The following theorem, which is the infinite-dimensional analogue of Theorem 6.3, shows that we can again get an equation to sample from $\varphi\,d\nu$ by adding $\nabla \log \varphi$ to the drift of the linear equation.

Theorem 6.7 *Let $U: E \to \mathbb{R}$ be bounded from above and Fréchet differentiable. Assume that $\mathcal{L}$ and $F = U'$ satisfy assumptions (A1)–(A5), let $\nu = \mathcal{N}(0, -\mathcal{L}^{-1})$. Then the probability measure μ given by*

$$\mathrm{d}\mu(x) = c\,\mathrm{e}^{U(x)}\,\mathrm{d}\nu(x),$$

where c is a normalization constant, is the unique invariant measure for (6.5).

The following result helps to convert the preceding theorem into useful numerical methods: properties of the target distribution μ can be found by considering ergodic averages of the solution of the SDE (6.5).

Theorem 6.8 *Assume that (A1)–(A5) hold and let μ be the invariant measure for (6.5). Then one has*

$$\lim_{T \to \infty} \frac{1}{T} \int_0^T \varphi(x_t)\,\mathrm{d}t = \int_E \varphi(x)\,\mu(\mathrm{d}x), \quad \text{almost surely}$$

for every initial condition x_0 in the support of μ and for every bounded measurable function $\varphi: E \to \mathbb{R}$.

While these theorems are formulated in a way that helps to identify the stationary distribution of a given stochastic evolution equation, we will use the equations in the reverse way: starting with a target distribution μ with a known density $\varphi = e^U$ w.r.t. a Gaussian measure ν we will construct semilinear SDEs with invariant measure μ. From Theorem 6.7 we know that a possible choice for the drift is $F = (\log \varphi)'$, in direct analogy with the finite-dimensional result from Theorem 6.3. This procedure is illustrated in the following example.

Example 6.2. Consider the $\mathbb{R}^d$-valued SDE

$$dX_u = AX_u\,du + f(X_u)\,du + B\,dW_u, \quad X_0 = x^- \qquad (6.7)$$

on the time interval $[0,1]$, where $A, B \in \mathbb{R}^{d \times d}$ are matrices, $x^- \in \mathbb{R}^d$ is the starting point and W is a standard Brownian motion on $\mathbb{R}^d$. In this situation we can apply the following form of the Girsanov formula.

Lemma 6.9 *Let $\nu = \mathcal{L}(Z)$ be the distribution of the solution of the linear SDE (6.2) and $\mu = \mathcal{L}(X)$ be the distribution of the solution of (6.7). Assume that (6.7) has a.s. no explosions until time 1 and that B is invertible. Then μ has a density φ w.r.t. ν on $C([0,1], \mathbb{R}^d)$ which is given by*

$$\varphi(X) = \exp\Big(\int_0^1 (BB^*)^{-1} f(X_u)\,dX_u$$
$$- \int_0^1 \Big\langle AX_u + \frac{1}{2} f(X_u), (BB^*)^{-1} f(X_u) \Big\rangle\,du \Big).$$

If $f = -BB^* \nabla V$ for some potential $V : \mathbb{R}^d \to \mathbb{R}$, then φ can be written as

$$\varphi(X) = \exp\Big(V(X_0) - V(X_1)$$
$$- \int_0^1 \Big\langle AX_u + \frac{1}{2} f(X_u), (BB^*)^{-1} f(X_u) \Big\rangle + \frac{1}{2} \operatorname{div} f(X_u)\,du \Big).$$

Proof. Since X (by assumption) and Z (since it solves a linear SDE) have no explosions, we can apply Girsanov's theorem [7, Theorem 11A] to find the densities of $\mathcal{L}(X)$ and $\mathcal{L}(Z)$ w.r.t. the distribution of the Brownian motion $\mathcal{L}(BW)$. Taking the ratio of these two densities gives the first expression for φ. The second form of φ can be found by applying Ito's formula to $V(X)$ and substituting the result into the first part. $\square$

In the following we will assume that f has the required gradient form so we can use the second form of φ from the lemma (without the stochastic integral). From Example 6.1 we obtain a second-order differential operator $\mathcal{L}$ on $L^2\big([0,1],\mathbb{R}^d\big)$ such that

$$\mathrm{d}z_t = \mathcal{L}(z_t - m)\,\mathrm{d}t + \sqrt{2}\,\mathrm{d}w_t,$$

where m is the mean of Z, has stationary distribution ν. From Theorem 6.7 we see, assuming (A1)–(A5) are satisfied, that we can add the drift $F = (\log\varphi)'$ to this equation to obtain a $C\big([0,1],\mathbb{R}^d\big)$-valued SDE with stationary distribution μ. A simple calculation shows

$$F(x) = (BB^*)^{-1}f(x_1)\delta_1 - \nabla\Psi(x), \qquad \forall x \in C\big([0,1],\mathbb{R}^d\big),$$

where δ_1 is a Dirac mass at $u = 1$ and Ψ is given by

$$\Psi(\xi) = \Big\langle A\xi + \frac{1}{2}f(\xi), (BB^*)^{-1}f(\xi)\Big\rangle + \frac{1}{2}\operatorname{div}f(\xi) \qquad \forall \xi \in \mathbb{R}^d. \quad (6.8)$$

Under mild assumptions on A, B and f, the conditions for Theorems 6.6, 6.7 and 6.8 are satisfied and the stationary distribution of

$$\mathrm{d}x_t = \mathcal{L}(x_t - m)\,\mathrm{d}t + F(x_t)\,\mathrm{d}t + \sqrt{2}\,\mathrm{d}w_t$$

coincides with the distribution of the process X. An explicit set of assumptions on A, B, and f for the result to hold can be found in [8].

Again, we would like to write this equation as a stochastic partial differential equation. In order to do so, we should just add the drift F to the SPDE from Example 6.1. One complication is the presence of the Dirac-term in F. Since, assuming smooth w for this argument, the source term $(BB^*)^{-1}f(x_1)\delta_1$ will lead to a jump of size $f(x_1)$ in the u-derivative of the solution, we can incorporate the Dirac term in the boundary condition by formally writing the SPDE as

$$\partial_t x(t,u) = Lx(t,u) - \nabla\Psi(x(t,u)) + \sqrt{2}\,\partial_t w(t,u)$$
$$\forall (t,u) \in (0,\infty) \times (0,1),$$
$$x(t,0) = x^-, \quad \partial_u x(t,1) = Ax(t,1) + f(x(t,1)) \qquad \forall t \in (0,\infty).$$

6.4 Conditioned diffusions

In the previous sections we have seen how the Langevin sampling method can be generalized to infinite-dimensional situations and how this can be used to construct SPDEs which sample from the distribution of a finite-dimensional diffusion process. In this section we focus on our main interest of this text, namely on applying the presented techniques to sample from conditioned diffusion processes.

Consider the following $\mathbb{R}^d$-valued SDE on the time interval $[0, 1]$:

$$\mathrm{d}X_u = AX_u \,\mathrm{d}u + f(X_u)\,\mathrm{d}u + B\,\mathrm{d}W_u, \quad X_0 = x^-. \qquad (6.9)$$

As before, $A, B \in \mathbb{R}^{d \times d}$ are matrices, $x^- \in \mathbb{R}^d$ is the starting point, W is a standard Brownian motion on $\mathbb{R}^d$ and we assume that $f = -BB^* \nabla V$ for some potential $V : \mathbb{R}^d \to \mathbb{R}$ and that B is invertible. Our aim is to construct an SPDE which has the distribution of X, conditioned on some event C, as its stationary distribution.

Let Z be the solution of the linear SDE

$$\mathrm{d}Z_u = AZ_u \,\mathrm{d}u + B\,\mathrm{d}W_u, \quad Z_0 = x^-, \qquad (6.10)$$

and set $m(u) = \mathbb{E}(Z(u)|C)$ for all $u \in [0, 1]$. In the cases we consider here, the event C is such that $\mathcal{L}(Z|C)$ is still Gaussian. The general idea is to perform a construction consisting of the following steps.

(i) Use the results of Section 6.3.1 to obtain an L^2-valued SDE which has the centred Gaussian measure $\mathcal{L}(Z - m|C)$ as its stationary distribution.

(ii) Use the Girsanov formula and results about conditional distributions to derive the density of the conditional distribution $\mathcal{L}(X|C)$ w.r.t. $\mathcal{L}(Z|C)$. Using substitution, this gives the density of the shifted distribution $\mathcal{L}(X - m|C)$ w.r.t. the centred measure $\mathcal{L}(Z - m|C)$.

(iii) Use the results of Section 6.3.2 and the density from step 2 to derive a $C([0, 1], \mathbb{R}^d)$-valued SDE with stationary distribution $\mathcal{L}(X - m|C)$. Shifting the process by m reverses the centring from step 2 and gives the required sampling equation. Optionally write the L^2-valued SDE as an SPDE.

Combining all these steps leads to an SPDE which samples from the conditional distribution $\mathcal{L}(X|C)$ in its stationary measure. The details of the above steps depend on the specific situation under consideration. We will study one special case in detail in the next section, where we develop a method for nonlinear filtering by using the Langevin method to sample from the distribution of some signal given the observations. In the remainder of this section we illustrate the technique in a simpler setting.

Example 6.3. We can use the technique described above to construct an SPDE which samples bridges from

$$\mathrm{d}X_u = AX_u \,\mathrm{d}u + f(X_u)\,\mathrm{d}u + B\,\mathrm{d}W_u, \quad X_0 = x^-, \quad X_1 = x^+, \qquad (6.11)$$

that is, the stationary distribution of the SPDE coincides with the distribution of solutions of the SDE (6.9), conditioned on $X_1 = x^+$.

Step 1: We need to find an SPDE with stationary distribution $\mathcal{L}(Z|Z_1 = x^+)$. Mean and covariance of the conditioned process can be found by conditioning the random variable $(Z(u), Z(v), Z(1))$ for $u \le v \le 1$ on the value of $Z(1)$. Since this is a finite-dimensional Gaussian random variable, mean and covariance of the conditional distribution can be explicitly calculated. Let m and C be the mean and covariance function of $\mathcal{L}(Z)$. Then $\mathcal{L}(Z|Z_1 = x^+)$ is a Gaussian measure with mean

$$\tilde{m}(u) = m(u) + C(u,1)C(1,1)^{-1}(x^+ - m(1))$$

and covariance operator $\tilde{C}$ with $\tilde{C}x = \int \tilde{C}(\cdot, v)x(v)\, dv$ where the covariance function is given by

$$\tilde{C}(u,v) = C(u,v) - C(u,1)C(1,1)^{-1}C(1,v).$$

A simple calculation shows that $\tilde{\mathcal{L}} = -\tilde{C}^{-1}$ is again the differential operator L from (6.3), but this time on the domain

$$\mathcal{D}(\tilde{\mathcal{L}}) = \{f \in H^2([0,1], \mathbb{R}^d) \mid f(0) = 0, f(1) = 0\}.$$

Thus the stationary distribution of

$$dz_t = \tilde{\mathcal{L}}z_t\, dt + \sqrt{2}\, dw_t$$

is $\mathcal{L}(Z - \tilde{m}|Z_1 = x^+)$ by Theorem 6.4.

Step 2: We have already seen in Example 6.2 that the density of $\mathcal{L}(X)$ w.r.t. $\mathcal{L}(Z)$ is given by

$$\varphi(X) = \exp\left(V(x^-) - V(X_1) - \int_0^1 \Psi(X_u)\, du\right)$$

with the Ψ from equation (6.8). The following lemma shows that the density of $\mathcal{L}(X|X_1 = x^+)$ w.r.t. $\mathcal{L}(Z|Z_1 = x^+)$ coincides, up to a multiplicative constant, with φ.

Lemma 6.10 *Let P, Q be probability measures on $S \times T$ where $(S, \mathcal{A})$ and $(T, \mathcal{B})$ are measurable spaces and let $X: S \times T \to S$ and $Y: S \times T \to T$ be the canonical projections. Assume that P has a density φ w.r.t. Q and that the conditional distribution $Q_{X|Y=y}$ exists. Then the conditional distribution $P_{X|Y=y}$ exists and is given by*

$$\frac{dP_{X|Y=y}}{dQ_{X|Y=y}}(x) = \begin{cases} \frac{1}{c(y)}\varphi(x,y) & \text{if } c(y) > 0, \\ 1 & \text{otherwise} \end{cases}$$

where $c(y) = \int_S \varphi(x, y) \, dQ_{X|Y=y}(x)$ for all $y \in T$.

Thus, the density of $\mathcal{L}(X - m|X_1 = x^+)$ w.r.t. $\mathcal{L}(Z - m|Z_1 = x^+)$ is

$$\tilde{\varphi}(X) = c \exp\left(\int_0^1 \Psi(X_u + m_u) \, du\right)$$

for some normalization constant c where Ψ is given by (6.8).

Step 3: Assuming that the conditions for Theorems 6.6, 6.7 and 6.8 are satisfied, the stationary distribution of

$$d\tilde{x} = \mathcal{L}\tilde{x} \, dt - \nabla\Psi(\tilde{x} + \tilde{m}) \, dt + \sqrt{2} \, dw_t$$

is then $\mathcal{L}(X - \tilde{m}|X_1 = x^+)$. Thus the process $x = \tilde{x} + \tilde{m}$, solving

$$dx_t = \mathcal{L}(x_t - \tilde{m}) \, dt - \nabla\Psi(x_t) \, dt + \sqrt{2} \, dw_t, \tag{6.12}$$

can be used to sample from the target distribution $\mathcal{L}(X|X_1 = x^+)$.

Finally, we can rewrite this evolution equation as an SPDE: since the mean $\tilde{m}$ satisfies $\tilde{m}(0) = x^-$, $\tilde{m}(1) = x^+$ and $L\tilde{m} = 0$ on $(0, 1)$, we can formally write (6.12) in the form

$$\partial_t x(t, u) = Lx(t, u) - \nabla\Psi(x(t, u)) + \sqrt{2} \, \partial_t w(t, u),$$
$$\forall(t, u) \in (0, \infty) \times (0, 1),$$
$$x(t, 0) = x^-, \quad x(t, 1) = x^+ \quad \forall t \in (0, \infty).$$

Note that use of this formulation no longer requires knowledge of the conditioned mean $\tilde{m}$.

Figure 6.1 shows the result of a numerical simulation which implements the method derived in Example 6.3 to sample bridges of the process (6.11). For the simulation we use the drift

$$f(x) = -\left(\frac{(x-1)^2(x+1)^2}{1+x^2}\right)' = x\left(\frac{8}{(1+x^2)^2} - 2\right), \tag{6.13}$$

$A = 0$, $B = I$ and the end-points $x^- = -1$ and $x^+ = +1$. To allow the process to transition a few times between the stable equilibrium points, we chose $u \in [0, 100]$. The upper panel illustrates how one can get an approximation to a typical sample path of (6.11): it displays $u \mapsto x(t, u)$ for a big value of t. Assuming that the sampling process is already close to equilibrium, this path should closely resemble a typical bridge path. The second panel illustrates how statistical properties of the bridges can be approximated by taking ergodic averages using Theorem 6.8. The

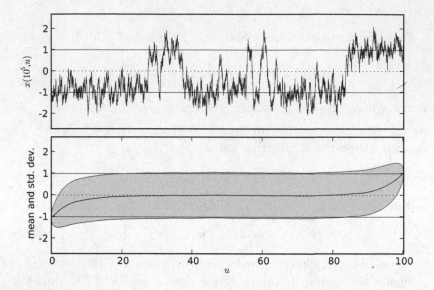

Fig. 6.1. *Illustration of the bridge sampling method from Example 6.3. The drift f in (6.11) is chosen to be the gradient of a double-well potential with stable equilibrium points at −1 and 1 and an unstable equilibrium point at 0 (see (6.13)), the process starts in $x^- = -1$ and is conditioned on ending up in $x^+ = +1$. The upper panel shows the value of the Langevin SPDE at time $t = 10^5$ as a function of u. This is an approximation to a typical bridge path. The lower panel shows a one-standard-deviation band around the mean of the solution as a function of u, obtained by taking averages over the interval $t \in [0, 10^5]$. This gives an approximation for the mean and standard deviation of the bridge process (6.11).*

line in the centre of the shaded band shows

$$\bar{m}(u) = \frac{1}{T} \int_0^T x(t, u) \, \mathrm{d}u$$

as a function of u for a big value of T. By Theorem 6.8 we have $\bar{m}(u) \approx \tilde{m}(u)$. The width of the band is given by

$$\bar{\sigma}(u) = \left(\frac{1}{T} \int_0^T \left(x(t, u) - \bar{m}(u) \right)^2 \mathrm{d}u \right)^{1/2}.$$

Again by Theorem 6.8, $\bar{\sigma}(u)$ is approximately equal to the standard deviation of the bridge at position u.

6.5 Nonlinear smoothing

In this section we will give a more challenging application of the method developed in the previous sections: we will describe how nonlinear smoothing problems can be formulated as a problem of sampling conditioned diffusions and how it can be solved using Langevin sampling.

Let $d = m + n$ with $m, n \in \mathbb{N}$ and consider a d-dimensional diffusion process given by

$$\mathrm{d}X_u = AX_u\,\mathrm{d}u + f(X_u)\,\mathrm{d}u + B\,\mathrm{d}W_u, \quad X_0 = x^-,$$

where B is invertible and $(BB^*)^{-1}f$ is a gradient. Assume that only the last n components of this process can be observed and that we want to gain as much knowledge as possible about the unobserved m components from one observation of the last n components. We write $X_u = (X_u^{(1)}, X_u^{(2)}) \in \mathbb{R}^m \times \mathbb{R}^n$ and call $X^{(1)}$ the 'signal' and $X^{(2)}$ the 'observation'.

While the problem is formally very easy to solve, the solution is just the conditional distribution $\mathcal{L}(X^{(1)}|X^{(2)})$, the task of actually algorithmically computing this solution is quite challenging. There are two commonly used ways of solving this problem: the traditional method, employed for example in particle filters, is to use the Zakai equation to construct an approximation to the density of $\mathcal{L}(X_u^{(1)}|X_v^{(2)}, 0 \leq v \leq u)$. The solution we propose here is to construct an SPDE which samples from the distribution $\mathcal{L}(X^{(1)}|X^{(2)})$. Questions about this conditional distribution can then be answered by considering ergodic averages. It transpires that this way of solving the smoothing problem can be derived as a special case of the general technique of sampling from conditioned diffusions which we presented in section 6.4.

Commonly, finding $\mathcal{L}(X_u^{(1)}|X_v^{(2)}, 0 \leq v \leq u)$ is called 'filtering' and finding $\mathcal{L}(X^{(1)}|X^{(2)})$ is called 'smoothing'. The standard methods, like the Kalman filter and particle filter based approaches, proceed by first solving the filtering problem and then, optionally, solving the smoothing problem in a second, backward sweep over the data. The method we propose here directly solves the smoothing problem and thus all observations must be present from the start of the computation.

6.5.1 Construction of the smoothing SPDE

The construction of the SPDE to sample from the conditional distribution of $X^{(1)}$ given $X^{(2)}$ follows the steps outlined in Section 6.4. We start the construction by considering the linear, $\mathbb{R}^{m+n}$-valued SDE

$$dZ_u = AZ_u \, du + B \, dW_u, \qquad Z_0 = x^-, \qquad (6.14)$$

which will give our reference measure as before. Since this SDE is linear, its solution is a Gaussian process, and thus the distribution $\mathcal{L}(Z^{(1)}|Z^{(2)})$ is also Gaussian. First we have to identify the mean and covariance of this distribution. The abstract mechanism we use here is given in the following lemma.

Lemma 6.11 *Let $\mathcal{H} = \mathcal{H}_1 \oplus \mathcal{H}_2$ be a separable Hilbert space with projectors $\Pi_i \colon \mathcal{H} \to \mathcal{H}_i$. Let $(Z^{(1)}, Z^{(2)})$ be an $\mathcal{H}$-valued Gaussian random variable with mean $m = (m_1, m_2)$ and positive definite covariance operator $\mathcal{C}$ and define $\mathcal{C}_{ij} = \Pi_i \mathcal{C} \Pi_j^*$. Then the conditional distribution of $Z^{(1)}$ given $Z^{(2)}$ is Gaussian with mean*

$$m_{1|2} = m_1 + \mathcal{C}_{12} \mathcal{C}_{22}^{-1} \big(Z^{(2)} - m_2 \big)$$

and covariance operator

$$\mathcal{C}_{1|2} = \mathcal{C}_{11} - \mathcal{C}_{12} \mathcal{C}_{22}^{-1} \mathcal{C}_{21}.$$

If we define as above $\mathcal{L} = (-\mathcal{C})^{-1}$ and formally define $\mathcal{L}_{ij} = \Pi_i \mathcal{L} \Pi_j^*$, then a simple formal calculation shows that $m_{1|2}$ and $\mathcal{C}_{1|2}$ are expected to be given by

$$m_{1|2} = m_1 - \mathcal{L}_{11}^{-1} \mathcal{L}_{12} \big(Z^{(2)} - m_2 \big), \qquad \mathcal{C}_{1|2} = -\mathcal{L}_{11}^{-1}. \qquad (6.15)$$

In contrast to the lemma above, the relations (6.15) do not hold in general (consider for example the case $\mathcal{C}_{1|2} = 0$), but in our situation it can be shown that domains for the operators $\mathcal{L}_{ij}$ can be chosen so that all of the given expressions are defined and that the conditional mean and expectation really have the form given in (6.15). Details of this construction can be found in [9, lemma 4.6]. By Theorem 6.4, the $L^2([0,1], \mathbb{R}^d)$-valued SDE

$$dz_t = \mathcal{L}_{11} z_t \, dt + \sqrt{2} \, dw_t$$

has $\mathcal{L}(Z^{(1)} - m_{1|2}|Z^{(2)})$ as its stationary distribution. We have already identified the differential operator $\mathcal{L}$ in Section 6.3.1.

Now we can just follow the programme outlined in Section 6.4: the version of Girsanov formula from Lemma 6.9 gives the density φ of $\mathcal{L}(X)$ w.r.t. $\mathcal{L}(Z)$. From Lemma 6.10 we know that the conditional density $\varphi_{1|2}$ of $X^{(1)}$ given $X^{(2)}$ differs from $x \mapsto \varphi(x, X^{(2)})$ only by a multiplicative constant which depends only on $X^{(2)}$. Thus we have

$\nabla \log \varphi_{1|2} = \nabla_1 \log \varphi(\,\cdot\,, X^{(2)})$ where ∇ denotes the Fréchet derivative on $C([0,1], \mathbb{R}^d)$ and ∇_1 denotes the Fréchet derivative w.r.t. the first m components. By Theorem 6.7 the equation

$$\mathrm{d}x_t = \mathcal{L}_{11}(x_t - m_{1|2})\, \mathrm{d}t + \nabla_1 \log \varphi(x_t, X^{(2)})\, \mathrm{d}t + \sqrt{2}\mathrm{d}w_t$$

has $\mathcal{L}(X^{(1)}|X^{(2)})$ as its stationary distribution and thus can be used as a Monte Carlo method to solve the smoothing problem.

Example 6.4. In the standard smoothing setup the signal $X^{(1)}$ evolves on its own without reference to the observation. The observation depends both on the signal and on additional noise. To fit this situation in the framework described above we consider the following case:

$$A = \begin{pmatrix} 0 & 0 \\ A_{21} & 0 \end{pmatrix}, \qquad B = \begin{pmatrix} B_{11} & 0 \\ 0 & B_{22} \end{pmatrix},$$

with $A_{21} \in \mathbb{R}^{n \times m}$, $B_{11} \in \mathbb{R}^{m \times m}$ and $B_{22} \in \mathbb{R}^{n \times n}$. Furthermore let $V(x,y) = V_1(x) + V_2(y)$ and $f = -BB^*\nabla V$. In this situation, equation (6.14) can be written as

$$\begin{aligned} \mathrm{d}X^{(1)} &= f_1(X^{(1)})\, \mathrm{d}u + & B_{11}\, \mathrm{d}W^{(1)} \\ \mathrm{d}X^{(2)} &= f_2(X^{(2)})\, \mathrm{d}u + A_{21}X^{(1)}\, \mathrm{d}u + B_{22}\, \mathrm{d}W^{(2)} \end{aligned} \tag{6.16}$$

with $f_1 = -B_{11}B_{11}^*\nabla V_1$ and $f_2 = -B_{22}B_{22}^*\nabla V_2$.

For this choice of the matrices A and B the differential operator $\mathcal{L}$ is

$$\begin{aligned} \begin{pmatrix} L_{11} & L_{12} \\ L_{21} & L_{22} \end{pmatrix} &= \begin{pmatrix} \partial_u & A_{21}^* \\ 0 & \partial_u \end{pmatrix} \begin{pmatrix} B_{11}B_{11}^* & 0 \\ 0 & B_{22}B_{22}^* \end{pmatrix}^{-1} \begin{pmatrix} \partial_u & 0 \\ -A_{21} & \partial_u \end{pmatrix} \\ &= \begin{pmatrix} \partial_u (B_{11}B_{11}^*)^{-1}\partial_u - A_{21}^*(B_{22}B_{22}^*)^{-1}A_{21} & A_{21}^*(B_{22}B_{22}^*)^{-1}\partial_u \\ -\partial_u(B_{22}B_{22}^*)^{-1}A_{21} & \partial_u(B_{22}B_{22}^*)^{-1}\partial_u \end{pmatrix}, \end{aligned}$$

defined on some appropriate domain. A more detailed analysis, as found in [9], Section 4, shows that $\mathcal{L}_{11}$ in (6.15) can be taken to be L_{11} on the domain

$$\mathcal{D}(\mathcal{L}_{11}) = \{f \in H^2([0,1], \mathbb{R}^d) \mid f(0) = 0, \partial_u f(1) = 0\}.$$

From (6.15) we find that $m_{1|2}$ is the solution of

$$\mathcal{L}_{11}(m_{1|2} - m_1) = -A_{21}^*(B_{22}B_{22}^*)^{-1}\left(\frac{\mathrm{d}Z^{(2)}}{\mathrm{d}u} - m_2\right).$$

Here $\frac{\mathrm{d}Z^{(2)}}{\mathrm{d}u}$ only exists as a distribution, but since $\mathcal{L}_{11}$ is a second-order differential operator, the solution $m_{1|2}$ is a smooth function.

The density of $\mathcal{L}(X^{(1)}|X^{(2)})$ w.r.t. $\mathcal{L}(Z^{(1)}|Z^{(2)})$ can be simplified because of the simple structure of the matrices A and B: we get

$$\varphi(X^{(1)}|X^{(2)}) = c\exp\Big(-V_1(X_1^{(1)}) - \frac{1}{2}\int_0^1 |B_{11}^{-1}f_1(X_u^{(1)})|^2$$
$$+ \operatorname{div} f_1(X_u^{(1)})\,\mathrm{d}u$$
$$- \int_0^1 \langle X_u^{(1)}, A_{21}^*(B_{22}B_{22}^*)^{-1}f_2(X_u^{(2)})\rangle\,\mathrm{d}u\Big)$$

for some normalization constant c and the density of the target distribution $\mu = \mathcal{L}(X^{(1)} - m_{1|2}|X^{(2)})$ w.r.t. $\nu = \mathcal{L}(Z^{(1)} - m_{1|2}|Z^{(2)})$ is $\varphi(X - m_{1|2}|Y)$. Thus, for given $X^{(2)}$, the Fréchet derivative of $\log\varphi(X^{(1)}|X^{(2)})$ is

$$F(x) = \nabla_1 \log\varphi(x|X^{(2)})$$
$$= -\nabla V_1(x_1)\delta_1 - \nabla\Phi(x) - A_{21}^*(B_{22}B_{22}^*)^{-1}f_2(X^{(2)})$$

for all $x \in C([0,1], \mathbb{R}^m)$, where δ_1 is a Dirac mass at $u = 1$ and

$$\Phi(\xi) = \frac{1}{2}\big(|B_{11}^{-1}f_1(\xi)|^2 + \operatorname{div} f_1(\xi)\big) \qquad \forall \xi \in \mathbb{R}^m.$$

With F we have found the drift to be used in Theorem 6.7: the equation

$$\mathrm{d}\tilde{x}_t = \mathcal{L}_{11}\tilde{x}_t\,\mathrm{d}t - \nabla\Phi(\tilde{x}_t + m_{1|2})\,\mathrm{d}t - A_{21}^*(B_{22}B_{22}^*)^{-1}f_2(X^{(2)})\,\mathrm{d}t$$
$$- \nabla V_1(\tilde{x}_t(1) + m_{1|2}(1))\delta_1\,\mathrm{d}t + \sqrt{2}\,\mathrm{d}w_t$$

is ergodic and has $\mathcal{L}(X^{(1)} - m_{1|2}|X^{(2)})$ as its stationary distribution. Defining $x_t = \tilde{x}_t + m_{1|2}$ for all $t \geq 0$ and formally writing the equation for x as an SPDE again, we find that the SPDE

$$\partial_t x(t,u) = (B_{11}B_{11}^*)^{-1}\partial_u^2 x(t,u) - \nabla\Phi\big(x(t,u)\big)$$
$$+ A_{21}^*(B_{22}B_{22}^*)^{-1}\Big(\frac{\mathrm{d}X^{(2)}}{\mathrm{d}u}(u) - f_2(X^{(2)}(u)) - A_{21}x(t,u)\Big)$$
$$+ \sqrt{2}\,\partial_t w(t,u) \tag{6.17}$$

with boundary conditions

$$x(t,0) = 0, \quad \partial_u x(t,1) = f_1\big(x(t,1)\big)$$

for all $t \geq 0$ is the Langevin equation on $C([0,1], \mathbb{R}^m)$ to sample from the distribution $\mathcal{L}(X^{(1)}|X^{(2)})$. In the derivation above we did not check whether the conditions (A1), ..., (A5), which are required for our sampling method, are satisfied. In general this depends on the specific choice

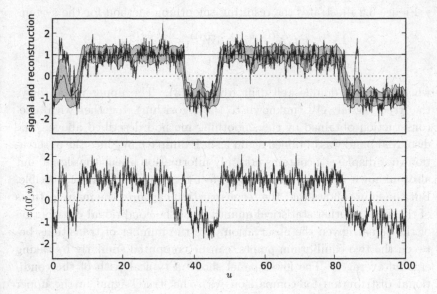

Fig. 6.2. *Illustration of the smoothing method from Example 6.4. The upper panel shows the true signal (unknown to the algorithm) together with a one-standard-deviation band around the mean of the sampling SPDE. This band can be seen as a reconstruction of the signal, but since the observation (not displayed) incorporates additional noise, a perfect reconstruction is not possible. The lower panel shows a typical path of the conditional distribution of the signal, given the observation, obtained by taking the value of the sampling SPDE at large t.*

of f, A and B. A (quite technical) set of conditions such that the theorems apply can be found in [8].

A comparison between the sampling equation derived here and the equation derived in Example 6.2 to sample from the unconditional distribution $\mathcal{L}(X^{(1)})$ reveals that the only difference caused by the conditioning is the presence of the term

$$A_{21}^*(B_{22}B_{22}^*)^{-1}\Big(\frac{\mathrm{d}X^{(2)}}{\mathrm{d}u}(u) - f_2(X^{(2)}(u)) - A_{21}x(t,u)\Big).$$

The presence of this additional drift term moves the solution of the sampling SPDE towards paths $X^{(1)}$ which minimize the 'energy' of the noise required for the second equation in (6.16) to hold.

Figure 6.2 illustrates the resulting smoothing method for the system

$$dX_u^{(1)} = f(X_u^{(1)}) \, du + dW_u^{(1)}, \qquad X_0^{(1)} = -1$$
$$dX_u^{(2)} = X_u^{(1)} \, du + dW_u^{(2)}, \qquad X_0^{(2)} = 0$$

where f is the double-well drift from (6.13). The upper panel shows the 'true' signal $X^{(1)}$ (unknown to the algorithm), together with a reconstruction obtained by the smoothing method described above. The displayed band was obtained again as in Example 6.3. Since the observation (not displayed) contains not only information about the signal, but also unknown additional noise, a perfect reconstruction is not possible. But the figure shows that the reconstruction captures the main features of the signal. Other statistical quantities of the conditional distribution of the signal, given the observation, like the number of transitions between the two equilibrium points, can be computed similarly by taking ergodic averages. The lower panel shows a typical path of the conditional distribution for comparison with the 'true' signal in the upper panel.

6.5.2 Some remarks about smoothing

While the sampling technique developed in the previous section solves the same problem as traditional filters/smoothers do, it does so in a very different way: instead of trying to obtain the density of the conditional distribution, our method constructs samples from the conditional distribution which can be used as the basis of an MCMC algorithm.

Filtering and smoothing are sometimes used in high-dimensional situations. For example, applications in weather prediction, where filtering is used to incorporate the observed weather data into a model, now use values of d which are as big as 10^7 or 10^8. When d is big, a map from $\mathbb{R}^d$ to $\mathbb{R}$ like the density of $\mathcal{L}(X_u^{(1)} | X_v^{(2)}, 0 \le v \le u)$ is a complex object which is very hard to accurately represent in a computer. A standard way to deal with this problem, used in particle filter methods, is to approximate the conditional distribution as a sum of weighted Dirac masses. Another approach is to approximate the conditional distribution by a Gaussian, but in high-dimensional situations even storing the covariance matrix of this Gaussian has a non-negligible cost and sometimes even further approximations are necessary. In comparison, a map from $\mathbb{R}$ to $\mathbb{R}^d$, like the paths obtained by the smoothing method discussed here, is a much more manageable object. Thus the discussed method might be advantageous in high dimensions when smoothing is required and not just filtering.

Another observation to note is that the situation considered in Example 6.4 is just one of many possible situations where a Langevin sampling based filtering method can be derived. Similar constructions are possible in many situations, for example it is easy to derive a sampling SPDE to sample from a diffusion conditioned on discrete noisy observations. See [2] for further examples.

More information about filtering and pointers into the literature can be found in [1].

6.6 Metropolis–Hastings algorithm on path space

In the previous sections we showed how an infinite-dimensional analogue of the Langevin equation can be used to sample from the distribution of conditioned diffusions. One of the main motivations behind this approach is that it directly translates into an implementable algorithm to solve these sampling problems. In this section we will discuss some issues which arise in this context. When implementing the method for practical use one has to numerically solve the sampling SPDE (6.5) and thus one has to discretise this equation in both 'space' u and time t. The two kinds of discretization raise different issues and here we will mostly focus on the effects of discretizing time.

There are two constraints which affect the choice of time step size Δt. Firstly, we are only interested in the stationary distribution of the sampling SPDE and thus, for our purposes, it doesn't matter if the numerical simulation accurately represents the trajectories of the solution but we require the invariant measure of the discretized equation to be close to the invariant measure of the exact equation. And, secondly, we will use the numerical solution to approximate ergodic averages as in Theorem 6.8 and thus we need to simulate the solution over long time intervals. This leads to a trade-off in the choice of the step size Δt: small Δt requires many steps to cover big time intervals and thus makes the resulting method computationally expensive whereas big Δt leads to big discretization error and makes the results less accurate.

One solution to this dilemma is the following idea, described in more detail in [4]: one can use a discretisation with a big step size Δt, but then use a rejection mechanism to compensate for the resulting discretisation error. More specifically, given an approximation $\hat{x}(t)$ to the exact solution x_t, a discretized version of the evolution equation gives an approximation to the solution at time $t + \Delta t$. But instead of directly using the computed value $\hat{y}(t + \Delta t)$ for the numerical solution, one can use it as the proposal in a Metropolis–Hastings algorithm and either accept or reject it as described in Theorem 6.1.

A (partially implicit) Euler method for solving the equation

$$\mathrm{d}x_t = \mathcal{L}x_t\,\mathrm{d}t + F(x_t)\,\mathrm{d}t + \sqrt{2}\,\mathrm{d}w_t \tag{6.18}$$

from Section 6.3.2 can be formulated as

$$X_{n+1} = X_n + \mathcal{L}\big(\theta X_{n+1} + (1-\theta)X_n\big)\,\Delta t + F(X_n)\,\Delta t + \sqrt{2}\,\xi_n,$$

where the ξ_n have the same distribution as the increments of the cylindrical Wiener process w. The parameter $\theta \in [0,1]$ controls the implicitness of the method. We did not include implicitness in the evaluation of the nonlinear part F of the drift, to make it easy to solve the iteration equation for X_{n+1}: one gets

$$X_{n+1} = \big(I - \Delta t\,\theta\mathcal{L}\big)^{-1}\big(I + \Delta t\,(1-\theta)\mathcal{L}\big)X_n$$
$$+ \Delta t\,\big(I - \Delta t\,\theta\mathcal{L}\big)^{-1}F(X_n) + \sqrt{2}\big(I - \Delta t\,\theta\mathcal{L}\big)^{-1}\xi_n. \tag{6.19}$$

It is not *a priori* clear what space this equation takes values in, since the cylindrical Wiener process w, and thus its increments, do not live in the Hilbert space $\mathcal{H}$. However, since $-\mathcal{L}^{-1}$ is trace class (it is the covariance of a Gaussian measure, see Section 6.3.1), for $\theta > 0$ the operator $A = (I - \Delta t\,\theta\mathcal{L})^{-1}$ is Hilbert–Schmidt and thus the random increments $A\xi_n$ take values in $\mathcal{H}$. For this reason we restrict ourselves to the case $\theta > 0$ here.

When trying to use X_{n+1} as the proposal in a Metropolis algorithm, there is the following surprising dichotomy.

Theorem 6.12 *Let $\mathcal{H} = L^2([0,1], \mathbb{R}^d)$ and let $\mathcal{L}$ be a symmetric, negative definite operator on $\mathcal{H}$ as in Section 6.3.2. Let μ be the invariant measure of (6.18). Let $\theta > 0$ and define the transition kernel P on $\mathcal{H}$ by*

$$P(x, \cdot) = \mathcal{L}(X_{n+1}|X_n = x) \qquad \forall x \in \mathcal{H},$$

where X_{n+1} is defined by equation (6.19). Then there are two cases:

(a) If $\theta \neq 1/2$, then the distributions $\mu(\mathrm{d}y)P(y, \mathrm{d}x)$ and $\mu(\mathrm{d}x)P(x, \mathrm{d}y)$ on $\mathcal{H} \times \mathcal{H}$ are singular w.r.t. each other and thus the Metropolis algorithm cannot be used.

(b) If $\theta = 1/2$, then the distributions $\mu(\mathrm{d}y)P(y, \mathrm{d}x)$ and $\mu(\mathrm{d}x)P(x, \mathrm{d}y)$ on $\mathcal{H} \times \mathcal{H}$ are equivalent and thus the Metropolis algorithm can be used.

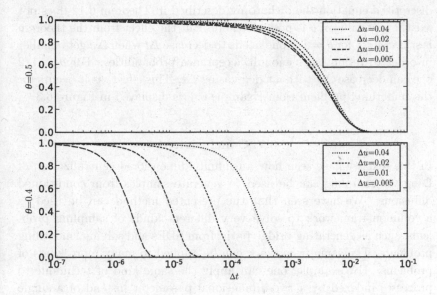

Fig. 6.3. *This figure illustrates how the acceptance rates of the Metropolised algorithm for a discretized version of the smoothing problem from Example 6.4 depend on the time discretization step size Δt. The different curves correspond to different space discretizations Δu. The upper panel gives the average acceptance probabilities in equilibrium for $\theta = 1/2$. In this case the Metropolis-Hastings algorithm can also be applied to the infinite-dimensional problem. The lower panel illustrates the case $\theta = 0.4$, which only makes sense for the discretized equation. One can see that the method degenerates as $\Delta u \to 0$.*

Proof. For $X \in \mathcal{H}$ let $\langle X \rangle_u$ be the quadratic variation of X until time u. Then, by imitating the proof of [4], Proposition 4.1, for $(X, Y) \sim \mu(dx)P(x, dy)$ we have

$$\langle Y \rangle_u = \frac{(1 - \theta)^2}{\theta^2} \langle X \rangle_u \qquad \forall u \in [0, 1]$$

almost surely. Since under μ the quadratic variation is a.s. constant, this shows that the measures in part (a) are singular whenever $(1 - \theta)^2/\theta^2 \neq 1$, i.e. when $\theta \neq 1/2$. A proof for part (b) when $\mathcal{L}$ is a second derivative operator with Dirichlet boundary conditions can be found in [4], Theorem 4.1. An inspection of this proof reveals that it still holds in the more general situation considered here. □

To implement the methods described in this text, the Langevin SPDE needs to be discretized in 'space' as well as in time. Some remarks about the required space discretization can be found in [4]. For the space-discretized equation the dichotomy described in Theorem 6.12 does not exist, every value of θ is possible there. But the effect from the theorem is still visible: for $\theta \neq 1/2$ one needs to decrease Δt when Δu gets smaller in order to retain large enough acceptance probabilities. For $\theta = 1/2$ one can decrease Δu without decreasing Δt. This effect, as it occurs for the smoothing problem from Example 6.4, is displayed in Figure 6.3.

6.7 Conclusion

In this text we have seen how an infinite-dimensional generalization of Langevin sampling can be used to generate samples from conditioned diffusions. We have seen that the presented method can be used as a common framework to solve very different kinds of sampling problems, such as generating bridge paths from SDEs and solving smoothing problems. The same framework can be applied to many more kinds of problems. For example, one can apply the same kind of technique to processes indexed by a two-dimensional parameter instead of a single time variable. This might give rise to techniques which could be applied in image analysis, for example. It will be interesting to see what future applications will be developed based on this.

Throughout this text, we concentrated on sampling techniques which were direct generalizations of the finite-dimensional result from Theorem 6.3. But of course, since we are only interested in the stationary distribution, the sampling equation is not uniquely determined; many choices are possible. For example, in the finite-dimensional case the SDE

$$dX_t = LX_t \, dt + \nabla \log \varphi(X_t) \, dt + \sqrt{2} \, dW_t$$

and the 'preconditioned' SDE

$$dX_t = GLX_t \, dt + G\nabla \log \varphi(X_t) \, dt + \sqrt{2G} \, dW_t,$$

where G is a symmetric, positive matrix, share the same invariant measure. This relation carries over to the infinite-dimensional situation. By taking e.g. $G = -L^{-1}$ one obtains a new equation with very different properties: the cylindrical noise is now replaced by a significantly more regular noise, but the smoothing effect from the operator L is no longer present. This technique is discussed in [8] and [4]. Other choices of sampling equations, including second-order equations, are discussed in [1].

In the further development of the presented sampling techniques, several open problems remain. For example, in this text we always assumed that the densities we obtained from the Girsanov formula can be rewritten without resorting to a stochastic integral. This restricted the choice of drift functions for the underlying diffusion processes to functions which are a gradient plus a linear function. It transpires that this restriction is not easily lifted: the theorems presented here no longer apply and, while it is easy to formally derive sampling equations, it is very difficult to even give sense to the resulting equations. A conjecture about the results in the non-gradient case can be found in [8].

Other open problems include questions about efficient implementation of the method. This requires numerical solutions of the resulting SPDEs and a careful choice of step sizes for discretisation is required.

Bibliography

[1] A. Apte, M. Hairer, A. M. Stuart, and J. Voss. Sampling the posterior: An approach to non-gaussian data assimilation. *Physica D: Nonlinear Phenomena*, 230(1–2):50–64, 2007.

[2] A. Apte, C. K. R. T. Jones, A. M. Stuart, and J. Voss. Data assimilation: Mathematical and statistical perspectives. To appear in the *International Journal for Numerical Methods in Fluids*, 2007.

[3] V. I. Bogachev. *Gaussian Measures*, volume 62 of *Mathematical Surveys and Monographs*. American Mathematical Society, Providence, RI, 1998.

[4] A. Beskos, G. Roberts, A. M. Stuart, and J. Voss. MCMC methods for diffusion bridges. Submitted, 2007.

[5] S. Chib, M. Pitt, and N. Shephard. Likelihood based inference for diffusion driven models. Working paper, 2004.

[6] G. Da Prato and J. Zabczyk. *Stochastic Equations in Infinite Dimensions*, volume 44 of *Encyclopedia of Mathematics and its Applications*. Cambridge University Press, 1992.

[7] K. D. Elworthy. *Stochastic Differential Equations on Manifolds*, volume 70 of *London Mathematical Society Lecture Note Series*. Cambridge University Press, 1982.

[8] M. Hairer, A. M. Stuart, and J. Voss. Analysis of SPDEs arising in path sampling, part II: The nonlinear case. *Ann Appl Prob*, 17(5/6):1657–1706, 2007.

[9] M. Hairer, A. M. Stuart, J. Voss, and P. Wiberg. Analysis of SPDEs arising in path sampling, part I: The Gaussian case. *Communications in Mathematical Sciences*, 3(4):587–603, 2005.

[10] I. Karatzas and S. E. Shreve. *Brownian Motion and Stochastic Calculus*. Berlin: Springer, second edition, 1991.

7

Asymptotic formulae for coding problems and intermediate optimization problems: a review

Steffen Dereich

Department of Mathematical Sciences
University of Bath
Bath BA2 7AY, United Kingdom

Abstract

In this paper we review asymptotic formulae for the three coding problems induced by a range constraint (*quantization*), an entropy constraint (*entropy coding*) and a mutual information constraint (*distortion rate function*) on the approximation $\hat{X}$ to an original signal X. We consider finite-dimensional random variables as well as stochastic processes as original signal. A main objective of the paper is to explain relationships between the original problems and certain intermediate convex optimization problems (the *point* or *rate allocation problem*). These intermediate optimization problems often build the basis for the proof of asymptotic formulae.

7.1 Introduction

In practice, discretization problems appeared for the first time in the context of Pulse-Code-Modulation (*PCM*). Provided that a source signal has limited bandwidth, it is possible to reconstruct it perfectly, when knowing its values on an appropriate grid of time points. However, in order to digitize the data, it still remains to *quantize* the signal at these time points.

Today the motivation for studying discretization problems is twofold. The *information theoretic* approach asks for a digital representation for an incoming random signal X. The digital representation is typically a finite number of binary digits, which is thought of as the basis for

Trends in Stochastic Analysis, ed. J. Blath, P. Mörters and M. Scheutzow.
Published by Cambridge University Press. ©Cambridge University Press 2008.

further reconstructions of the original signal. The digital representation is to be stored on a computer or to be sent over a channel, and one needs to find a good trade-off between the quality of the reconstruction and the complexity of the describing representation. When adopting the information-theoretic viewpoint, the signal X describes a real-world object that is to be translated into an approximation $\hat{X}$ on the basis of measurements or a first very fine discretization.

A second viewpoint is adopted in the context of *simulation* of random vectors. Often a random object X is described implicitly via a probability distribution and one asks for discrete simulations $\hat{X}$ of X. In that case, one wants to approximate the distribution $\mathbb{P}_X$ of X (by some discrete distribution $\mathbb{P}_{\hat{X}}$) rather than the random element X itself. Recently, the finding of finitely supported approximations has attracted much attention since it can be used to define cubature formulae: for a real function f defined on the state space of X, one approximates the expectation $\mathbb{E}f(X)$ by the expectation $\mathbb{E}f(\hat{X})$. Provided one knows all possible outcomes of $\hat{X}$ including the corresponding probability weights, one can explicitly express the latter expectation as a finite sum. Now the aim is to find $\mathbb{P}_{\hat{X}}$ that is supported on a small set and that is close to $\mathbb{P}_X$ in a Wasserstein metric.

Although both problems share a lot of similarities, there are also important differences: when adopting the coding perspective, one is interested to have easily implementable mechanisms that map X into its digital representation, and that reconstruct $\hat{X}$ on the basis of the representation. In this case the algorithm does not need to incorporate the exact probability distribution of $\hat{X}$. Conversely, if the approximation is to be used as a cubature formula, one needs to get hold of the probability weights of the approximation $\hat{X}$. However, it is not necessary to couple the simulated random element $\hat{X}$ to an original signal, since one only asks for an approximating distribution in that case.

The information-theoretic viewpoint has been treated in an abundance of monographs and articles. The most influential contributions were Shannon's works on the Source Coding Theorem [44]. General accounts on this topic are for instance the monographs by Cover and Thomas [8] and by Ihara [29]. Surveys on the development of source coding theory and quantization were published by Kieffer [31] and by Gray and Neuhoff [27] in the engineering literature. Moreover, a mathematical account on finite-dimensional quantization was provided by Graf and Luschgy [26].

Applications of discretization schemes in finance are described, for instance, in Pagès and Printems [43] and further relations to quadrature problems are established in Creutzig et al. [10].

In this paper, the discretization problem is formalized as a minimization problem of certain objective functions (representing the approximation error) under given complexity constraints on the approximation. Our main focus lies on the development of asymptotic formulae for the best-achievable approximation error when the information content of the approximation tends to infinity. Our problems will be described via nonconvex optimization problems, and an explicit evaluation of minimizers is typically not feasible. However, it is often possible to relate the original problem to simpler convex optimization problems. Solving these intermediate optimization problems often leads to asymptotic formulae in the associated approximation problems. Moreover, the solutions give a crude description of "good" approximation schemes, and they can be used for initializing numerical methods or to define close to optimal approximation schemes. In this paper we try to convey intuition on how intermediate optimization problems can be derived and used to solve the asymptotic coding problems. In doing so we survey several classical and recent results in that direction. Our complexity constraints are given through constraints on the disk space needed to save the binary representation or the number of points maximally attained by the approximation. Note that these constraints do not incorporate the computational complexity of finding a closest representation or computing the probability weights of the approximating random variable, so that the results presented here can only serve as benchmark (which may show the optimality of certain schemes) or as a basis for the development of feasible approximation schemes.

To be more precise, we think of a problem constituted of a *source signal* (the *original*), modelled as a random variable X taking values in a measurable space E, and an error criterion given as a product measurable map $\rho : E \times E \to [0, \infty]$ (the *distortion measure*). Then the objective is to minimize the expectation

$$\mathbb{E}\rho(X, \hat{X}) \tag{7.1}$$

over a set of E-valued random variables $\hat{X}$ (the *reconstruction*) satisfying a *complexity constraint* (also called *information constraint*). Often we shall also consider certain moments $s > 0$ of the approximation error as objective function. In that case (7.1) is replaced by

$$\mathbb{E}[\rho(X, \hat{X})^s]^{1/s} = \left\| \rho(X, \hat{X}) \right\|_{L^s(\mathbb{P})}. \tag{7.2}$$

Mainly, we shall be working with *norm-based distortion measures*, i.e. distortions ρ admitting the representation $\rho(x, \hat{x}) = \|x - \hat{x}\|$.

7.1.1 The information constraints

We work with three different complexity constraints that all depend on a parameter $r \geq 0$, the *rate* (see for instance [32] for further motivation):

- $\log |\text{range}(\hat{X})| \leq r$ (*quantization constraint*)
- $H(\hat{X}) \leq r$, where H denotes the entropy of $\hat{X}$ (*entropy constraint*)
- $I(X; \hat{X}) \leq r$, where I denotes the Shannon mutual information of X and $\hat{X}$ (*mutual information constraint*).

Here and elsewhere, we use the standard notations for entropy and mutual information:

$$H(\hat{X}) = \begin{cases} -\sum_x p_x \log p_x & \text{if } \hat{X} \text{ is discrete with weights } (p_x), \\ \infty & \text{otherwise} \end{cases}$$

and

$$I(X, \hat{X}) = \begin{cases} \int \log \frac{\mathrm{d}\mathbb{P}_{X,\hat{X}}}{\mathrm{d}\mathbb{P}_X \otimes \mathbb{P}_{\hat{X}}} \, \mathrm{d}\mathbb{P}_{X,\hat{X}} & \text{if } \mathbb{P}_{X,\hat{X}} \ll \mathbb{P}_X \otimes \mathbb{P}_{\hat{X}}, \\ \infty & \text{otherwise.} \end{cases}$$

In general, we denote by $\mathbb{P}_Z$ the distribution function of a random variable Z indicated in the subscript. As we shall explain later, the rigorous treatment of the minimization problem often leads to simpler intermediate convex optimization problems. The solution to such an intermediate optimization problem represents a rough description of how much information good coding schemes assign to different locations. Its information can be used to appropriately initialize numerical schemes used to find close to optimal codebooks. Alternatively, one may locally apply (not necessarily optimal) quantizers with rates adjusted to the solution of the corresponding convex optimization problem, and thus enhance coding schemes.

In order to code a countable set I, one typically uses *prefix-free codes*; these are maps $\psi : I \rightarrow \{0,1\}^*$ ($\{0,1\}^*$ denoting the strings of binary digits of finite length) such that for any $i \neq j$ in I the code $\psi(i)$ is not a prefix of $\psi(j)$. That means a prefix-free code is naturally related to a binary tree, where the leaves of the tree correspond to the elements of I and the code describes the path from the root to the leaf (0 meaning left child, 1 meaning right child). Clearly, such a code allows us to decode the original message uniquely. Moreover, a concatenation of a fixed number of prefix-free codes leads to a new prefix-free code.

Suppose for now that the logarithms are taken to the base 2. Then the quantization constraint implies that the set $\mathcal{C} = \text{range}(\hat{X}) \subset E$ (called *codebook*) is of size $\log|\mathcal{C}| \leq r$. Hence, there exists a prefix-free representation for $\hat{X}$ (or for $\mathcal{C}$) of length at most $\lceil r \rceil$. The quantization constraint is a worst-case constraint on the code length. Note that any finite set $\mathcal{C} \subset E$ with $\log|\mathcal{C}| \leq r$ induces a *quantizer* $\hat{X}$ for X of rate r via

$$\hat{X} = \operatorname*{argmin}_{\hat{x} \in \mathcal{C}} \rho(X, \hat{X}).$$

Next, suppose that $\hat{X}$ satisfies the entropy constraint. Using Lempel–Ziv coding there exists a code ψ for the range of $\hat{X}$ such that

$$\mathbb{E}\,\text{length}(\psi(\hat{X})) < r + 1.$$

Thus the entropy constraint is an average-case constraint on the bit length needed. The mutual information constraint, is motivated by Shannon's celebrated source coding theorem which will be stated later (see Theorem 7.11). For a general account on information theory we refer the reader to the monographs by Cover and Thomas [8] and by Ihara [29].

For a reconstruction $\hat{X}$ the constraints are ordered as follows

$$I(X; \hat{X}) \leq H(\hat{X}) \leq \log|\text{range}(\hat{X})|,$$

so that the mutual information constraint is the least restrictive. Moreover, our notions of information satisfy the following *additivity property*: suppose that $\hat{X}_1$ and $\hat{X}_2$ are random vectors attaining values in some Borel space and suppose that $\hat{X}$ is given as $\varphi(\hat{X}_1, \hat{X}_2)$ where φ is some measurable function; then

- $\log|\text{range}(\hat{X})| \leq \log|\text{range}(\hat{X}_1)| + \log|\text{range}(\hat{X}_2)|$
- $H(\hat{X}) \leq H(\hat{X}_1) + H(\hat{X}_2)$
- $I(X; \hat{X}) \leq I(X; \hat{X}_1) + H(\hat{X}_2)$.

Essentially, these estimates state that when combining the information contained in two random vectors $\hat{X}_1$ and $\hat{X}_2$ the resulting random vector has information content less than the sum of the single information contents. Note that the mutual information is special in the sense that the property $I(X; \hat{X}) \leq I(X; \hat{X}_1) + I(X; \hat{X}_2)$ does not hold in general!

Now let us introduce the notation used for the minimal values in the minimization problems. When minimizing (7.2) under the quantization constraint for a given rate r and moment s, the minimal value will be denoted by $D^{(q)}(r,s)$:

$$D^{(q)}(r,s) = \inf\{\|\rho(X,\hat{X})\|_{L^s(\mathbb{P})} : \log|\operatorname{range}(\hat{X})| \le r\}.$$

Here the (q) in the upper index refers to the quantization constraint. Similarly, we write $D^{(e)}(r,s)$ and $D(r,s)$ for the minimal values induced by the entropy and mutual information constraint, respectively.

Strictly speaking, the quantities $D^{(e)}(\cdot)$ and $D(\cdot)$ depend on the underlying probability space. Since we prefer a notion of complexity that only depends on the distribution μ of the underlying signal X, we allow extensions of the probability space when minimizing over $\hat{X}$ so that the pair $(X, \hat{X})$ can assume any probability distribution on the product space having first marginal μ.

Sometimes the source signal or the distortion measure might not be clear from the context. In that case we include this information in the notation. For instance, we write $D(r, s|\mu, \|\cdot\|)$ when considering a μ-distributed original under the objective function

$$\mathbb{E}[\|X - \hat{X}\|^s]^{1/s},$$

and we write $D(r|\mu, \rho)$ in order to refer to the objective function

$$\mathbb{E}[\rho(X, \hat{X})].$$

Thereafter we write $f \sim g$ iff $\lim \frac{f}{g} = 1$, while $f \lesssim g$ stands for $\limsup \frac{f}{g} \le 1$. Finally, $f \approx g$ means

$$0 < \liminf \frac{f}{g} \le \limsup \frac{f}{g} < \infty,$$

and $f \precsim g$ means

$$\limsup \frac{f}{g} < \infty.$$

Moreover, we use the Landau symbols o and $\mathcal{O}$.

7.1.2 Synopsis

The paper is outlined as follows. In the next section, we provide asymptotic formulae for finite-dimensional quantization problems. Historically the concept of an intermediate convex optimization problem appeared there for the first time. We proceed in Section 7.3 with a treatment of Banach space-valued Gaussian signals under norm-based distortion measures. In Section 7.4 we encounter the next intermediate optimization problem when considering Hilbert space valued Gaussian signals. Then it follows a treatment of 1-dimensional diffusions in Section 7.5. Here again convex minimization problems play a crucial role in the analysis. Finally, we conclude the paper in Section 7.6 with two further approaches for deriving (weak) asymptotic formulae. One is applicable for 1-dimensional stochastic processes and the other one for Lévy processes.

7.2 Finite-dimensional signals

This section is devoted to some classical and some recent results on finite-dimensional quantization. First we will introduce the concept of an intermediate optimization problem in the classical setting. The following section then proceeds with a treatment of Orlicz-norm distortions.

7.2.1 Classical setting

Suppose now that X is a $\mathbb{R}^d$-valued original, where d denotes an arbitrary integer. Moreover, fix a norm $|\cdot|$ on $\mathbb{R}^d$, let ρ be the corresponding norm-based distortion and fix a moment $s > 0$. We shall consider the asymptotic quantization problem under the additional assumptions that the absolutely continuous part μ_c of $\mu = \mathcal{L}(X)$ (w.r.t. Lebesgue measure) does not vanish and that $\mathbb{E}[|X|^{\tilde{s}}] < \infty$ for some $\tilde{s} > s$.

Originally, this problem has been addressed in [47] and [6]. More recently, Graf and Luschgy [26] gave a rigorous proof of a general high-resolution formula in their mathematical account on finite-dimensional quantization. Let us state the asymptotic formula:

Theorem 7.1

$$\lim_{n \to \infty} n^{1/d} D^{(q)}(\log n, s) = c(|\cdot|, s) \left\| \frac{d\mu_c}{d\lambda^d} \right\|_{L^{d/(d+s)}(\lambda^d)}^{1/s}.$$

Here, $c = c(|\cdot|, s) \in (0, \infty)$ is a constant depending on the norm $|\cdot|$ and the moment s only.

Note that the singular part of μ has no influence on the asymptotic quantization problem. Unfortunately, the constant c is known explicitly only in a few cases (e.g. when $|\cdot|$ denotes a 2-dimensional Euclidean norm or a d-dimensional supremum norm). Notice that it is convenient to state the result in the number of approximating points rather than in the rate.

A heuristic derivation of the point allocation problem

Let us shortly give the main ideas along which one can prove the theorem. First one considers a $[0,1)^d$-uniformly distributed random variable X as original. Based on the self similarity of the uniform distribution one can show that

$$\lim_{n \to \infty} n^{1/d} \, D^{(q)}(\log n, s) = \inf_{n \in \mathbb{N}} n^{1/d} \, D^{(q)}(\log n, s) = c, \qquad (7.3)$$

where the constant c lies in $(0, \infty)$ and agrees with the c in the asymptotic formula above.

In order to explain the findings for the general case, first assume that μ is absolutely continuous w.r.t. Lebesgue measure and that its density $h = \frac{\mathrm{d}\mu}{\mathrm{d}\lambda^d}$ can be represented as

$$h(x) = \sum_{i=1}^{m} \mathbf{1}_{B_i}(x) \, h_i, \qquad (7.4)$$

where $m \in \mathbb{N}$ and $B_i \subset \mathbb{R}^d$ $(i = 1, \dots, m)$ denote disjoint cuboids having side length l_i and $(h_i)_{i=1,\dots,m}$ is a $\mathbb{R}_+^m$-valued vector. We want to analyze a quantization scheme based on a combination of optimal codebooks for the measures $\mathcal{U}(B_i)$, where in general $\mathcal{U}(B)$ denotes the uniform distribution on the set B. We fix a function $\xi : \mathbb{R}^d \to [0, \infty)$ with $\int \xi = 1$ of the form

$$\xi(x) = \sum_{i=1}^{m} \mathbf{1}_{B_i}(x) \, \xi_i \qquad (7.5)$$

and denote by $\mathcal{C}_i$ an $\mathcal{U}(B_i)$-optimal codebook of size $n \, \xi_i \, \lambda^d(B_i) = n \int_{B_i} \xi$. Here, n parameterizes the size constraint of the global codebook $\mathcal{C}$ defined as $\mathcal{C} = \bigcup_{i=1}^{m} \mathcal{C}_i$. Notice that the inferred error $\mathbb{E} \min_{\hat{x} \in \mathcal{C}} |X - \hat{x}|^s$ is for large n approximately equal to

$$\sum_{i=1}^{m} \underbrace{l_i^s c^s (n\xi_i \lambda^d(B_i))^{-s/d}}_{\text{av. error in box } i} \underbrace{h_i \lambda^d(B_i)}_{\mathbb{P}(X \in B_i)} = \left(\frac{c}{n^{1/d}}\right)^s \sum_{i=1}^{m} \xi_i^{-s/d} \, h_i \lambda^d(B_i)$$

$$= \left(\frac{c}{n^{1/d}}\right)^s \int \xi(x)^{-s/d} h(x) \, \mathrm{d}x.$$

Here the average error in box i can be explained as follows: the terms $c^s (n\xi_i \lambda^d(B_i))^{-s/d}$ represent the error obtained for the $\mathcal{U}([0,1)^d)$ distribution when applying an optimal codebook of size $n\xi_i \lambda^d(B_i)$ (due to (7.3)). The term l_i^s arises from the scaling needed to get from $\mathcal{U}([0,1)^d)$ to $\mathcal{U}(B_i)$ and the fact that we consider the sth moment.

We arrive at the minimization problem

$$\int \xi(x)^{-s/d} h(x) \, \mathrm{d}x = \min!$$ (7.6)

where the infimum is taken over all non-negative ξ with $\int \xi = 1$. This convex optimization problem has a unique solution that will be given explicitly below. In particular, its solution is of the form (7.5) in our particular setting.

On the other hand, accepting that for "good" codebooks a typical realization of X that has fallen into B_i is also approximated by an element of B_i allows one to conclude that the above construction is close to optimal. In particular, one might correctly guess that the minimal value of the minimization problem (7.6) leads to the optimal asymptotic rate in the quantization problem.

We refer to the function ξ as *point density* and to the minimization problem as the *point allocation problem*.

Rigorous results related to the point allocation problem

The importance of the point allocation problem in the classical setting was firstly conjectured by Lloyd and Gersho (see [24]). First rigorous proofs are due to Bucklew [5] (see also [25]). In the general setting (as introduced in the first lines of this section) one can prove rigorously the following statement.

Theorem 7.2 *For each $n \in \mathbb{N}$ fix a codebook $\mathcal{C}(n) \subset \mathbb{R}^d$ with at most n elements and consider the associated empirical measure*

$$\nu_n = \frac{1}{n} \sum_{\hat{x} \in \mathcal{C}(n)} \delta_{\hat{x}}.$$

Supposing that for some infinite index set $I \subset \mathbb{N}$ the measures $(\nu_n)_{n \in I}$ converge vaguely to a measure ν, one has

$$\liminf_{\substack{n \to \infty \\ n \in I}} n^{s/d} \, \mathbb{E}[\min_{\hat{x} \in \mathcal{C}(n)} |X - \hat{x}|^s] \geq c^s \int_{\mathbb{R}^d} \xi(x)^{-s/d} \, h(x) \, \mathrm{d}x,$$ (7.7)

where $\xi = \frac{\mathrm{d}\nu_c}{\mathrm{d}\lambda^d}$ and $h = \frac{\mathrm{d}\mu_c}{\mathrm{d}\lambda^d}$. On the other hand, one has for an arbitrary measurable function $\xi : \mathbb{R}^d \to [0, \infty)$ with $\int \xi \leq 1$,

$$\limsup_{n \to \infty} n^{s/d} \, D^{(q)}(\log n, s)^s \leq c^s \int_{\mathbb{R}^d} \xi(x)^{-s/d} \, h(x) \, \mathrm{d}x.$$ (7.8)

Consequently, the solution of the point allocation problem leads to the asymptotics of the quantization error given in Theorem 7.1. Equations (7.7) and (7.8) are even more powerful: they show that for an asymptotically optimal family $(\mathcal{C}(n))_{n\in\mathbb{N}}$ of codebooks, in the sense that

$$|\mathcal{C}(n)| \leq n \quad \text{and} \quad \mathbb{E}[\min_{\hat{x}\in\mathcal{C}(n)} |X-\hat{x}|^s]^{1/s} \lesssim D^{(q)}(\log n, s),$$

any accumulation point of $(\nu_n)_{n\in\mathbb{N}}$ is a minimizer of the point allocation problem (more explicitly $\xi = \frac{d\nu_c}{d\lambda^d}$ with ν_c denoting the continuous part of the accumulation point is a minimizer). Since the objective function is even *strictly convex* in ξ, the minimizer is unique up to Lebesgue null-sets. Together with the property that the optimal ξ satisfies $\int \xi = 1$, one concludes that $(\nu_n)_{n\in\mathbb{N}}$ converges in the weak topology to a measure having as density the solution ξ to the point allocation problem.

It remains to solve the point allocation problem. Applying the reverse Hölder inequality with adjoint indices $p = -d/s$ and $q = d/(s+d)$ one gets for ξ with $\int \xi \leq 1$

$$\int_{\mathbb{R}^d} \xi(x)^{-s/d} h(x)\,dx \geq \|\xi^{-s/d}\|_{L^p(\mathbb{R}^d)} \|h\|_{L^q(\mathbb{R}^d)}$$

$$= \|\xi\|_{L^1(\mathbb{R}^d)}^{-s/d} \|h\|_{L^q(\mathbb{R}^d)} \geq \left\|\frac{d\mu_c}{d\lambda^d}\right\|_{L^{d/(d+s)}(\mathbb{R}^d)}. \tag{7.9}$$

On the other hand, the inequalities can be replaced by equalities when choosing ξ as the optimal point density

$$\xi(x) = \frac{1}{\int h^{d/(d+s)}} h(x)^{d/(d+s)}. \tag{7.10}$$

Consequently, it follows that for an asymptotically optimal family of codebooks $(\mathcal{C}(n))_{n\in\mathbb{N}}$ (in the sense mentioned above) the empirical measures ν_n ($n\in\mathbb{N}$) converge to a probability measure having density ξ given by (7.10).

Moreover, estimate (7.7) gives immediately a lower bound for the efficiency of *mismatched codebooks*: when using asymptotically optimal codebooks $(\mathcal{C}(n))_{n\in\mathbb{N}}$ for the moment s in the case where the underlying distortion is taken to a different moment $s' > 0$, one has:

$$\liminf_{n\to\infty} n^{1/d}\, \mathbb{E}[\min_{\hat{x}\in\mathcal{C}(n)} |X-\hat{x}|^{s'}]^{1/s'}$$

$$\geq c(|\cdot|, s') \left(\int h^{d/(d+s)}\right)^{1/d} \left(\int h^{1-\frac{s'}{d+s}}\right)^{1/s'}.$$

Interestingly, the latter integral is infinite when $\lambda^d(\{h > 0\}) = \infty$ and $s' > d + s$, so that in that case the rate of convergence to zero is of a different order. For further reading concerning mismatched codebooks (in particular, corresponding upper bounds), we refer to the paper by Graf, Luschgy and Pagès [25].

We have presented this classical quantization result in detail, since it represents a prototype of a coding result. Typically, the optimization problem (related to a coding problem) is non-convex and it is not possible to give explicit solutions. In practice, one needs to apply numerical or probabilistic methods to obtain good solutions. In order to analyze the problem, a powerful tool is to relate the original problem to an intermediate convex minimization problem. Such a relation then typically allows us to derive asymptotic formulae. The intermediate problem is also of practical interest: for instance, in the above example, the optimal point density can be used to initialize procedures used for generating close to optimal codebooks.

7.2.2 Orlicz-norm distortion

In the classical setting, the objective function for the approximation loss is given as

$$\mathbb{E}[|X - \hat{X}|^s]^{1/s}.$$

Let now $f : [0, \infty) \to [0, \infty)$ be an increasing left continuous function with $\lim_{t \downarrow 0} f(t) = 0$. It is natural to pose the question of what happens when replacing the objective function by

$$\mathbb{E}[f(|X - \hat{X}|)].$$

This problem has been treated by Delattre et al. in [11] in the case where f behaves like a polynomial at 0, which means

$$\lim_{\varepsilon \downarrow 0} \frac{f(\varepsilon)}{\kappa \varepsilon^\alpha} = 1,$$

for two parameters $\alpha, \kappa > 0$. They found that, under certain concentration assumptions on X, the asymptotic quantization problem behaves as in the classical case with $s = \alpha$. In particular, the optimal point density does not differ from the one derived before.

However, if the rate is too small such that typical approximation deviations fall into a range where $\kappa \varepsilon^\alpha$ is not a good approximation to $f(\varepsilon)$, then the corresponding optimal point density does not give a reasonable description of good codebooks. Thus the optimal point density may not be useful for moderate rates. A possible remedy is to consider Orlicz-norm distortions instead. Let us first introduce the necessary notation.

Let $\varphi : [0, \infty) \to [0, \infty)$ be a monotonically increasing, left continuous function that satisfies $\lim_{t \downarrow 0} \varphi(t) = 0$. Note that this implies that φ is lower semicontinuous. We assume that $\varphi \neq 0$ and let $|\cdot|$ denote an arbitrary norm on $\mathbb{R}^d$. For any $\mathbb{R}^d$-valued random variable Z, the Orlicz norm $\|\cdot\|_\varphi$ is defined as

$$\|Z\|_\varphi = \inf\Big\{ t \geq 0 : \mathbb{E}\,\varphi\Big(\frac{|Z|}{t}\Big) \leq 1 \Big\},$$

with the convention that the infimum of the empty set is equal to infinity. Actually, the left continuity of φ together with monotone convergence implies that the infimum is attained, whenever the set is nonempty. We set

$$L^\varphi(\mathbb{P}) = \{ Z : Z\ \mathbb{R}^d\text{-valued r.v. with } \|Z\|_\varphi < \infty \}.$$

Then $\|\cdot\|_\varphi$ defines a norm on $L^\varphi(\mathbb{P})$ if φ is convex (otherwise the triangle inequality fails to hold). We do not assume convexity for φ. Nevertheless, with a slight abuse of notation, we allow ourselves to call $\|\cdot\|_\varphi$ an Orlicz norm. Choosing $\varphi(t) = t^p$, $p \geq 1$, yields the usual $L^p(\mathbb{P})$-norm.

Now we choose as objective function

$$\|X - \hat{X}\|_\varphi$$

and we denote by $D^{(q)}(r)$ the corresponding minimal quantization error of rate $r \geq 0$. The concept of introducing an intermediate optimization problem is also applicable in the Orlicz-norm setting. Let us quote the main results taken from [21].

The constant c of the classical setting is now replaced by a convex decreasing function $g : (0, \infty) \to [0, \infty)$ defined via

$$g(\zeta) = \lim_{n \to \infty} \inf_{\mathcal{C}(n)} \mathbb{E}\varphi((n/\zeta)^{1/d} \operatorname{dist}(U, \mathcal{C}(n))), \qquad (7.11)$$

where the infima are taken over all finite sets $\mathcal{C}(n) \subset \mathbb{R}^d$ with $|\mathcal{C}(n)| \leq n$ and U denotes a uniformly distributed r.v. on the unit cube $[0, 1)^d$.

Theorem 7.3 *Suppose that* $\mathbb{E}\psi(|X|) < \infty$ *for some function ψ satisfying a growth condition (G) depending on φ. Then*

$$\lim_{n \to \infty} n^{1/d}\, D^{(q)}(\log n) = J^{1/d}$$

where J is the minimal value in the optimization problem

$$\int_{\mathbb{R}^d} \xi(x)\, dx = \min!$$

among all non-negative ξ with

$$\int_{\mathbb{R}^d} g\big(\xi(x)\big)\, h(x)\, \mathrm{d}x \le 1,$$

where h denotes again the density of μ_c.

The point allocation problem in the Orlicz-norm setting

In the Orlicz-norm setting, the analogs of (7.7) and (7.8) are summarized in the following statement.

Theorem 7.4 *Let $I \subset (0, \infty)$ denote an index set with $\sup I = \infty$, and denote by $(\mathcal{C}(\eta))_{\eta \in I}$ a family of codebooks such that the associated measures*

$$\nu_\eta = \frac{1}{\eta} \sum_{\hat{x} \in \mathcal{C}(\eta)} \delta_{\hat{x}}$$

converge vaguely to a finite measure ν. Then one has

$$\liminf_{\substack{\eta \to \infty \\ \eta \in I}} \mathbb{E}\, \varphi\big(\eta^{1/d} \operatorname{dist}(X, \mathcal{C}(\eta))\big) \ge \int g\big(\xi(x)\big)\, \mathrm{d}\mu_c(x) \qquad (7.12)$$

for $\xi = \frac{\mathrm{d}\nu_c}{\mathrm{d}\lambda^d}$ where ν_c denotes again the absolutely continuous part of ν. On the other hand, for any non-negative $\xi : \mathbb{R}^d \to [0, \infty)$ with $J := \int \xi < \infty$ there exists a family of codebooks $(\mathcal{C}(\eta))_{\eta \ge 1}$ such that $\limsup_{\eta \to \infty} |\mathcal{C}(\eta)|/\eta \le J$ and

$$\limsup_{\eta \to \infty} \mathbb{E}\, \varphi\big(\eta^{1/d} \operatorname{dist}(X, \mathcal{C}(\eta))\big) \le \int g\big(\xi(x)\big)\, \mathrm{d}\mu_c(x). \qquad (7.13)$$

Similar as in the classical setting one can heuristically verify these two estimates for μ with density h of the form (7.4).

Let us show how the estimate (7.13) can be used to prove the upper bound in the asymptotic formula. Recall that J is the minimal value of

$$\int_{\mathbb{R}^d} \xi(x)\, \mathrm{d}x = \min!$$

where the infimum is taken among all non-negative ξ with

$$\int_{\mathbb{R}^d} g\big(\xi(x)\big)\, h(x)\, \mathrm{d}x \le 1,$$

and where h denotes again the density of μ_c.

Fix $\varepsilon > 0$. As one can easily derive from (7.13) there exists a family of codebooks $(\mathcal{C}(\eta))_{\eta \geq 1}$ such that for sufficiently large η

$$|\mathcal{C}_\eta| \leq (J + \varepsilon)\eta \quad \text{and} \quad \mathbb{E}\,\varphi(\eta^{1/d}\,\mathrm{dist}(X, \mathcal{C}(\eta))) \leq 1.$$

Note that the latter estimate is equivalent to $\|\mathrm{dist}(X, \mathcal{C}(\eta))\|_\varphi \leq \eta^{-1/d}$. Consequently, for sufficiently large η one has

$$D^{(q)}\big(\log((J + \varepsilon)\eta)\big) \leq \eta^{-1/d}$$

or equivalently switching from η to $\bar{\eta} = (J + \varepsilon)\,\eta$:

$$D^{(q)}(\log \bar{\eta}) \leq (J + \varepsilon)^{1/d}\bar{\eta}^{-1/d}.$$

This proves the upper bound. The proof of the lower bound is similar and therefore omitted.

Solutions to the point allocation problem

Solutions to the point allocation problem can be represented in terms of the conjugate $\bar{g} : [0, \infty) \to [0, \infty)$ defined as

$$\bar{g}(a) = \inf_{\eta \geq 0}[a\eta + g(\eta)], \qquad a \geq 0.$$

The function $\bar{g}$ is continuous, monotonically increasing and concave, and it satisfies $\bar{g}(0) = 0$.

Theorem 7.5 *We suppose that*

$$\mu_c(\mathbb{R}^d)\sup_{t \geq 0}\varphi(t) > 1.$$

(Otherwise $J = 0$ and $\xi = 0$ is an optimal point density.) The point allocation problem has an integrable solution iff the integral

$$\int \bar{g}\left|\frac{\vartheta}{h(x)}\right| \mathrm{d}\mu_c(x) \tag{7.14}$$

is finite for some $\vartheta > 0$. In such a case there exists a parameter $\zeta > 0$ such that the optimal point density ξ satisfies

$$\int g(\xi(x))\,\mathrm{d}\mu_c(x) = 1 \quad \text{and} \quad \bar{g}'_+\left|\frac{\zeta}{h(x)}\right| \leq \xi(x) \leq \bar{g}'_-\left|\frac{\zeta}{h(x)}\right|, \qquad x \in \mathbb{R}^d. \tag{7.15}$$

Here, the functions $\bar{g}'_+$ and $\bar{g}'_-$ denote the right-hand side and left-hand side derivative of $\bar{g}$, respectively, and we denote $\bar{g}'_+(\infty) = \bar{g}'_-(\infty) = 0$. In particular, the optimal point density is unique, whenever $\bar{g}$ is differentiable or – expressed in terms of g – whenever g is strictly convex.

Back to the original problem

For simplicity we assume that $\bar{g}$ is continuous, and, for any given $\zeta > 0$, we let ξ^ζ be the point density $\xi^\zeta(x) = \bar{g}'\left(\frac{\zeta}{h(x)}\right)$ $(x \in \mathbb{R}^d)$ and denote by $\bar{\xi}^\zeta(x) = \xi^\zeta(x)/\|\xi\|_{L^1(\lambda^d)}$ its normalized version. Suppose now that $\hat{X}^{(n)}$ is a quantizer of rate $\log n$ (assuming at most n different values) that minimizes the objective function

$$\mathbb{E}f(|X - \hat{X}^{(n)}|).$$

We denote by δ its minimal value and set $\bar{\varphi} = \frac{1}{\delta}f$. Then $\hat{X}^{(n)}$ is also an optimal quantizer for the Orlicz-norm objective function

$$\|X - \hat{X}^{(n)}\|_{\bar{\varphi}}.$$

Using the ideas above we can link the problem to a point allocation problem and find an normalized optimal point density $\bar{\xi}$. Recall that the definition of $\bar{\varphi}$ still depends on δ. However, straightforward analysis shows that $\bar{\xi}$ is also contained in the family $(\bar{\xi}^\zeta)_{\zeta>0}$ when taking $\varphi = f$.

Hence, one can relate, for given $n \in \mathbb{N}$, the original quantization problem to a normalized point density of the family $(\bar{\xi}^\zeta)_{\zeta>0}$. We believe that this approach leads to more reasonable descriptions of optimal codebooks for moderate n.

7.3 Gaussian signals

In this section we summarize results on the asymptotic coding problems for general Banach space-valued (centered) Gaussian signals. Let us be more precise about our setting: we fix a separable Banach space $(E, \|\cdot\|)$ and call an E-valued (Borel measurable) random variable X a *Gaussian signal* if for any f in the topological dual E' of E the real-valued random variable $f(X)$ is a zero-mean normally distributed random variable. In this context, the Dirac measure in 0 is also conceived as a normal distribution. A typical example is, for instance, X being a (fractional) Wiener process considered in the space of continuous functions $E = C[0,1]$ endowed with supremum norm. In general, we look at the norm-based distortion measure $\rho(x, \hat{x}) = \|x - \hat{x}\|$ so that our objective function is again

$$\mathbb{E}[\|X - \hat{X}\|^s]^{1/s}.$$

7.3.1 Asymptotic estimates

As was first observed in the dissertation by Fehringer [23] the quantization problem is linked to the behavior of the so-called *small ball function*, that is the function

$$\varphi(\varepsilon) = -\log \mathbb{P}(\|X\| \leq \varepsilon) \qquad (\varepsilon > 0).$$

We summarize the results in the following theorem.

Theorem 7.6 *Suppose that the small ball function satisfies*

$$\varphi^{-1}(\varepsilon) \approx \varphi^{-1}(2\varepsilon) \quad as \quad \varepsilon \downarrow 0,$$

where φ^{-1} denotes the inverse of φ. Then for all moments $s \geq 1$ one has

$$\varphi^{-1}(r) \lesssim D(r,s) \leq D^{(q)}(r,s) \lesssim 2\varphi^{-1}(r/2), \qquad r \to \infty.$$

The lower and upper bounds were first proved for the quantization error in [19], and later complemented by the remaining lower bound for the distortion rate function [16].

Remark 7.7

(i) *The upper and lower bounds do not depend on s. This suggests that "good" codebooks lead to a random approximation error that is concentrated around a typical value or interval. The problem of proving such a result is still open in the general setting. However, as we will see below, one can get stronger results in several particular settings.*

(ii) *The small ball function is a well-studied object. Mostly, the small ball function of a functional signal satisfies the assumptions of the theorem and the estimates provided by the theorem agree asymptotically up to some factor. Thus one can immediately infer the rate of convergence in the coding problems for several Gaussian processes. For a general account on small deviations one might consult [35].*

(iii) *The asymptotic behavior of the small ball function at zero is related to other quantities describing the complexity such as entropy numbers (see [34], [36]), Kolmogorov width and average Kolmogorov width. A general treatment of such quantities together with the quantization problem can be found in the dissertation by Creutzig [9] (see also [7]).*

The proof of the main result relies heavily on the measure concentration features of Gaussian measures exhibited by the isoperimetric inequality and the Ehrhard inequality. Further important tools are the Cameron–Martin formula and the Anderson inequality.

7.3.2 Some examples

We summarize some results that can be extracted from the link to the small ball function. We only give the results for the approximation problems. For a general account we refer the reader to Li and Shao [35]. First let $X = (X_t)_{t \in [0,1]}$ denote a Wiener process.

- When E is chosen to be $C[0,1]$ endowed with supremum norm, one gets

$$\frac{\pi}{\sqrt{8r}} \lesssim D(r,s) \leq D^{(q)}(r,s) \lesssim \frac{\pi}{\sqrt{r}}$$

as $r \to \infty$.
- If $E = L_p[0,1]$ $(p \geq 1)$, then one has

$$\frac{c_p}{\sqrt{r}} \lesssim D(r,s) \leq D^{(q)}(r,s) \lesssim \frac{\sqrt{8}c_p}{\sqrt{r}}$$

where

$$c_p = 2^{1/p} \sqrt{p} \Big(\frac{\lambda_1(p)}{2+p}\Big)^{(2+p)/2p}$$

and

$$\lambda_1(p) = \inf\Big\{ \int_{-\infty}^{\infty} |x|^p f(x)^2 \, dx + \frac{1}{2} \int_{-\infty}^{\infty} f'(x)^2 \, dx \Big\}$$

where the infimum is taken over all differentiable $f \in L^2(\mathbb{R})$ with unit-norm.
- If $E = C^\alpha$ $(\alpha \in (0,1/2))$ is the space of α-Hölder continuous functions over the time $[0,1]$ endowed with the standard Hölder norm

$$\|f\|_{C^\alpha} := \sup_{0 \leq s < t \leq 1} \frac{|f(t) - f(s)|}{|t - s|^\alpha},$$

then

$$\frac{c_\alpha}{r^{(1-2\alpha)/2}} \lesssim D(r,s) \leq D^{(q)}(r,s) \lesssim 2^{(3-2\alpha)/2} \frac{c_\alpha}{r^{(1-2\alpha)/2}}$$

for a constant $c_\alpha > 0$ not known explicitly.

Next, let us consider a fractional Brownian sheet $X = (X_t)_{t \in [0,1]^d}$ with parameter $\gamma = (\gamma_1, \ldots, \gamma_d) \in (0,2)^d$ as original signal. The process is characterized as the centered continuous Gaussian process on $[0,1]^d$ with covariance kernel

$$\mathbb{E}[X_t X_u] = \frac{1}{2^d} \prod_{j=1}^{d} \left[|t_j|^{\gamma_j} + |u_j|^{\gamma_j} - |t_j - u_j|^{\gamma_j} \right], \qquad t, u \in [0,1]^d.$$

As underlying space we consider $E = C([0,1]^d)$ the space of continuous functions endowed with supremum norm.

If there is a unique minimum, say γ_1, in $\gamma = (\gamma_1, \ldots, \gamma_d)$, one has

$$D(r,s) \approx D^{(q)}(r,s) \approx r^{-\gamma_1/2}, \qquad r \to \infty$$

due to [42]. However, if there are two minimal coordinates, say γ_1 and γ_2, then

$$D(r,s) \approx D^{(q)}(r,s) \approx r^{-\gamma_1/2}(\log r)^{1+\gamma_1/2}$$

owing to [3, 46]. For the case that there are more than two minimal elements in the vector γ, it is still an open problem to find the weak asymptotic order of the small ball function.

7.3.3 A particular random coding strategy

We now introduce a random coding strategy that has been originally used to prove the upper bound in Theorem 7.6. Let $(Y_i)_{i \in \mathbb{N}}$ be a sequence of independent random vectors with the same law as X. We consider the random set $\mathcal{C}(r) = \{Y_1, \ldots, Y_{\lfloor e^r \rfloor}\}$ ($r \geq 0$ indicating again the rate) as codebook for X, and set

$$D^{(r)}(r,s) = \mathbb{E}[\min_{i=1,\ldots,\lfloor e^r \rfloor} \|X - Y_i\|^s]^{1/s}.$$

A detailed analysis of this approximation error was carried out in [13]:

Theorem 7.8 *Assume that there exists $\kappa < \infty$ such that*

$$\left(1 + \frac{1}{\kappa}\right)\varphi(2\varepsilon) \leq \varphi(\varepsilon) \leq \kappa\varphi(2\varepsilon)$$

for all sufficiently small $\varepsilon > 0$. Then there exists a continuous, strictly decreasing function $\varphi_* : (0,\infty) \to (0,\infty)$ such that

$$D^{(r)}(r,s) \sim \varphi_*^{-1}(r)$$

for any $s > 0$.

The function φ_* can be represented in terms of a random small ball function. Let

$$\ell_\varepsilon(x) = -\log \mathbb{P}(\|X - x\| \leq \varepsilon) = -\log \mu(B(x, \varepsilon)) \qquad (x \in E, \varepsilon > 0);$$

then φ_* can be chosen as $\varphi_*(\varepsilon) = \mathbb{E}\ell_\varepsilon(X)$.

The proof of the theorem relies on the following strong limit theorem: assuming that there is a constant $\kappa < \infty$ such that $\varphi(\varepsilon) \leq \kappa\varphi(2\varepsilon)$ for sufficiently small $\varepsilon > 0$, one has

$$\lim_{\varepsilon \downarrow 0} \frac{\ell_\varepsilon(X)}{\varphi_*(\varepsilon)} = 1, \text{ a.s.}$$

In information theory, the concept of proving a strong limit theorem in order to control the efficiency of a coding procedure is quite common. For instance, the proof of Shannon's Source Coding Theorem can be based on such a result. Since $\ell_\varepsilon(X)$ is concentrated around a typical value, the conditional probability $\mathbb{P}(d(X, \mathcal{C}(r)) \leq \varepsilon|X)$ almost does not depend on the realization of X when $\varepsilon > 0$ is small. Moreover, for large rate r, there is a critical value ε_c for which the probability decays fast from almost 1 to almost 0. Around this critical value the approximation error is highly concentrated, and the moment s does not have an influence on the asymptotics. Such a strong limit theorem is often referred to as *asymptotic equipartition property*. For further reading concerning the asymptotic equipartition problem in the context of Shannon's Source Coding Theorem, we refer the reader to [12].

Open problem Can one prove the equivalence of moments in the quantization problem under weak assumption on the small ball function φ?

7.3.4 The fractional Brownian motion

In this subsection we consider fractional Brownian motion $X = (X_t)_{t \in [0,1]}$ with Hurst index $H \in (0, 1)$ as underlying signal. Its distribution is characterized as the unique Gaussian measure on $C[0, 1]$ with covariance kernel:

$$\mathbb{E}[X_u \, X_v] = \frac{1}{2}\left[u^{2H} + v^{2H} - |u - v|^{2H}\right].$$

We state the main result.

Theorem 7.9 *If $E = L^p[0,1]$ for some $p \geq 1$, there exists a constant $\kappa_p > 0$ such that for any $s > 0$*

$$\lim_{r \to \infty} r^H D^{(q)}(r,s) = \lim_{r \to \infty} r^H D(r,s) = \kappa_p. \qquad (7.16)$$

Additionally, if $E = C[0,1]$, there exists a constant $\kappa_\infty > 0$ such that for any $s > 0$

$$\lim_{r \to \infty} r^H D^{(q)}(r,s) = \lim_{r \to \infty} r^H D^{(e)}(r,s) = \kappa_\infty.$$

The results are proved for the entropy and quantization constraint in [20]. The extension to the distortion rate function is established in [18].

Remark 7.10

- *In [20] the result is proved for one-dimensional processes only. However, having a careful look at the proof, the result remains true for the multi-dimensional Wiener process, too.*
- *The moment s and the choice of the complexity constraint do not influence the asymptotics of the approximation error. Thus for good approximation schemes the approximation error is concentrated around a typical value: for fixed $s > 0$ let $(\mathcal{C}(r))_{r \geq 0}$ be a family of asymptotically optimal codebooks in the sense that*

$$\log |\mathcal{C}(r)| \leq r \quad and \quad \mathbb{E}[\min_{\hat{x} \in \mathcal{C}(r)} \|X - \hat{x}\|^s]^{1/s} \sim D^{(q)}(r,s);$$

then one has

$$\min_{\hat{x} \in \mathcal{C}(r)} \|X - \hat{x}\| \sim D^{(q)}(r,s), \quad in\ probability,$$

in the sense that, for any $\varepsilon > 0$,

$$\lim_{r \to \infty} \mathbb{P}\big((1-\varepsilon)D^{(q)}(r,s) \leq \min_{\hat{x} \in \mathcal{C}(r)} \|X - \hat{x}\| \leq (1+\varepsilon)D^{(q)}(r,s)\big) = 1.$$

- *The equivalence of moments and the equivalence in the coding quantities are very special, and, as we will see below, more heterogeneous processes as the diffusion processes will not share these features. In [20], the equivalence of moments is used to prove the equivalence of the entropy and quantization constraint. In particular both features (equivalences) seem to be related in a general way.*

Open problem It is still an open problem whether one can prove (7.16) in the case where $E = C[0,1]$.

7.4 Hilbert space-valued Gaussian signals

Let now $E = H$ denote a (separable) Hilbert space and X be an H-valued Gaussian signal. In that case one can represent X in the Karhunen–Loève expansion, that is as the a.s. limit

$$X = \sum_i \sqrt{\lambda_i} \xi_i e_i, \tag{7.17}$$

where

- (λ_i) is an $\mathbb{R}_+$-valued sequence (the eigenvalues of the corresponding covariance operator)
- (e_i) is a orthonormal system in H (the corresponding normalized eigenvalues)
- (ξ_i) is an i.i.d. sequence of standard normals.

Then there is an isometric isomorphism π mapping the range of X (i.e. the smallest closed set in H containing X a.s.) to l^2 such that

$$\pi(X) = (\sqrt{\lambda_1}\xi_1, \sqrt{\lambda_2}\xi_2, \dots).$$

One can prove that applying a contraction on the original signal does not increase its coding complexity under either information constraint. Therefore, the coding quantities are the same for X in H as for $(\sqrt{\lambda_i}\xi_i)$ in l^2. Thus we can and will assume without loss of generality that X is given in the form $(\sqrt{\lambda_i}\xi_i)$.

Before we treat the general coding problem, we first restrict our attention to the distortion rate function $D(r, 2)$. It is one of the few examples that can be given explicitly in terms of a solution to a rate allocation problem. We start by providing some elementary results on mutual information and distortion rate functions.

7.4.1 *The mutual information and Shannon's Source Coding Theorem*

Let us introduce the notion of conditional mutual information. For random variables A, B and C taking values in standard Borel spaces, we denote

$$I(A; B|C = c) = H(\mathbb{P}_{A,B|C=c} \| \mathbb{P}_{A|C=c} \otimes \mathbb{P}_{B|C=c}),$$

where in general

$$H(P\|Q) = \begin{cases} \int \log \frac{dP}{dQ} \, dP & \text{if } P \ll Q, \\ \infty & \text{otherwise} \end{cases}$$

denotes the *relative entropy*. Then one denotes by

$$I(A; B|C) = \int I(A; B|C = c) \, d\mathbb{P}_C(c)$$

the *mutual information of A and B conditional on C*.

The mutual information can be verified to satisfy the following properties (see for instance [29]):

(i) $I(A; B|C) \geq 0$ (*positivity*); $I(A; B|C) = 0$ iff A and B are independent given C

(ii) $I(A; B|C) = I(B; A|C)$ (*symmetry*)

(iii) $I(A_1, A_2; B|C) = I(A_1; B|C) + I(A_2; B|A_1, C)$;
in particular, $I(A_1; B|C) \leq I(A_1, A_2; B|C)$ (*monotonicity*)

(iv) $I(A; B|C) \geq I(\varphi(A); B|C)$ for a Borel measurable map φ between Borel spaces.

All above results remain valid for the (unconditional) mutual information.

The mutual information constraint has its origin in Shannon's celebrated Source Coding Theorem (see [44] for the original version):

Theorem 7.11 (Shannon's Source Coding Theorem) *Let μ be a probability measure on a Borel space E and let $\rho : E \times E \to [0, \infty]$ be a product measurable map. If there exist $\bar{x} \in E$ and $s > 1$ with $\int \rho(x, \bar{x})^s \, d\mu(x) < \infty$, then one has for any continuity point r of the associated distortion rate function $D(\cdot|\mu, \rho)$:*

$$\lim_{m \to \infty} \frac{1}{m} D^{(q)}(mr|\mu^{\otimes m}, \rho_m) = D(r|\mu, \rho),$$

where $\rho_m : E^m \times E^m \to [0, \infty]$ is defined as

$$\rho_m(x, \hat{x}) = \sum_{i=1}^{m} \rho(x_i, \hat{x}_i).$$

Remark 7.12

- *As a consequence of the convexity of the relative entropy, the distortion rate function is convex. Moreover, it is decreasing and bounded from below by 0. Consequently, it has at most one discontinuity in which it jumps from ∞ to some finite value.*

- *For a given $m \in \mathbb{N}$, the distortion measure ρ_m is called* single letter distortion measure, *since its value can be expressed as a* sum of the errors inferred in each single letter (here the term letter refers to the single E-valued entries).

- The distortion rate function can be evaluated explicitly in a few cases only. One of these is the case where the original is normally distributed and the error criterion is given by the mean squared error:

$$D(r|\mathcal{N}(0,\sigma^2), |\cdot|^2) = \sigma^2 e^{-2r}.$$

In that case, the joint distribution of the minimizing pair $(X; \hat{X})$ is explicitly known. It is centered Gaussian with

$$\begin{aligned} \mathbb{E}[X\,\hat{X}] &= \mathbb{E}[\hat{X}^2] \\ &= \sigma^2(1 - e^{-2r}) \end{aligned}$$

7.4.2 The derivation of the distortion rate function $D(r,2)$

Let us denote by $D(\cdot)$ the distortion rate function of the Gaussian original X under the objective function

$$\mathbb{E}[\|X - \hat{X}\|^2],$$

that is $D(r) = D(r,2)^2$ $(r \geq 0)$. The distortion rate function will be given as a solution to a rate allocation problem.

The lower bound

Suppose that $\hat{X}$ is a reconstruction for X with $I(X; \hat{X}) \leq r$ for a given rate r. For $a, b \in \mathbb{N}$ we let $X_a^b = (X_i)_{i=a,\dots,b}$ with the natural extension when $b = \infty$ or $a > b$. With property (iii) one gets

$$I(X; \hat{X}) = I(X_1; \hat{X}) + I(X_2^\infty; \hat{X}|X_1) \geq I(X_1; \hat{X}_1) + I(X_2^\infty; \hat{X}|X_1).$$

Repeating the argument infinitely often yields

$$I(X; \hat{X}) \geq \sum_{i=1}^{\infty} I(X_i; \hat{X}_i|X_1^{i-1}). \tag{7.18}$$

For each $i \in \mathbb{N}$ we conceive $r_i = I(X_i; \hat{X}_i|X_1^{i-1})$ as the rate allocated to the ith coordinate. Note that one has $\sum_i r_i \leq r$.

Let us now fix $i \in \mathbb{N}$ and analyze $\mathbb{E}|X_i - \hat{X}_i|^2$. It can be rewritten as

$$\mathbb{E}|X_i - \hat{X}_i|^2 = \mathbb{E}\big[\mathbb{E}[|X_i - \hat{X}_i|^2|X_1^{i-1}]\big]. \tag{7.19}$$

Note that conditional upon X_1^{i-1} the random variable X_i is $\mathcal{N}(0, \lambda_i)$-distributed so that

$$\mathbb{E}[|X_i - \hat{X}_i|^2 | X_1^{i-1} = x_1^{i-1}] \geq D\big(I(X_i; \hat{X}_i | X_1^{i-1} = x_1^{i-1}) | \mathcal{N}(0, \lambda_i), | \cdot |^2\big).$$

$$(7.20)$$

Together with (7.19) and the convexity of the distortion rate functions one gets

$$\mathbb{E}|X_i - \hat{X}_i|^2 \geq \int D\big(I(X_i; \hat{X}_i | X_1^{i-1} = x_1^{i-1}) | \mathcal{N}(0, \lambda_i), | \cdot |^2\big) \, d\mathbb{P}_{X_1^{i-1}}(x_1^{i-1})$$

$$\geq D(r_i | \mathcal{N}(0, \lambda_i), | \cdot |^2) = \lambda_i e^{-2r_i}.$$

$$(7.21)$$

Altogether, we arrive at

$$\mathbb{E}\|X - \hat{X}\|^2 \geq \sum_{i=1}^{\infty} \lambda_i e^{-2r_i} \quad \text{and} \quad \sum_{i=1}^{\infty} r_i \leq r.$$

The upper bound

Now we fix a non-negative sequence (r_i) with $\sum_i r_i \leq r$. As mentioned in Remark 7.12, for each $i \in \mathbb{N}$, there exists a pair $(X_i, \hat{X}_i)$ (on a possibly enlarged probability space) satisfying

$$\mathbb{E}[|X_i - \hat{X}_i|^2] = \lambda_i e^{-2r_i} \quad \text{and} \quad I(X_i; \hat{X}_i) = r_i.$$

These pairs can be chosen in such a way that $(X_i, \hat{X}_i)_{i \in \mathbb{N}}$ form an independent sequence of random variables. Fix $N \in \mathbb{N}$ and note that due to property (iii) of mutual information one has

$$I(X_1^N; \hat{X}_1^N) = I(X_2^N; \hat{X}_1^N) + I(X_1^N; \hat{X}_1^N | X_2^N)$$

$$= I(X_2^N; \hat{X}_2^N) + \underbrace{I(X_2^N; \hat{X}_1 | \hat{X}_2^N)}_{=0}$$

$$+ \underbrace{I(X_1; \hat{X}_2^N | X_2^N)}_{=0} + \underbrace{I(X_1; \hat{X}_1 | X_2^N, \hat{X}_2^N)}_{=I(X_1; \hat{X}_1)}.$$

Here the second and third term vanish due to the independence of the conditional distributions (property (i)). Moreover, one can remove the conditioning in the last term, since $(X_1, \hat{X}_1)$ is independent of $(X_2^N, \hat{X}_2^N)$. Repeating the argument now gives

$$I(X_1^N; \hat{X}_1^N) = \sum_{i=1}^{N} I(X_i; \hat{X}_i)$$

and a further argument (based on the lower semicontinuity of the relative entropy) leads to

$$I(X; \hat{X}) = \lim_{N \to \infty} I(X_1^N; \hat{X}_1^N) = \sum_{i=1}^{\infty} I(X_i; \hat{X}_i) = \sum_{i=1}^{\infty} r_i \leq r.$$

Moreover, $\mathbb{E}\|X - \hat{X}\|^2 = \sum_i \lambda_i e^{-2r_i}$.

Kolmogorov's inverse water filling principle

As a result of the computations above, the value $D(r)$ can be represented as the minimal value of the strictly convex optimization problem

$$\sum_i \lambda_i e^{-2r_i} = \min!$$

where the infimum is taken over all non-negative sequences (r_i) with $\sum_i r_i = r$. It is common to restate the minimization problem in terms of the errors $d_i = \lambda_i e^{-2r_i}$ inferred in the single coordinates as

$$\sum_i d_i = \min!$$

where (d_i) is an arbitrary non-negative sequence with

$$\sum_i \frac{1}{2} \log_+ \frac{\lambda_i}{d_i} = r. \tag{7.22}$$

Using Lagrange multipliers one finds that the unique minimizer is of the form

$$d_i = \kappa \wedge \lambda_i,$$

where $\kappa > 0$ is a parameter that needs to be chosen such that (7.22) is valid. This link between $r \geq 0$ and $\kappa \in (0, \lambda_1]$ provides a one-to-one correspondence. This result was originally derived by Kolmogorov [33] and the formula is often referred to as *Kolmogorov's inverse water filling principle*. As illustrated in Figure 7.1, one can represent each coordinate as a box of size λ_i; then one fills water into the boxes until a level κ is reached for which the corresponding total rate is equal to r. The white area of the boxes represents the coding error. As we will see later the striped area also has an information-theoretic meaning.

Résumé

Let us recall the main properties that enabled us to relate the coding problem of the process X to that of a single normal random variable.

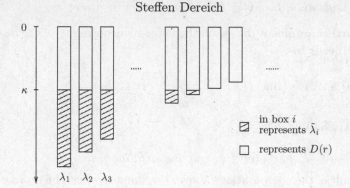

Fig. 7.1. Inverse water filling principle

In order to derive the lower estimate of the error in one coordinate (see (7.20) and (7.21)) we used the independence of a single coordinate and the remaining coordinates of X. Then in order to conclude back on the total approximation error, we used that it is given as the sum of the approximation errors in the single coordinates (*single letter distortion measure*), so we could exchange sum and expectation to derive the result. We shall see later that the above ideas can be applied in similar settings provided one considers independent letters under a single letter distortion measure.

Mostly, the eigenvalues (λ_i) are not known explicitly and thus the minimization problem cannot be solved explicitly. However, the asymptotics of $D(\cdot)$ depend only on the asymptotic behavior of (λ_i), and in the case where (λ_i) is regularly varying, both expressions can be linked by an asymptotic formula: one has

$$D(r) \sim \frac{\alpha^\alpha}{2^{\alpha-1}(\alpha-1)} r \lambda_{\lceil r \rceil}, \qquad (7.23)$$

if the sequence (λ_i) is regularly varying with index $-\alpha < -1$, i.e. there exists a continuous function $\ell : [1, \infty) \to \mathbb{R}$ such that

$$\lambda_i \sim i^{-\alpha} \ell(i) \text{ as } i \to \infty$$

and for any $\eta > 1$ one has $\ell(\eta t) \sim \ell(t)$ as $t \to \infty$.

7.4.3 Generalizations to other coding quantities (first approach)

Let us now look at the quantization problem for $s = 2$. In this section, we want to explain an approach taken by Luschgy and Pagès [39] (see also [38]) to prove the asymptotic equivalence of the quantization error $D^{(q)}(\cdot, 2)$ and the distortion rate function $D(\cdot, 2)$. We adopt a slightly more general setting than that of the original work. However, the generalization can be carried out easily, and we prefer to give the more general assumptions since we believe that these are the natural (most general) ones under which the following arguments work.

Theorem 7.13 *Assume that the eigenvalues are ordered by their size and that there is an increasing $\mathbb{N}_0$-valued sequence $(j_k)_{k \in \mathbb{N}}$ with $j_1 = 0$ such that*

(i) $\lim_{k \to \infty} j_{k+1} - j_k = \infty$ *and*
(ii) $\lim_{k \to \infty} \lambda_{j_k+1}/\lambda_{j_{k+1}} = 1$.

Then one has

$$D(r, 2) \sim D^{(q)}(r, 2).$$

Since the quantization error is always larger than the distortion rate function, it suffices to prove an appropriate upper bound for the quantization error.

Sketch of the proof

In the following, we still consider the single letter distortion measure $\rho(x, \hat{x}) = \|x - \hat{x}\|^2$ and we call the minimal value of the objective function $\mathbb{E}[\|X - \hat{X}\|^2]$ the quantization error of an approximation $\hat{X}$.

Based on the sequence (j_k) we decompose the process X into subbands

$$X_{j_k+1}^{j_{k+1}} = (X_{j_k+1}, \ldots, X_{j_{k+1}}) \qquad (k \in \mathbb{N}).$$

Note that owing to property (ii), the eigenvalues corresponding to a subband only differ by a factor tending to 1 as k tends to ∞. By replacing the eigenvalues of each subband by their largest value, one ends up with a process having a larger quantization error than the original one. Actually, one can show rigorously that the approximation error on each subband increases at most by the factor $\frac{\lambda_{j_k+1}}{\lambda_{j_{k+1}}}$. Since the values of finitely many eigenvalues do not affect the strong asymptotics of the distortion rate function, it remains to show asymptotic equivalence of both coding quantities for the modified process.

Each subband consists of an i.i.d. sequence of growing size and the distortion is just the sum of the distortions in the single letters. Thus we are in a situation where the Source Coding Theorem can be applied: similar to Proposition 4.4 in [39] one can prove that

$$\eta(m) := \sup_{\substack{r \geq 0, \\ \sigma^2 > 0}} \frac{D^{(q)}(r|\mathcal{N}(0,\sigma^2)^{\otimes m}, |\cdot|^2)}{D(r|\mathcal{N}(0,\sigma^2)^{\otimes m}, |\cdot|^2)} \tag{7.24}$$

is finite for all $m \in \mathbb{N}$ and satisfies $\lim_{m \to \infty} \eta(m) = 1$.

Let us sketch the remaining part of the proof of Theorem 7.13. For a given rate $r \geq 0$ we denote by (r_i) the corresponding solution to the rate allocation problem. Then choose for each subband $X_{j_k+1}^{j_{k+1}}$ an optimal codebook $\mathcal{C}_k \subset \mathbb{R}^{j_{k+1}-j_k}$ of size $\exp(\sum_{i=j_k+1}^{j_{k+1}} r_i)$ and denote by $\mathcal{C}$ the product codebook $\mathcal{C} = \prod_{k=1}^{\infty} \mathcal{C}_k$. It contains at most $\exp(\sum_{i=1}^{\infty} r_i) = \exp(r)$ elements. Moreover, the inferred coding error satisfies

$$\mathbb{E}[\min_{\hat{x} \in \mathcal{C}} \|X - \hat{x}\|^2] = \sum_{k \in \mathbb{N}} \mathbb{E}[\min_{\hat{x} \in \mathcal{C}_k} |X_{j_k+1}^{j_{k+1}} - \hat{x}|^2]$$

$$\leq \sum_{k \in \mathbb{N}} \eta(j_{k+1} - j_k) \, D(\sum_{i=1}^{\infty} r_i | X_{j_k+1}^{j_{k+1}}, |\cdot|^2).$$

For arbitrary $\varepsilon > 0$ one can now fix k_0 such that for all $k \geq k_0$ one has $\eta(j_{k+1} - j_k) \leq 1 + \varepsilon$; then

$$\sum_{k \geq k_0} \eta(j_{k+1} - j_k) \, D\Big(\sum_{i=j_k+1}^{j_{k+1}} r_i \Big| X_{j_k+1}^{j_{k+1}}, |\cdot|^2 \Big) \leq (1+\varepsilon) \, D(r|X, |\cdot|^2).$$

The remaining first $k_0 - 1$ summands are of lower order (as $r \to \infty$) and one retrieves the result for the slightly modified process.

7.4.4 Generalizations to other coding quantities (second approach)

A second approach in the analysis of the asymptotic coding errors has been undertaken in [14]. The results can be stated as follows:

Theorem 7.14 *If the eigenvalues satisfy*

$$\lim_{n \to \infty} \frac{\log \log 1/\lambda_n}{n} = 0,$$

then, for any $s > 0$,

$$D^{(q)}(r,s) \sim D(r,s) \sim D(r,2).$$

The analysis of this result is more elaborate and we only want to give the very basic ideas of the proof of the upper bound of the quantization error. A central role is played by an asymptotic equipartition property.

The underlying asymptotic equipartition property

For a fixed rate $r \geq 0$, the solution to the rate allocation problem is linked to a unique parameter $\kappa \in (0, \lambda_1]$ (as explained above). Now set $\tilde{\lambda}_i = (\lambda_i - \kappa) \wedge 0$ and denote by $\tilde{X}^{(r)} = (\tilde{X}_i^{(r)})_{i \in \mathbb{N}}$ an l^2-valued random variable having as entries independent $\mathcal{N}(0, \tilde{\lambda}_i)$-distributed r.v. $\tilde{X}_i^{(r)}$ (the variances $\tilde{\lambda}_i$ ($i \in \mathbb{N}$) are visualized as striped boxes in Figure 7.1). The asymptotic equipartition property states that for $\varepsilon > 0$ fixed and r going to infinity, the probability that

$$-\log \mathbb{P}\big(\|X - \tilde{X}^{(r)}\|^2 \leq (1 + \varepsilon)D(r)\big|X\big) \leq r + \frac{D(r)}{-D'(r)}\varepsilon$$

tends to one. Here $D'(\cdot)$ denotes the right-hand side derivative of the convex function $D(\cdot)$.

The proof of the equipartition property relies on a detailed analysis of the random logarithmic moment generating function

$$\Lambda_X(\theta) := \log \mathbb{E}\big[e^{\theta\|X - \tilde{X}^{(r)}\|^2}\big|X\big]$$

$$= \sum_{i \in \mathbb{N}}\Big[-\frac{1}{2}\log(1 - 2\theta\tilde{\lambda}_i) + \frac{\theta\lambda_i}{1 - 2\theta\tilde{\lambda}_i}\xi_i^2\Big]$$

(where (ξ_i) is the sequence of independent standard normals from representation (7.17)), and the close relationship between $-\log \mathbb{P}(\|X - \tilde{X}^{(r)}\|^2 \leq \zeta|X)$ and the random Legendre transform

$$\Lambda_X^*(\zeta) := \sup_{\theta \leq 0}[\theta\zeta - \Lambda_X(\theta)]$$

given in [14, Theorem 3.4.1].

The asymptotic equipartition property implies that the random codebooks $\mathcal{C}(r)$ ($r \geq 0$) consisting of $\lfloor\exp(r + 2\frac{D(r)}{-D'(r)}\varepsilon)\rfloor$ independent random copies of $\tilde{X}^{(r)}$ satisfy

$$\lim_{r \to \infty} \mathbb{P}\big(\min_{\hat{x} \in \mathcal{C}(r)} \|X - \hat{x}\|^2 \leq (1 + \varepsilon)D(r)\big) = 1.$$

Moreover, $D\big(r + 2\frac{D(r)}{-D'(r)}\varepsilon\big) \geq (1 - 2\varepsilon)D(r)$ (due to the convexity of $D(\cdot)$) so that typically the random approximation error satisfies

$$\min_{\hat{x} \in \mathcal{C}(r)} \|X - \hat{x}\|^2 \leq \frac{1 + \varepsilon}{1 - 2\varepsilon} D(\log |\mathcal{C}(r)|) \qquad (7.25)$$

for large r and fixed $\varepsilon \in (0, 1/2)$. It remains to control the error inferred in the case when estimate (7.25) is not valid.

7.4.5 A particular random quantization procedure

As before we want to compare the coding results quoted so far with the efficiency of the particular random quantization strategy introduced above. Let $(Y_i)_{i \in \mathbb{N}}$ denote a sequence of independent random vectors with the same law as X. We consider the quantization error inferred by the random codebooks $\mathcal{C}(r) = \{Y_1, \ldots, Y_{\lfloor e^r \rfloor}\}$. Let

$$D^{(r)}(r, s) = \mathbb{E}[\min_{i=1,\ldots,\lfloor e^r \rfloor} \|X - Y_i\|^s]^{1/s}.$$

As mentioned above the asymptotics of $D^{(r)}(\cdot, s)$ are related to a randomly centered small ball function. Let us again denote $\ell_\varepsilon(x) = -\log \mathbb{P}(\|X - x\| \leq \varepsilon)$ $(x \in H, \varepsilon > 0)$. We quote the main result of [15].

Theorem 7.15 *One has*

$$\lim_{\varepsilon \downarrow 0} \frac{\ell_\varepsilon(X)}{\varphi_*(\varepsilon)} = 1, \ a.s., \tag{7.26}$$

where $\varphi_*(\varepsilon) = \Lambda^*(\varepsilon^2) = \sup_{\theta \in \mathbb{R}}[\varepsilon^2 \theta - \Lambda(\theta)]$ *is the Legendre transform of*

$$\Lambda(\theta) = \sum_i \left[-\frac{1}{2} \log(1 - 2\theta \lambda_i) + \frac{\theta \lambda_i}{1 - 2\theta \lambda_i} \right]$$

and $\log(z) = -\infty$ *for* $z \leq 0$.

Note that the theorem does not assume any assumptions on the eigenvalues. Given that the eigenvalues are regularly varying it is possible to directly relate the function $\varphi_*(\varepsilon)$ to the standard small ball function $\varphi(\varepsilon) = \ell_\varepsilon(0)$ $(\varepsilon > 0)$:

Theorem 7.16 *If the eigenvalues are regularly varying with index* $-\alpha < -1$ *(in the sense described above), one has*

$$\lim_{\varepsilon \downarrow 0} \frac{\varphi_*(\varepsilon)}{\varphi(\varepsilon)} = \left(\frac{\alpha + 1}{\alpha} \right)^{\frac{\alpha}{\alpha - 1}}.$$

Let us now compare the error induced by the random coding strategy with the optimal quantization error in the case where the eigenvalues satisfy

$$\lambda_i \sim c\, i^{-\alpha}$$

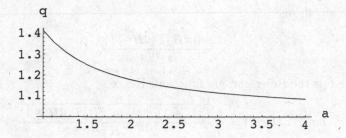

Fig. 7.2. Comparison of $D^{(r)}(\cdot, s)$ and $D^{(q)}(\cdot, s)$ in dependence on α.

for two constants $c > 0$ and $\alpha > 1$. Starting from the standard small ball function (see [45] or [37] for a general treatment) one can deduce the asymptotics of the random quantization error $D^{(r)}(\cdot, s)$. A comparison with $D^{(q)}(\cdot, s)$ then gives

$$\lim_{r \to \infty} \frac{D^{(r)}(r, s)}{D^{(q)}(r, s)} = \left[\frac{(\alpha^2 - 1)\pi}{\alpha^3 \sin(\frac{\pi}{\alpha})} \right]^{\alpha/2}.$$

The limiting value on the right-hand side is plotted in Figure 7.2.

7.4.6 Examples

- The most prominent example is $X = (X_t)_{t \in [0,1]}$ being a Wiener process in $L^2[0, 1]$. In that case the Karhunen–Loève expansion is known explicitly: the eigenvalues are given by $\lambda_i = (\pi(i - 1/2))^{-2}$ and the corresponding normalized eigenfunctions are

$$e_i(t) = \sqrt{2} \sin(\pi(n - 1/2)t) \qquad (t \in [0, 1]).$$

Thus the eigenvalues are regularly varying with index -2 and one gets

$$\lim_{r \to \infty} \sqrt{r} \, D(r, 2) = \frac{\sqrt{2}}{\pi}.$$

- Next, let $X = (X_t)_{t \in [0,1]}$ be a fractional Brownian motion with Hurst parameter $H \in (0, 1)$ (see Section 7.3.4 for the defining properties) As underlying Hilbert space we consider again $L^2[0, 1]$.

In that case the Karhunen–Loéve expansion is not known explicitly and there exist only suboptimal representations (see [22],[28]). However, the asymptotic behavior of the ordered sequence of eigenvalues is known [4]: one has

$$\lambda_i \sim \frac{\sin(\pi H)\,\Gamma(2H+1)}{(i\pi)^{2H+1}},$$

where Γ is the Euler gamma function. Hence,

$$\lim_{r\to\infty} r^H\, D(r,2) = \sqrt{\frac{(2H+1)^{2H+1}\sin(\pi H)\,\Gamma(2H+1)}{2^{2H+1}\,H\,\pi^{2H+1}}}.$$

as $r \to \infty$.

- Next, let $X = (X_t)_{t\in[0,1]^d}$ denote a d-dimensional fractional Brownian sheet with parameter $H = (H_1, \dots, H_1)$. Then the sequence of ordered eigenvalues satisfies

$$\lambda_i \sim \left(\frac{\sin(\pi H_1)\Gamma(2H_1+1)}{\pi^{2H_1+1}}\right)^d ((d-1)!)^{-(2H_1+1)}\left(\frac{(\log i)^{d-1}}{i}\right)^{2H_1+1}$$

so that

$$\lim_{r\to\infty} \frac{r^{H_1}}{(\log r)^{(d-1)(2H_1+1)/2}}\, D(r,2)$$

$$= \sqrt{\frac{(2H_1+1)^{2H_1+1}}{2^{2H_1+1}H_1((d-1)!)^{2H_1+1}}\left(\frac{\sin(\pi H_1)\Gamma(2H_1+1)}{\pi^{2H_1+1}}\right)^d}$$

(see [39] or [14]).

7.5 Diffusions

The quantization complexity of diffusion processes was first treated in an article by Luschgy and Pagès [40]. There weak asymptotic estimates for the quantization problem were derived for a class of 1-dimensional diffusions.

In the following we want to focus on one approach that has been developed in two papers by the author [17, 18]. It leads to asymptotic formulae for several coding quantities. Moreover, it is based on a rate allocation problem and we believe that it fits best into the context of this paper.

We consider as original signal an $\mathbb{R}$-valued process $X = (X_t)_{t\in[0,1]}$ that solves the integral equation

$$X_t = x_0 + \int_0^t \sigma(X_u, u)\, \mathrm{d}W_u + \int_0^t b(X_u, u)\, \mathrm{d}u, \qquad (7.27)$$

where $W = (W_t)_{t \in [0,1]}$ denotes a standard Wiener process, $x_0 \in \mathbb{R}$ denotes an arbitrary starting point and $\sigma, b : \mathbb{R} \times [0,1] \to \mathbb{R}$ are continuous functions.

We impose the following regularity assumptions on the functions σ and b: there exist constants $\beta \in (0,1]$ and $L < \infty$ such that for $x, x' \in \mathbb{R}$ and $t, t' \in [0,1]$:

$$|b(x,t)| \le L(|x| + 1) \quad \text{and}$$
$$|\sigma(x,t) - \sigma(x',t')| \le L[|x - x'|^\beta + |x - x'| + |t - t'|^\beta]. \tag{7.28}$$

Note that the assumptions imply neither existence nor uniqueness of the solution. However, the analysis does not rely on the uniqueness of the solution and we only need to assume the existence of one solution which will be fixed for the rest of this section.

To simplify notation, we denote by $(\sigma_t)_{t \in [0,1]}$ and $(b_t)_{t \in [0,1]}$ the stochastic processes $(\sigma(X_t, t))$ and $(b(X_t, t))$, respectively. Let us first state the main results:

Theorem 7.17 *(Dereich [17]) If $E = C[0,1]$, then for each $s > 0$ one has*

$$\lim_{r \to \infty} \sqrt{r} \, D^{(q)}(r,s) = \kappa_\infty \left\| \|\sigma.\|_{L^2[0,1]} \right\|_{L^s(\mathbb{P})}$$

and

$$\lim_{r \to \infty} \sqrt{r} \, D^{(e)}(r,s) = \kappa_\infty \left\| \|\sigma.\|_{L^2[0,1]} \right\|_{L^{2s/(s+2)}(\mathbb{P})},$$

where κ_∞ is the real constant appearing in Theorem 7.9.

Theorem 7.18 *(Dereich [18]) If $E = L^p[0,1]$ for $p \ge 1$, then for every $s > 0$ one has*

$$\lim_{r \to \infty} \sqrt{r} \, D^{(q)}(r,s) = \kappa_p \left\| \|\sigma.\|_{L^{2p/(2+p)}[0,1]} \right\|_{L^s(\mathbb{P})}$$

and

$$\lim_{r \to \infty} \sqrt{r} \, D(r,s) = \lim_{r \to \infty} \sqrt{r} \, D^{(e)}(r,s) = \kappa_p \left\| \|\sigma.\|_{L^{2p/(2+p)}[0,1]} \right\|_{L^{2s/(s+2)}(\mathbb{P})},$$

where κ_p is the constant from Theorem 7.9.

The analysis of the asymptotic coding problem is based on a decoupling argument. The decoupling argument allows us to connect the complexity of the diffusion to that of the Wiener process. After one has applied the decoupling techniques one can prove the asymptotic formulae by considering certain rate allocation problems. In the next section we give a heuristic explanation of the decoupling method. Then we use these results to derive heuristically the lower bound. The solution of the corresponding rate allocation problem can also be used to define asymptotically optimal quantizers or entropy coding schemes.

7.5.1 The decoupling method

Let us start with an intuitive explanation of the decoupling method. We fix $s \geq 1$ and let the underlying norm $\| \cdot \|$ be supremum norm over the interval $[0, 1]$.

The main idea is to write the diffusion as the sum of an easily analyzable process and a remaining term of negligible complexity. Let us first explain what we understand by negligible terms.

Negligible terms

The approximation error inferred from either complexity constraint leads to an asymptotic error of order $r^{-1/2}$, where r is and will be used to indicate the rate of the approximation. In order to simplify the following discussion, we assume that changing the rate r by a term of order $o(r)$ does not have an influence on the strong asymptotics of the coding quantity of interest. This property is valid for the diffusion under either information constraint as one can easily infer from Theorems 7.17 and 7.18. However, the proof of this fact is non-trivial.

Suppose now that $(Y^{(r)})_{r \geq 0}$ is a family of processes such that there exist discrete approximations $\hat{A}^{(r)}$ to $A^{(r)} := X - Y^{(r)}$ satisfying

$$\log |\operatorname{range} \hat{A}^{(r)}| = o(r) \quad \text{and} \quad \mathbb{E}[\|A^{(r)} - \hat{A}^{(r)}\|^s]^{1/s} = o(r^{-1/2}).$$

Then one can replace the process X by $Y^{(r)}$ (actually by the family $(Y^{(r)})_{r \geq 0}$) without changing the strong asymptotics of either of the approximation errors. Indeed, one can relate approximations $\hat{Y}^{(r)}$ of $Y^{(r)}$ to approximations $\hat{X}^{(r)}$ of X via $\hat{X}^{(r)} = \hat{Y}^{(r)} + \hat{A}^{(r)}$ and get

$$\left| \mathbb{E}[\|X - \hat{X}^{(r)}\|^s]^{1/s} - \mathbb{E}[\|Y^{(r)} - \hat{Y}^{(r)}\|^s]^{1/s} \right| = o(r^{-1/2}). \tag{7.29}$$

Moreover, reconstructions $\hat{X}^{(r)}$ of rate r give rise to reconstructions $\hat{Y}^{(r)}$ of rate $(1 + o(1))r$ and vice versa (see the additivity properties listed in Section 7.1. Here it does not matter which complexity constraint we use. Moreover, the supremum norm is the strongest norm under consideration and (7.29) remains valid for the same family $(Y^{(r)})_{r \geq 0}$ when $\| \cdot \|$ denotes an L^p-norm. We call the term $A^{(r)}$ asymptotically negligible.

Relating X to a decoupled time change of a Wiener process

We represent X in its Doob–Meyer decomposition as a sum of a martingale

$$M_t = \int_0^t \underbrace{\sigma(X_u, u)}_{\sigma_u} \, \mathrm{d}W_u$$

and a drift term

$$A_t = x_0 + \int_0^t \underbrace{b(X_u, u)}_{b_u} \, \mathrm{d}u.$$

The drift term is significantly more regular than the martingale term and one can verify that it is asymptotically negligible: we only need to consider the martingale part. It can be viewed as a time change of a Wiener process:

$$(M_t) = (\tilde{W}_{\varphi(t)}),$$

where $(\tilde{W}_t)_{t \geq 0}$ denotes a Wiener process and $(\varphi(t))_{t \in [0,1]}$ is given by $\varphi(t) = \int_0^t \sigma_u^2 \, du$. Unfortunately, the time change and the Wiener process are not independent so that one cannot immediately apply the asymptotic formulae for the Wiener process. On the other hand, conditional upon $(\varphi(t))$, the process $(\tilde{W}_t)_{t \geq 0}$ is no longer a Wiener process and the process might even be deterministic up to time $\varphi(1)$.

In order to bypass this problem we introduce an approximation $\hat{\varphi} = (\hat{\varphi}(t))_{t \in [0,1]}$ (depending on the rate r) to φ such that the approximation error

$$\mathbb{E}[\|M. - \tilde{W}_{\hat{\varphi}(\cdot)}\|^s]^{1/s} = o(r^{-1/2})$$

is negligible compared to the total coding error (which is of order $r^{-1/2}$). On the other hand, conditional upon $\hat{\varphi}$ the process $(\tilde{W}_t)_{t \geq 0}$ should not significantly deviate from a Wiener process. For controlling the influence of the conditioning, we view the additional information induced by the approximation $\hat{\varphi}$ as an initial enlargement of the canonical filtration $(\mathcal{F}_t^{\tilde{W}})$ induced by the process $\tilde{W}$. Let us be more precise about the estimates used here.

For an absolutely continuous function $f : [0, \infty) \to \mathbb{R}$ with differential $\dot{f}$ we set

$$\|f\|_{\mathcal{H}} = \left(\int_0^\infty |\dot{f}(u)|^2 \, \mathrm{d}u \right)^{1/2}.$$

The Wiener process $\tilde{W}$ is now represented in its $(\mathcal{G}_t) = (\mathcal{F}_t^{\tilde{W}} \vee \sigma(\hat{\varphi}))$-Doob–Meyer decomposition as

$$\tilde{W}_t = \bar{W}_t + \bar{A}_t.$$

Again recall that $\hat{\varphi}$ and thus also $\bar{W}$ and $\bar{A}$ depend on r. Now $(\bar{W}_t)$ is a $(\mathcal{G}_t)$-Wiener process that is independent of $(\hat{\varphi}(t))$. Supposing that the process $(\hat{\varphi}(t))$ has finite range, one can conclude that $(\tilde{W}_t)$ is indeed a $(\mathcal{G}_t)$-semi martingale which justifies the Doob–Meyer decomposition above. Moreover, there exists a constant c depending on $s \geq 1$ only, such that

$$\mathbb{E}[\|\bar{A}\|_{\mathcal{H}}^{2s}]^{1/s} \leq c\,(1 + \log|\,\mathrm{range}\,(\hat{\varphi})|)$$

(see [30], [1]). On the other hand, the Sobolev space $\mathcal{H}$ (the space of absolutely continuous functions with finite $\|\cdot\|_{\mathcal{H}}$-norm) is compactly embedded into $C[0,T]$ for a finite time horizon $T > 0$ and one can use analytic results on this embedding to show the asymptotic negligibility of the term $(\bar{A}_{\hat{\varphi}(t)})_{t \in [0,1]}$. Altogether, one gets that the coding complexity of X and $(\bar{W}_{\hat{\varphi}(t)})_{t \in [0,1]}$ coincide. Here one needs to be aware of the fact that the definition of $(\hat{\varphi}(t))$ and thus of $(\bar{W}_t)$ depend on the parameter r.

The main result

Let us state the precise decoupling result. Fix $\alpha \in (0, \beta/2)$ and denote by $\hat{\varphi}^{(n)} = (\hat{\varphi}_t^{(n)})_{t \in [0,1]}$ ($n \in \mathbb{N}$) a random increasing and continuous function that is linear on each interval $[i/n, (i+1)/n]$ ($i = 0, \ldots, n-1$) and satisfies

$$\hat{\varphi}^{(n)}(i/n) = \arg\min_{y \in I(n)} |\varphi(i/n) - y| \qquad (i = 0, \ldots, n),$$

where $I(n)$ is defined as

$$I(n) = \Big\{ j\,\frac{1}{n^{1+\alpha}} : j \in \mathbb{N}_0, j \leq n^{2(1+\alpha)} \Big\}.$$

Theorem 7.19 *Fix* $\zeta \in ((1+\alpha)^{-1}, 1)$, *choose* $n \in \mathbb{N}$ *in dependence on* $r > 0$ *as* $n = n(r) = \lceil r^\zeta \rceil$, *and denote by* $\tilde{W} = \bar{W}^{(n)} + \bar{Y}^{(n)}$ *the* $(\mathcal{F}_t^{\tilde{W}} \vee \sigma(\hat{\varphi}^{(n)}))$-*Doob–Meyer decomposition of* $\tilde{W}$. *For arbitrarily fixed* $s > 0$ *there exist* $C[0,1]$-*valued r.v.'s* $\bar{R}^{(n)}$ *and* $\hat{R}^{(r)}$ *such that*

- $X = \bar{W}^{(n)}_{\hat{\varphi}^{(n)}(\cdot)} + \bar{R}^{(n)}$
- $\bar{W}^{(n)}$ *is a Wiener process that is independent of* $\hat{\varphi}^{(n)}$
- $\mathbb{E}[\|\bar{R}^{(n)} - \hat{R}^{(r)}\|^s]^{1/s} = \mathcal{O}(r^{-\frac{1}{2}-\delta})$, *for some* $\delta > 0$
- $\log|\,\mathrm{range}\,(\hat{R}^{(r)}, \hat{\varphi}^{(n)})| = \mathcal{O}(r^\gamma)$, *for some* $\gamma \in (0,1)$.

7.5.2 The corresponding rate allocation problem

We only consider the case where $E = L^p[0,1]$ for some $p \geq 1$. Theorem 7.19 states that the process X and the process $(\bar{W}^{(n)}_{\hat{\varphi}^{(n)}(t)})_{t \in [0,1]}$ (actually the family of processes) have the same asymptotic complexity. The number n is still related to r via $\lceil r^\varsigma \rceil$.

We adopt the notation of Theorem 7.19, and let for $i \in \{0, \ldots, n-1\}$ and $t \in [0, 1/n)$

$$\bar{X}^{(n)}_{\frac{i}{n}+t} := \bar{X}^{(n,i)}_t := \bar{W}^{(n)}_{\hat{\varphi}^{(n)}(\frac{i}{n}+t)} - \bar{W}^{(n)}_{\hat{\varphi}^{(n)}(\frac{i}{n})}.$$

Note that r/n tends to infinity and one can verify that the difference between $\bar{X}^{(n)} = (\bar{X}^{(n)}_t)_{t \in [0,1)}$ and $\bar{W}^{(n)}_{\hat{\varphi}^{(n)}(\cdot)}$ is negligible in the sense explained above.

Conditional on $\hat{\varphi}^{(n)}$ the process $\bar{X}^{(n)}$ is a concatenation of n independent scaled Wiener processes $\bar{X}^{(n,0)}, \ldots, \bar{X}^{(n,n-1)}$ in the sense that each of the components equals $(\hat{\sigma}_i W_t)_{t \in [0,1/n)}$ in law, where $\hat{\sigma}_i \geq 0$ is given via

$$\hat{\sigma}^2_i := n(\hat{\varphi}(i/n) - \hat{\varphi}((i-1)/n)).$$

Concatenations of Wiener processes

First we suppose that $\hat{\varphi}^{(n)}$ is deterministic and that $s = p$. The *letters* $\bar{X}^{(n,0)}, \ldots, \bar{X}^{(n,n-1)}$ are independent and the objective function

$$\mathbb{E}[\|\bar{X}^{(n)} - \hat{X}^{(r)}\|^p] = \mathbb{E}\Big[\int_0^1 |\bar{X}^{(n)}_t - \hat{X}^{(r)}_t|^p \, dt\Big]$$

can be understood as single letter distortion measure. Thus the discussion of Section 7.4.2 yields that for an arbitrary rate $\bar{r} \geq 0$ the DRF $D(\bar{r}|\bar{X}^{(n)}, \|\cdot\|^p)$ is naturally related to a rate allocation problem. One has

$$D(\bar{r}|\bar{X}^{(n)}, \|\cdot\|^p) = \inf_{(r_i)} \sum_{i=0}^{n-1} \underbrace{D(r_i|\hat{\sigma}_i W, \|\cdot\|^p_{L^p[0,1/n)})}_{=\hat{\sigma}^p_i D(r_i|W, \|\cdot\|_{L^p[0,1/n)}, p)^p}, \qquad (7.30)$$

where the infimum is taken over all non-negative vectors $(r_i)_{i=0,\ldots,n-1}$ with $\sum_{i=0}^{n-1} r_i = \bar{r}$. Moreover, the map

$$\pi : L^p[0, 1/n) \to L^p[0,1), \quad (x_t)_{t \in [0,1/n)} \mapsto (n^{-1/p} x_{t/n})_{t \in [0,1)}$$

is an isometric isomorphism so that

$$D(r_i|W, \| \cdot \|_{L^p[0,1/n)}, p) = D(r_i|n^{-\frac{1}{p}-\frac{1}{2}}\sqrt{n}W_{\cdot/n}, \| \cdot \|_{L^p[0,1)}, p)$$
$$= n^{-\frac{1}{p}-\frac{1}{2}} \underbrace{D(r_i|W, \| \cdot \|_{L^p[0,1)}, p)}_{\sim \kappa_p r_i^{-1/2}}.$$

Supposing that the rates r_i $(i = 0, \ldots, n-1)$ are large so that $\kappa_p r_i^{-1/2}$ is a reasonable approximation to the latter DRF, one concludes together with (7.30) that $D(\bar{r}|\bar{X}^{(n)}, \| \cdot \|, p)$ is approximately equal to

$$\kappa_p \inf_{(r_i)} \Big(\frac{1}{n} \sum_{i=0}^{n-1} \frac{\hat{\sigma}_i^p}{(nr_i)^{p/2}} \Big)^{1/p}. \tag{7.31}$$

The infimum can be evaluated explicitly by applying the reverse Hölder inequality analogously to the computations in (7.9). It is equal to

$$\Big(\frac{1}{n} \sum_{i=0}^{n-1} |\hat{\sigma}_i|^{2p/(p+2)} \Big)^{(p+2)/2p} \frac{1}{\sqrt{\bar{r}}}. \tag{7.32}$$

Next, let $\hat{\sigma}_t := \hat{\sigma}_i$ for $i = 0, \ldots, n-1$ and $t \in [i/n, (i+1)/n)$ and observe that (7.32) can be rewritten as

$$\Big(\int_0^1 |\hat{\sigma}_t|^{2p/(p+2)} \, dt \Big)^{(p+2)/2p} \frac{1}{\sqrt{\bar{r}}}. \tag{7.33}$$

Rigorously one can prove that one infers (asymptotically) the same error for the other information constraints and for all other moments s. We quote the exact result:

Lemma 7.20 *For fixed $s \in (0, \infty)$ there exists a real valued function $h = h_{s.p} : \mathbb{R}_+ \to \mathbb{R}_+$ with $\lim_{\bar{r} \to \infty} h(\bar{r}) = 1$ such that the following statements are valid.*

Suppose that $\hat{\varphi}^{(n)}$ is deterministic and let $\hat{X}$ be a reconstruction for $\bar{X}^{(n)}$ of rate $\bar{r} > 0$. Then there exists a $[0, \infty)$-valued sequence $(r_i)_{i=0,\ldots,n-1}$ with $\sum_{i=0}^{n-1} r_i \le \bar{r}$ such that for any $r_* > 0$:

$$\mathbb{E}[\| \bar{X}^{(n)} - \hat{X} \|^s]^{1/s} \ge h(r_*) \kappa_p \Big(\frac{1}{n} \sum_{i=0}^{n-1} \frac{|\hat{\sigma}_i|^p}{(n(r_i + r_*))^{p/2}} \Big)^{1/p}.$$

On the other hand, for any $\mathbb{R}_+$-valued vector $(r_i)_{i=0,\ldots,n-1}$, there exists a codebook $\mathcal{C} \subset L^p[0,1]$ with $\log |\mathcal{C}| \le \sum_{i=0}^{n-1} r_i$ and

$$\mathbb{E}[\min_{\hat{x} \in \mathcal{C}} \| \bar{X}^{(n)} - \hat{x} \|^s]^{1/s} \le h(r^*) \kappa_p \Big(\frac{1}{n} \sum_{i=0}^{n-1} \frac{|\hat{\sigma}_i|^p}{(nr_i)^{p/2}} \Big)^{1/p},$$

where $r^* = \min_{i=0,\dots,n-1} r_i$.

The lower bound

In the discussion above we have assumed that the time change $\hat{\varphi}^{(n)}$ is deterministic. Now we switch back to the original problem and suppose that $\hat{\varphi}^{(n)}$ and $\bar{X}^{(n)}$ ($n \in \mathbb{N}$) are as introduced in the beginning of this subsection. In particular, $\hat{\varphi}^{(n)}$ is again assumed to be random. Again the parameter n is linked to $r > 0$ via $n = \lceil r^\varsigma \rceil$.

We fix a family $(\hat{X}^{(r)})_{r\geq 0}$ of reconstructions such that each reconstruction has finite mutual information $I(\bar{X}^{(n)}; \hat{X}^{(r)})$ to be finite, and we set $r(\phi) = I(\bar{X}^{(n)}; \hat{X}^{(r)}|\varphi^{(n)} = \phi)$ and $R = r(\hat{\varphi}^{(n)})$. The random variable R is to be conceived as the random rate reserved for coding $\bar{X}^{(n)}$ given the time change $\hat{\varphi}^{(n)}$.

Using the lower bound (7.33) for the approximation error, one gets

$$\mathbb{E}[\|\bar{X}^{(n)} - \hat{X}^{(r)}\|^s]^{1/s} = \mathbb{E}\big[\mathbb{E}[\|\bar{X}^{(n)} - \hat{X}^{(r)}\|^s|\hat{\varphi}^{(n)}]\big]^{1/s}$$
$$\gtrsim \mathbb{E}\Big[\Big(\int_0^1 |\hat{\sigma}_t|^{2p/(p+2)}\,\mathrm{d}t\Big)^{s(p+2)/2p}\frac{1}{R^{s/2}}\Big]^{1/s}.$$

It remains to translate a given information constraint into a condition onto the random rate R.

The quantization constraint

Let us now assume that the family $(\hat{X}^{(r)})$ satisfies the quantization constraint $\log|\operatorname{range}\hat{X}^{(r)}| \leq r$. Then $R \leq r$ almost surely so that one gets the lower bound

$$\mathbb{E}[\|\bar{X}^{(n)} - \hat{X}^{(r)}\|^s]^{1/s} \gtrsim \mathbb{E}\Big[\Big(\int_0^1 |\hat{\sigma}_t|^{2p/(p+2)}\,\mathrm{d}t\Big)^{s(p+2)/2p}\Big]^{1/s}\frac{1}{\sqrt{r}}.$$

Since $\int_0^1 |\hat{\sigma}_t|^{2p/(p+2)}\,\mathrm{d}t$ converges to $\int_0^1 |\sigma_t|^{2p/(p+2)}\,\mathrm{d}t$ one finally obtains

$$\mathbb{E}[\|\bar{X}^{(n)} - \hat{X}^{(r)}\|^s]^{1/s} \gtrsim \big\|\|\sigma.\|_{L^{2p/(p+2)}[0,1]}\big\|_{L^s(\mathbb{P})}\frac{1}{\sqrt{r}}.$$

The mutual information constraint

Now we assume that $I(\bar{X}; \hat{X}^{(r)}) \leq r$. Then

$$I(X; \hat{X}|\hat{\varphi}) \leq I(X; \hat{X}, \hat{\varphi}^{(n)}) \leq I(\bar{X}; \hat{X}^{(r)}) + \log|\operatorname{range}\hat{\varphi}^{(n)}| \lesssim r.$$

Therefore,

$$\mathbb{E}R \lesssim r, \tag{7.34}$$

and one gets an asymptotic lower bound for the distortion rate function when minimizing

$$\mathbb{E}\left[\left(\int_0^1 |\hat{\sigma}_t|^{2p/(p+2)}\,\mathrm{d}t\right)^{s(p+2)/2p}\frac{1}{R^{s/2}}\right]^{1/s}$$

under the constraint (7.34). Also in this rate allocation problem the Hölder inequality for negative exponents solves the problem and one gets:

$$\mathbb{E}[\|\bar{X}^{(n)} - \hat{X}^{(r)}\|^s]^{1/s} \gtrsim \left\|\|\hat{\sigma}.\|_{L^{2p/(p+2)}[0,1]}\right\|_{L^{2s/(s+2)}(\mathbb{P})}\frac{1}{\sqrt{\mathbb{E}[R]}}$$

$$\gtrsim \left\|\|\sigma.\|_{L^{2p/(p+2)}[0,1]}\right\|_{L^{2s/(s+2)}(\mathbb{P})}\frac{1}{\sqrt{r}}.$$

7.6 Further results on asymptotic approximation of stochastic processes

As mentioned before, quantization of stochastic processes has been a vivid research area in recent years. So far we have restricted ourselves to surveying results that lead to strong asymptotic formulae and that are related to intermediate optimization problems. In this section we want to complete the paper by giving two further results which yield the correct weak asymptotics in many cases.

7.6.1 Estimates based on moment conditions on the increments

Let us next describe a very general approach undertaken in [41] to provide weak upper bounds for the quantization error. It is based on moment conditions on the increments of a stochastic process. Let $(X_t)_{t\in[0,1]}$ denote a real-valued stochastic process, that is $X : \Omega \times [0,1] \to \mathbb{R}$ is product measurable. We fix $\zeta > 0$ and assume that the marginals of X are in $L^\zeta(\mathbb{P})$.

We state the main result.

Theorem 7.21 Let $\varphi : [0,1] \to [0,\infty)$ be a regularly varying function at 0 with index $b > 0$, that is φ can be represented as $\varphi(t) = t^b \ell(t)$, where ℓ is continuous and satisfies $\ell(\alpha t) \sim \ell(t)$ as $t \downarrow 0$ for any $\alpha > 0$ (ℓ is called slowly varying). Moreover, we assume that for all $0 \leq u \leq t \leq 1$

$$\mathbb{E}[|X_t - X_u|^\zeta]^{1/\zeta} \leq \varphi(t-u),$$

in the case where $\zeta \geq 1$, and

$$\mathbb{E}[\sup_{v \in [u,t]} |X_v - X_s|^\zeta]^{1/\zeta} \leq \varphi(t - u),$$

otherwise. Then for $p, s \in (0, \zeta)$, one has

$$D^{(q)}(r, s|X, \| \cdot \|_{L^p[0,1]}) \precsim \varphi(1/r).$$

This theorem allows us to translate moment estimates for the increments into upper bounds for the quantization error. Let us demonstrate the power of this result by applying it to a class of diffusion processes. Let $X = (X_t)_{t \in [0,1]}$ satisfy

$$X_t = x_0 + \int_0^t G_u \, \mathrm{d}u + \int_0^t H_u \, \mathrm{d}W_u,$$

where $W = (W_t)_{t \in [0,1]}$ denotes a Wiener process, and $G = (G_t)_{t \in [0,1]}$ and $H = (H_t)_{t \in [0,1]}$ are assumed to be progressively measurable processes w.r.t. the canonical filtration induced by W. Supposing that for some $\zeta \geq 2$

$$\sup_{t \in [0,1]} \mathbb{E}[|G_t|^\zeta + |H_t|^\zeta] < \infty,$$

one can infer that for a constant $c \in \mathbb{R}_+$ one has

$$\|X_t - X_u\|_{L^\zeta(\mathbb{P})} \leq c \, (t - u)^{1/2}$$

for all $0 \leq u \leq t \leq 1$. Consequently,

$$D(r, s|X, \| \cdot \|_{L^p[0,1]}) \precsim \frac{1}{\sqrt{r}}.$$

As we have seen above this rate is optimal for the solutions to the autonomous stochastic differential equations studied above.

Moreover, the main result can be used to infer upper bounds for stationary processes and Lévy processes or to recover the weak asymptotics for fractional Brownian motions. For further details we refer the reader to [41].

7.6.2 Approximation of Lévy processes

Let now $X = (X_t)_{t \in [0,1]}$ denote a càdlàg Lévy process. Owing to the Lévy–Khintchine formula the marginals X_t ($t \in [0,1]$) admit a representation

$$\mathbb{E}e^{iuX_t} = e^{-t\psi(u)} \qquad (u \in \mathbb{R}),$$

where

$$\psi(u) = \frac{\sigma^2}{2}u^2 - i\beta u + \int_{\mathbb{R}\setminus\{0\}} (1 - e^{iux} + \mathbf{1}_{\{|x|\leq 1\}}iux)\,\nu(\mathrm{d}x),$$

for parameters $\sigma^2 \in [0,\infty)$, $\beta \in \mathbb{R}$ and a non-negative measure ν on $\mathbb{R}\setminus\{0\}$ with

$$\int_{\mathbb{R}\setminus\{0\}} 1 \wedge x^2\,\nu(\mathrm{d}x) < \infty.$$

The complexity of Lévy processes has been analyzed recently in [2] when the underlying distortion measure is induced by the $L^p[0,1]$-norm for a fixed $p \geq 1$. Its complexity is related to the function

$$F(\varepsilon) = \frac{\sigma^2}{\varepsilon^2} + \int_{\mathbb{R}\setminus\{0\}} \left[\left(\frac{x^2}{\varepsilon^2} \wedge 1\right) + \log_+ \frac{|x|}{\varepsilon}\right]\nu(\mathrm{d}x) \qquad (\varepsilon > 0).$$

In the following, we assume that $F(\varepsilon)$ is finite. Let us state the main results.

Theorem 7.22 (Upper bound) *There exist positive constants $c_1 = c_1(p)$ and c_2 such that for any $\varepsilon > 0$ and $s > 0$*

$$D^{(e)}(c_1 F(\varepsilon), s) \leq c_2\varepsilon.$$

The constants c_1 and c_2 can be chosen independently of the choice of the Lévy process. Additionally, if for $s > 0$ one has

- $\mathbb{E}\|X\|_{L^p[0,1]}^{s'} < \infty$ *for some $s' > s$*
- *for some $\zeta > 0$,*

$$\limsup_{\varepsilon \downarrow 0} \frac{\int_{|x|>\varepsilon}(|x|/\varepsilon)^\zeta\,\nu(\mathrm{d}x)}{\nu([-\varepsilon,\varepsilon]^c)} < \infty$$

then there exist constants c_1' and c_2' such that, for all $\varepsilon > 0$,

$$D^{(q)}(c_1' F(\varepsilon), s) \leq c_2'\varepsilon.$$

For stating the lower bound we will need the function

$$F_1(\varepsilon) = \frac{\sigma^2}{\varepsilon^2} + \int_{\mathbb{R}\setminus\{0\}} \left(\frac{x^2}{\varepsilon^2} \wedge 1\right)\nu(\mathrm{d}x) \qquad (\varepsilon > 0).$$

Theorem 7.23 (Lower bound) *There exist constants $c_1, c_2 > 0$ depending on $p \geq 1$ only such that the following holds. For every $\varepsilon > 0$ with $F_1(\varepsilon) \geq 18$ one has*

$$D(c_1 F(\varepsilon), p) \geq c_2\,\varepsilon.$$

Moreover, if $\nu(\mathbb{R}\backslash\{0\}) = \infty$ or $\sigma \neq 0$, one has for any $s > 0$,

$$D(c_1 F_1(\varepsilon), s) \gtrsim c_2 \varepsilon$$

as $\varepsilon \downarrow 0$.

The upper and lower bounds often are of the same weak asymptotic order. Let, for instance, X be an α-stable Lévy process (α being a parameter in $(0, 2]$). In that case, the bounds provided above are sharp, and one gets, for $s_1 > 0$ and $s_2 \in (0, \alpha)$, that

$$D(r, s_1) \approx D^{(e)}(r, s_1) \approx D^{(q)}(r, s_2) \approx \frac{1}{r^{1/\alpha}}.$$

The results of this subsection can also be related to a rate allocation problem. However, this link is weaker than the ones presented before, and we cannot derive the strong asymptotics of the coding quantities. The implication of the corresponding rate allocation problem can be roughly described as follows. When aiming at an accuracy of about ε, we need to assign rate of order $1 + \log(\ell/\varepsilon)$ to jumps of size ℓ if the jump is larger than ε. This information is used to code the position and direction of the jump. For jumps of size $\ell \leq \varepsilon$ one does not code each single jump separately. Instead one waits until the compensated cumulated sum of small jumps (plus diffusive component) has size of order ε, and then codes the corresponding exit time and direction. Since the compensated small jumps constitute a martingale the frequency at which such exits occur is of order

$$\frac{\sigma^2}{\varepsilon^2} + \int_{[-\varepsilon, \varepsilon]} \frac{x^2}{\varepsilon^2} \, \nu(\mathrm{d}x).$$

Combining the rates for the small and large jumps leads to the assertions of the above theorems. For more details we refer the reader to the original article.

Bibliography

[1] S. Ankirchner, S. Dereich, and P. Imkeller, Enlargement of filtrations and continuous Girsanov-type embeddings, In *Séminaire de Probabilités XL*, volume 1899 of Lectures Notes in Mathematics, 389–410, Springer, Berlin, 2007.

[2] F. Aurzada and S. Dereich. The coding complexity of Lévy processes. *Found. Comput. Math.*, to appear.

[3] E. Belinsky and W. Linde. Small ball probabilities of fractional Brownian sheets via fractional integration operators. *J. Theor. Prob.*, 15(3):589–612, 2002.

[4] J.C. Bronski. Small ball constants and tight eigenvalue asymptotics for fractional Brownian motions. *J. Theor. Prob.*, 16(1):87–100, 2003.

[5] J.A. Bucklew. Two results on the asymptotic performance of quantizers. *IEEE Trans. Inform. Theory*, 30(2, part 2):341–348, 1984.

[6] J.A. Bucklew and G.L. Wise. Multidimensional asymptotic quantization theory with rth power distortion measures. *IEEE Trans. Inf. Theory*, 28:239–247, 1982.

[7] B. Carl and I. Stephani. *Entropy, Compactness and the Approximation of Operators*, volume 98 of Cambridge Tracts in Mathematics. Cambridge University Press, 1990.

[8] Th.M. Cover and J.A. Thomas. *Elements of Information Theory*. Wiley Series in Telecommunications. New York: John Wiley, 1991.

[9] J. Creutzig. Approximation of Gaussian random vectors in Banach spaces. Ph.D. thesis, Friedrich-Schiller-Universität Jena, 2002.

[10] J. Creutzig, S. Dereich, Th. Müller-Gronbach, and K. Ritter. Infinite-dimensional quadrature and approximation of distributions. *Found. Comput. Math*, to appear.

[11] S. Delattre, S. Graf, H. Luschgy, and G. Pagès. Quantization of probability distributions under norm-based distortion measures. *Statist. Decision*, 22:261–282, 2004.

[12] A. Dembo and I. Kontoyiannis. Source coding, large deviations, and approximate pattern matching. *IEEE Trans. Inform. Theory*, 48(6):1590–1615, 2002. Special issue on Shannon theory: perspective, trends, and applications.

[13] S. Dereich and M.A. Lifshits. Probabilities of randomly centered small balls and quantization in Banach spaces. *Ann. Prob.*, 33(4):1397–1421, 2005.

[14] S. Dereich. High resolution coding of stochastic processes and small ball probabilities. Ph.D. thesis, TU Berlin, URL: http://edocs.tu-berlin.de/diss/2003/dereich_steffen.htm, 2003.

[15] S. Dereich. Small ball probabilities around random centers of Gaussian measures and applications to quantization. *J. Theor. Prob.*, 16(2):427–449, 2003.

[16] S. Dereich. Asymptotic behavior of the distortion-rate function for Gaussian processes in Banach spaces. *Bull. Sci. Math.*, 129(10):791–803, 2005.

[17] S. Dereich. The coding complexity of diffusion processes under supremum norm distortion. *Stochastic Process. Appl.*, vol. 118(6):917–937, 2008.

[18] S. Dereich. The coding complexity of diffusion processes under $L^p[0,1]$-norm distortion. *Stochastic Process. Appl.*, vol. 118(6):938–951, 2008.

[19] S. Dereich, F. Fehringer, A. Matoussi, and M. Scheutzow. On the link between small ball probabilities and the quantization problem for Gaussian measures on Banach spaces. *J. Theor. Prob.*, 16(1):249–265, 2003.

[20] S. Dereich and M. Scheutzow. High resolution quantization and entropy coding for fractional brownian motion. *Electron. J. Prob.*, 11:700–722, 2006.

[21] S. Dereich and Ch. Vormoor. The high resolution vector quantization problem with Orlicz norm distortion. Preprint, 2006.

[22] K. Dzhaparidze and H. van Zanten. A series expansion of fractional Brownian motion. *Prob. Theor. Related Fields*, 130(1):39–55, 2004.

[23] F. Fehringer. Kodierung von Gaußmaßen. Ph.D. thesis, TU Berlin, 2001.

[24] A. Gersho. Asymptotically optimal block quantization. *IEEE Trans. Inform.*

Theory, 25(4):373–380, 1979.

[25] S. Graf, H. Luschgy, and G. Pagès. Distortion mismatch in the quantization of probability measures. Preprint, arXiv:math/0602381v1 [math.PR], 2006.

[26] S. Graf and H. Luschgy. *Foundations of Quantization for Probability Distributions*. Lecture Notes in Mathematics 1730, Berlin: Springer, 2000.

[27] R.M. Gray and D.L. Neuhoff. Quantization. *IEEE Trans. Inf. Theory*, 44(6):2325–2383, 1998.

[28] E. Iglói. A rate-optimal trigonometric series expansion of the fractional Brownian motion. *Electron. J. Prob.*, 10, 2005.

[29] Sh. Ihara. *Information Theory for Continuous Systems*. Singapore: World Scientific, 1993.

[30] Th. Jeulin and M. Yor, editors. *Grossissements de filtrations: exemples et applications*, volume 1118 of Lecture Notes in Mathematics. Berlin: Springer-Verlag, 1985. Papers from the seminar on stochastic calculus held at the Université de Paris VI, Paris, 1982/1983.

[31] J.C. Kieffer. A survey of the theory of source coding with a fidelity criterion. *IEEE Trans. Inform. Theory*, 39(5):1473–1490, 1993.

[32] A.N. Kolmogorov. Three approaches to the quantitative definition of information. *Int. J. Comput. Math.*, 2:157–168, 1968.

[33] A.N. Kolmogorov. On the Shannon theory of information transmission in the case of continuous signals. *IRE Trans. Inform. Theory*, 2:102–108, 1956.

[34] J. Kuelbs and W.V. Li. Metric entropy and the small ball problem for Gaussian measures. *J. Funct. Anal.*, 116(1):133–157, 1993.

[35] W.V. Li and Q.-M. Shao. Gaussian processes: inequalities, small ball probabilities and applications. In *Stochastic Processes: Theory and Methods*, volume 19 of *Handbook of Statistics*, pages 533–597. Amsterdam: North-Holland, 2001.

[36] W.V. Li and W. Linde. Approximation, metric entropy and small ball estimates for Gaussian measures. *Ann. Probab.*, 27(3):1556–1578, 1999.

[37] M.A. Lifshits. On the lower tail probabilities of some random series. *Ann. Probab.*, 25(1):424–442, 1997.

[38] H. Luschgy and G. Pagès. Functional quantization of Gaussian processes. *J. Funct. Anal.*, 196(2):486–531, 2002.

[39] H. Luschgy and G. Pagès. Sharp asymptotics of the functional quantization problem for Gaussian processes. *Ann. Probab.*, 32(2):1574–1599, 2004.

[40] H. Luschgy and G. Pagès. Functional quantization of a class of Brownian diffusions: a constructive approach. *Stochastic Process. Appl.*, 116(2):310–336, 2006.

[41] H. Luschgy and G. Pagès. Functional quantization rate and mean pathwise regularity of processes with an application to Lévy processes. Preprint, arXiv:math/0601774v3 [math.PR], 2007.

[42] D.M. Mason and Zh. Shi. Small deviations for some multi-parameter Gaussian processes. *J. Theor. Probab.*, 14(1):213–239, 2001.

[43] G. Pagès and J. Printems. Functional quantization for numerics with an application to option pricing. *Monte Carlo Methods and Applications*, 11(4):407–446, 2005.

[44] C.E. Shannon. A mathematical theory of communication. *Bell System Tech. J.*, 27:379–423, 623–656, 1948.

[45] G.N. Sytaya. On some asymptotic representation of the Gaussian measure

in a Hilbert space. *Theor. Stoch. Proc.*, 2:94–104, 1974.

[46] M. Talagrand. The small ball problem for the Brownian sheet. *Ann. Prob.*, 22(3):1331–1354, 1994.

[47] P.L. Zador. Topics in the asymptotic quantization of continuous random variables. Bell Laboratories Technical Memorandum, 1966.

III. Stochastic analysis in mathematical physics

8

Intermittency on catalysts

Jürgen Gärtner

Institut für Mathematik
Technische Universität Berlin
Straße des 17. Juni 136
10623 Berlin, Germany

Frank den Hollander

Mathematical Institute
Leiden University
PO Box 9512
2300 RA Leiden, The Netherlands

Grégory Maillard

Section de Mathématiques
Ecole Polytechnique Fédérale de Lausanne
1015 Lausanne, Switzerland

Abstract

The present paper provides an overview of results obtained in four recent papers by the authors. These papers address the problem of intermittency for the Parabolic Anderson Model in a *time-dependent random medium*, describing the evolution of a "reactant" in the presence of a "catalyst". Three examples of catalysts are considered: (1) independent simple random walks; (2) symmetric exclusion process; (3) symmetric voter model. The focus is on the annealed Lyapunov exponents, i.e. the exponential growth rates of the successive moments of the reactant. It turns out that these exponents exhibit an interesting dependence on the dimension and on the diffusion constant.

Trends in Stochastic Analysis, ed. J. Blath, P. Mörters and M. Scheutzow.
Published by Cambridge University Press. ©Cambridge University Press 2008.

8.1 The Parabolic Anderson Model

8.1.1 Motivation

The *Parabolic Anderson Model* is the partial differential equation

$$\frac{\partial}{\partial t}u(x,t) = \kappa\Delta u(x,t) + \gamma\xi(x,t)u(x,t), \qquad x \in \mathbb{Z}^d, t \geq 0, \qquad (8.1)$$

for the $\mathbb{R}$-valued random field

$$u = \{u(x,t) \colon x \in \mathbb{Z}^d, t \geq 0\}, \qquad (8.2)$$

where $\kappa \in [0,\infty)$ is the diffusion constant, $\gamma \in [0,\infty)$ is the coupling constant, Δ is the discrete Laplacian, acting on u as

$$\Delta u(x,t) = \sum_{\substack{y \in \mathbb{Z}^d \\ \|y-x\|=1}} [u(y,t) - u(x,t)] \qquad (8.3)$$

($\|\cdot\|$ is the Euclidian norm), while

$$\xi = \{\xi(x,t) \colon x \in \mathbb{Z}^d, t \geq 0\} \qquad (8.4)$$

is an $\mathbb{R}$-valued random field that evolves with time and that drives the equation. As initial condition for (8.1) we take

$$u(\cdot,0) \equiv 1. \qquad (8.5)$$

One interpretation of (8.1) and (8.5) comes from *population dynamics*. Consider a spatially homogeneous system of two types of particles, A (catalyst) and B (reactant), subject to:

(i) A-particles evolve autonomously, according to a prescribed stationary dynamics given by the ξ-field, with $\xi(x,t)$ denoting the number of A-particles at site x at time t;

(ii) B-particles perform independent simple random walks with jump rate $2d\kappa$ and split into two at a rate that is equal to γ times the number of A-particles present at the same location;

(iii) the initial density of B-particles is 1.

Then

$$u(x,t) \quad = \quad \text{the average number of } B\text{-particles at site } x \text{ at time } t$$
$$\text{conditioned on the evolution of the } A\text{-particles.}$$
$$(8.6)$$

It is possible to add that B-particles die at rate $\delta \in (0,\infty)$. This amounts to the trivial transformation

$$u(x,t) \to u(x,t)e^{-\delta t}. \qquad (8.7)$$

What makes (8.1) particularly interesting is that the two terms on the right-hand side *compete with each other*: the diffusion (of B-particles) described by $\kappa\Delta$ tends to make u flat, while the branching (of B-particles caused by A-particles) described by ξ tends to make u irregular.

8.1.2 Intermittency

We will be interested in the presence or absence of *intermittency*. Intermittency means that for large t the branching dominates, i.e. the u-field develops sparse high peaks in such a way that u and its moments are each dominated by their own collection of peaks (see Gärtner and König [10], Section 1.3). In the *quenched* situation, i.e. conditional on ξ, this geometric picture of intermittency is well understood for several classes of *time-independent* random potentials ξ (see e.g. Sznitman [16] for Poisson clouds and Gärtner, König and Molchanov [11] for i.i.d. potentials with double-exponential and heavier upper tails; Gärtner and König [10] provides an overview). For *time-dependent* random potentials ξ, however, such a geometric picture is not yet available. Instead one restricts attention to understanding the phenomenon of intermittency indirectly by comparing the successive *annealed* Lyapunov exponents

$$\lambda_p = \lim_{t\to\infty} \Lambda_p(t), \qquad p \in \mathbb{N}, \tag{8.8}$$

with

$$\Lambda_p(t) = \frac{1}{t} \log \mathbb{E}\left([u(0,t)]^p\right)^{1/p}, \qquad p \in \mathbb{N}, t > 0, \tag{8.9}$$

where $\mathbb{E}$ denotes expectation w.r.t. ξ. One says that the solution u is *p-intermittent* if

$$\lambda_p > \lambda_{p-1}, \tag{8.10}$$

and *intermittent* if (8.10) holds for all $p \in \mathbb{N} \setminus \{1\}$.

Carmona and Molchanov [2] succeeded in investigating the annealed Lyapunov exponents, and in obtaining the qualitative picture of intermittency (in terms of these exponents), for potentials of the form

$$\xi(x,t) = \dot{W}_x(t), \tag{8.11}$$

where $\{W_x(t) \colon x \in \mathbb{Z}^d, t \geq 0\}$ denotes a collection of independent Brownian motions. (In this case, (8.1) corresponds to an infinite system of coupled Itô-diffusions.) They showed that for $d = 1, 2$ intermittency holds for all κ, whereas for $d \geq 3$ p-intermittency holds if and only if the diffusion constant κ is smaller than a critical threshold $\kappa_p = \kappa_p(d, \gamma)$

tending to infinity as $p \to \infty$. They also studied the asymptotics of the quenched Lyapunov exponent in the limit as $\kappa \downarrow 0$, which turns out to be singular. Subsequently, the latter was more thoroughly investigated in papers by Carmona, Molchanov and Viens [3], Carmona, Koralov and Molchanov [1], and Cranston, Mountford and Shiga [4].

In Sections 8.2–8.4 we consider three different choices for ξ, namely:

(1) Independent Simple Random Walks
(2) Symmetric Exclusion Process
(3) Symmetric Voter Model.

For each of these examples we study the annealed Lyapunov exponents as a function of d, κ and γ. Because of their *non-Gaussian* and *non-independent* spatial structure, these examples require techniques different from those developed for (8.11). Example (1) was studied earlier in Kesten and Sidoravicius [12]. We describe their work in Section 8.2.2.

By the Feynman–Kac formula, the solution of (8.1) and (8.5) reads

$$u(x,t) = E_x \left(\exp \left[\gamma \int_0^t ds \, \xi \left(X^\kappa(s), t-s \right) \right] \right), \qquad (8.12)$$

where X^κ is a simple random walk on $\mathbb{Z}^d$ with step rate $2d\kappa$ and E_x denotes expectation with respect to X^κ given $X^\kappa(0) = x$. This formula shows that understanding intermittency amounts to studying the *large deviation behaviour* of a random walk *sampling* a time-dependent random field.

8.2 Independent simple random walks

In this section we consider the case where ξ is a Poisson field of *independent simple random walks* (ISRW). We first describe the results obtained in Kesten and Sidoravicius [12]. After that we describe the refinements of these results obtained in Gärtner and den Hollander [6].

8.2.1 Model

ISRW is the Markov process with state space

$$\Omega = (\mathbb{N} \cup \{0\})^{\mathbb{Z}^d} \qquad (8.13)$$

whose generator acts on cylindrical functions f as

$$(Lf)(\eta) = \frac{1}{2d} \sum_{(x,y)} \eta(x)[f(\eta^{x \curvearrowright y}) - f(\eta)], \qquad (8.14)$$

where the sum runs over oriented bonds between neighboring sites, and

$$\eta^{x \curvearrowright y}(z) = \begin{cases} \eta(z) & \text{if } z \neq x, y, \\ \eta(x) - 1 & \text{if } z = x, \\ \eta(y) + 1 & \text{if } z = y, \end{cases} \qquad (8.15)$$

i.e. $\eta^{x \curvearrowright y}$ is the configuration obtained from η by moving a particle from x to y. We choose $\xi(\cdot, 0)$ according to the Poisson product measure with density $\rho \in (0, \infty)$, i.e. initially each site carries a number of particles that is Poisson distributed with mean ρ. For this choice, the ξ-field is stationary and reversible in time (see Kipnis and Landim [13]).

Under ISRW, particles move around independently as simple random walks, stepping at rate 1 and choosing from neighboring sites with probability $1/2d$ each.

8.2.2 Main theorems

Kesten and Sidoravicius [12] proved the following. They considered the language of A-particles and B-particles from population dynamics, as mentioned in Section 8.1.1, and included a death rate $\delta \in [0, \infty)$ for the B-particles (recall (8.7)).

(1) If $d = 1, 2$, then – for any choice of the parameters – the average number of B-particles per site tends to infinity at a rate that is faster than exponential.

(2) If $d \geq 3$, then – for γ sufficiently small and δ sufficiently large – the average number of B-particles per site tends to zero exponentially fast.

(3) If $d \geq 1$, then – conditional on the evolution of the A-particles – there is a phase transition: for small δ the B-particles locally survive, while for large δ they become locally extinct.

Properties (1) and (2) – which are annealed results – are implied by Theorems 8.2 and 8.3 below, while property (3) – which is a quenched result – is not. The main focus of [12] is on survival versus extinction. The approach in [12], being based on path estimates rather than on the Feynman-Kac representation, produces cruder results, but it is more robust against variations of the dynamics.

In Gärtner and den Hollander [6] the focus is on the annealed Lyapunov exponents. Theorems 8.1–8.3 below are taken from that paper.

Theorem 8.1 ([6, Theorem 24]) *Let $d \geq 1$, $\rho, \gamma \in (0, \infty)$ and $p \in \mathbb{N}$.*
(i) For all $\kappa \in [0, \infty)$, the limit in (8.8) exists.
(ii) If $\lambda_p(0) < \infty$, then $\kappa \to \lambda_p(\kappa)$ is finite, continuous, non-increasing and convex on $[0, \infty)$.

Let $p_t(x, y)$ denote the probability that simple random walk stepping at rate 1 moves from x to y in time t. Let

$$G_d = \int_0^\infty p_t(0,0) \, dt \qquad (8.16)$$

be the Green function at the origin of the simple random walk.

Theorem 8.2 ([6, Theorem 25]) *Let $d \geq 1$, $\rho, \gamma \in (0, \infty)$ and $p \in \mathbb{N}$. Then, for all $\kappa \in [0, \infty)$, $\lambda_p(\kappa) < \infty$ if and only if $p < 1/G_d\gamma$.*

It can be shown that, if $p > 1/G_d\gamma$, then $\Lambda_p(t)$ in (8.9) grows exponentially fast with t, i.e. the pth moment of $u(0, t)$ grows double exponentially fast with t. The constant in the exponent can be computed.

In the regime $p < 1/G_d\gamma$, $\kappa \mapsto \lambda_p(\kappa)$ has the following behaviour (see Fig. 8.1):

Theorem 8.3 ([6, Theorem 26]) *Let $d \geq 1$, $\rho, \gamma \in (0, \infty)$ and $p \in \mathbb{N}$ such that $p < 1/G_d\gamma$.*
(i) $\kappa \mapsto \lambda_p(\kappa)$ is continuous, strictly decreasing and convex on $[0, \infty)$.
(ii) For $\kappa = 0$,

$$\lambda_p(0) = \rho\gamma \, \frac{(1/G_d)}{(1/G_d) - p\gamma}. \qquad (8.17)$$

(iii) For $\kappa \to \infty$,

$$\lim_{\kappa \to \infty} 2d\kappa[\lambda_p(\kappa) - \rho\gamma] = \rho\gamma^2 G_d + 1_{d=3} \, (2d)^3 (\rho\gamma^2 p)^2 \, \mathcal{P}_3 \qquad (8.18)$$

with

$$\mathcal{P}_3 = \sup_{\substack{f \in H^1(\mathbb{R}^3) \\ \|f\|_2 = 1}} \left[\int_{\mathbb{R}^3} dx \, |f(x)|^2 \int_{\mathbb{R}^3} dy \, |f(y)|^2 \, \frac{1}{4\pi \|x - y\|} \right.$$
$$\left. - \int_{\mathbb{R}^3} dx \, |\nabla f(x)|^2 \right]. \qquad (8.19)$$

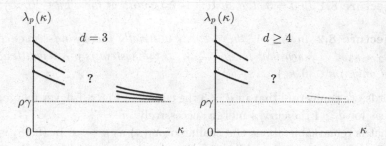

Fig. 8.1. $\kappa \mapsto \lambda_p(\kappa)$ for $p = 1, 2, 3$ when $p < 1/G_d\gamma$ for a simple random walk in $d = 3$ and $d \geq 4$.

8.2.3 Discussion

Theorem 8.2 says that if the catalyst is driven by a recurrent random walk ($G_d = \infty$) then it can pile up near the origin and make the reactant grow at an unbounded rate, while if the catalyst is driven by a transient random walk ($G_d < \infty$) then small enough moments of the reactant grow at a finite rate. We refer to this dichotomy as the *strongly catalytic*, respectively the *weakly catalytic*, regime.

Theorem 8.3(i) shows that, even in the weakly catalytic regime, some degree of *clumping* of the catalyst occurs, in that the growth rate of the reactant is $> \rho\gamma$, the average medium growth rate. As the diffusion constant κ of the reactant increases, the effect of the clumping of the catalyst on the reactant gradually diminishes, and the growth rate of the reactant gradually decreases to $\rho\gamma$.

Theorem 8.3(ii) shows that, again in the weakly catalytic regime, if the reactant stands still, then the system is intermittent. Apparently, the successive moments of the reactant are sensitive to *successive degrees of clumping*. By continuity, intermittency persists for small κ.

Theorem 8.3(iii) shows that all Lyapunov exponents decay to $\rho\gamma$ as $\kappa \to \infty$ in the same manner when $d \geq 4$ but not when $d = 3$. In fact, in $d = 3$ intermittency persists for large κ. It remains open whether the same is true for $d \geq 4$. To decide the latter, we need a finer asymptotics for $d \geq 4$. A large diffusion constant of the reactant hampers localization of the reactant around regions where the catalyst clumps, but it is not a priori clear whether this is able to destroy intermittency for $d \geq 4$. We conjecture:

Conjecture 8.1 *In $d = 3$, the system is intermittent for all $\kappa \in [0, \infty)$.*

Conjecture 8.2 *In $d \geq 4$, there exists a strictly increasing sequence $0 < \kappa_2 < \kappa_3 < \dots$ such that for $p = 2, 3, \dots$ the system is p-intermittent if and only if $\kappa \in [0, \kappa_p)$.*

In words, we conjecture that in $d = 3$ the curves in Fig. 8.1 never merge, whereas for $d \geq 4$ the curves merge successively.

What is remarkable about the scaling of $\lambda_p(\kappa)$ as $\kappa \to \infty$ in (8.18) is that $\mathcal{P}_3$ is the variational problem for the so-called *polaron model*. Here, one considers the quantity

$$\theta(t; \alpha) = \frac{1}{\alpha^2 t} \log E_0 \left(\exp \left[\alpha \int_0^t ds \int_s^t du \, \frac{e^{-(u-s)}}{|\beta(u) - \beta(s)|} \right] \right), \quad (8.20)$$

where $\alpha > 0$ and $(\beta(t))_{t \geq 0}$ is standard Brownian motion on $\mathbb{R}^3$ starting at $\beta(0) = 0$. Donsker and Varadhan [5] proved that

$$\lim_{\alpha \to \infty} \lim_{t \to \infty} \theta(t; \alpha) = 4\sqrt{\pi} \, \mathcal{P}_3. \quad (8.21)$$

Lieb [14] proved that (8.19) has a unique maximizer modulo translations and that the centered maximizer is radially symmetric, radially non-increasing, strictly positive and smooth. A deeper analysis shows that the link between the scaling of $\lambda_p(\kappa)$ for $\kappa \to \infty$ and the scaling of the polaron for $\alpha \to \infty$ comes from *moderate* deviation behaviour of ξ and *large* deviation behaviour of the occupation time measure of X^κ in (8.12). For details we refer the reader to Gärtner and den Hollander [6].

8.3 Symmetric exclusion process

In this section we consider the case where ξ is the *symmetric exclusion process* (SEP) in equilibrium. We summarize the results obtained in Gärtner, den Hollander and Maillard [7, 8].

8.3.1 Model

Let $p: \mathbb{Z}^d \times \mathbb{Z}^d \to [0, 1]$ be the transition kernel of an irreducible symmetric random walk. SEP is the Markov process with state space

$$\Omega = \{0, 1\}^{\mathbb{Z}^d} \quad (8.22)$$

whose generator L acts on cylindrical functions f as

$$(Lf)(\eta) = \sum_{\{x, y\} \subset \mathbb{Z}^d} p(x, y) \left[f(\eta^{x, y}) - f(\eta) \right], \quad (8.23)$$

where the sum runs over unoriented bonds between any pair of sites, and

$$\eta^{x,y}(z) = \begin{cases} \eta(z) & \text{if } z \neq x, y, \\ \eta(y) & \text{if } z = x, \\ \eta(x) & \text{if } z = y. \end{cases} \tag{8.24}$$

In words, the states of x and y are interchanged along the bond $\{x, y\}$ at rate $p(x, y)$. We choose $\xi(\cdot, 0)$ according to the Bernoulli product measure with density $\rho \in (0, 1)$. For this choice, the ξ-field is stationary and reversible in time (see Liggett [15]).

Under SEP, particles move around independently according to the symmetric random walk transition kernel $p(\cdot, \cdot)$, but subject to the restriction that no two particles can occupy the same site. A special case is simple random walk

$$p(x, y) = \begin{cases} \frac{1}{2d} & \text{if } \|x - y\| = 1, \\ 0 & \text{otherwise.} \end{cases} \tag{8.25}$$

8.3.2 Main theorems

Theorem 8.4 *Let* $d \geq 1$, $\rho \in (0, 1)$, $\gamma \in (0, \infty)$ *and* $p \in \mathbb{N}$.
(i) For all $\kappa \in [0, \infty)$, *the limit in* (8.8) *exists and is finite.*
(ii) On $[0, \infty)$, $\kappa \to \lambda_p(\kappa)$ *is continuous, non-increasing and convex.*

The following dichotomy holds (see Fig. 8.2):

Theorem 8.5 *Let* $d \geq 1$, $\rho \in (0, 1)$, $\gamma \in (0, \infty)$ *and* $p \in \mathbb{N}$.
(i) If $p(\cdot, \cdot)$ *is recurrent, then* $\lambda_p(\kappa) = \gamma$ *for all* $\kappa \in [0, \infty)$.
(ii) If $p(\cdot, \cdot)$ *is transient, then* $\rho\gamma < \lambda_p(\kappa) < \gamma$ *for all* $\kappa \in [0, \infty)$. *Moreover,* $\kappa \mapsto \lambda_p(\kappa)$ *is strictly decreasing with* $\lim_{\kappa \to \infty} \lambda_p(\kappa) = \rho\gamma$. *Furthermore,* $p \mapsto \lambda_p(0)$ *is strictly increasing.*

For transient simple random walk, $\kappa \mapsto \lambda_p(\kappa)$ has the following behaviour (similar to Fig. 8.1):

Theorem 8.6 *Let* $d \geq 3$, $\rho \in (0, 1)$, $\gamma \in (0, \infty)$ *and* $p \in \mathbb{N}$. *Assume* (8.25). *Then*

$$\lim_{\kappa \to \infty} 2d\kappa[\lambda_p(\kappa) - \rho\gamma] = \rho(1 - \rho)\gamma^2 G_d + 1_{\{d=3\}} (2d)^3 [\rho(1 - \rho)\gamma^2 p]^2 \mathcal{P}_3 \tag{8.26}$$

with G_d *and* $\mathcal{P}_3$ *as defined in* (8.16) *and* (8.19).

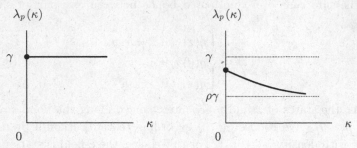

Fig. 8.2. Qualitative picture of $\kappa \mapsto \lambda_p(\kappa)$ for a recurrent (left) and a transient (right) random walk.

8.3.3 Discussion

The intuition behind Theorem 8.5 is the following. If the catalyst is driven by a recurrent random walk, then it suffers from "traffic jams", i.e. with not too small a probability there is a large region around the origin that the catalyst fully occupies for a long time. Since with not too small a probability the simple random walk (driving the reactant) can stay inside this large region for the same amount of time, the average growth rate of the reactant at the origin is maximal. This phenomenon may be expressed by saying that *for recurrent random walk clumping of the catalyst dominates the growth of the moments*. For transient random walk, on the other hand, clumping of the catalyst is present (the growth rate of the reactant is $> \rho\gamma$), but it is *not* dominant (the growth rate of the reactant is $< \gamma$). Again, when the reactant stands still or moves slowly, the successive moments of the reactant are sensitive to successive degrees of clumping of the catalyst. As the diffusion constant κ of the reactant increases, the effect of the clumping of the catalyst on the reactant gradually diminishes and the growth rate of the reactant gradually decreases to $\rho\gamma$.

Theorem 8.6 has the same interpretation as its analogue Theorem 8.3 (iii) for ISRW. We conjecture that the same behaviour occurs for SEP as in Conjectures 8.1–8.2 for ISRW.

8.4 Symmetric voter model

In this section we consider the case where ξ is the *symmetric voter model* (SVM) in equilibrium, or converging to equilibrium from a product measure. We summarize the results obtained in Gärtner, den Hollander and Maillard [9].

8.4.1 Model

As in Section 8.3, we abbreviate $\Omega = \{0,1\}^{\mathbb{Z}^d}$ and we let $p \colon \mathbb{Z}^d \times \mathbb{Z}^d \to [0,1]$ be the transition kernel of an irreducible symmetric random walk. The SVM is the Markov process on Ω whose generator L acts on cylindrical functions f as

$$(Lf)(\eta) = \sum_{x,y \in \mathbb{Z}^d} 1_{\{\eta(x) \neq \eta(y)\}} \, p(x,y) \, [f(\eta^y) - f(\eta)], \qquad (8.27)$$

where

$$\eta^y(z) = \begin{cases} \eta(z) & \text{if } z \neq y, \\ 1 - \eta(y) & \text{if } z = y. \end{cases} \qquad (8.28)$$

In words, site x imposes its state on site y at rate $p(x,y)$. The states 0 and 1 are referred to as opinions or, alternatively, as vacancy and particle. Contrary to ISRW and SEP, SVM is a non-conservative and non-reversible dynamics: opinions are not preserved.

We will consider two choices for the starting measure of ξ:

$$\left. \begin{array}{l} \nu_\rho, \text{ the Bernoulli product measure with density } \rho \in (0,1), \\ \mu_\rho, \text{ the equilibrium measure with density } \rho \in (0,1). \end{array} \right\} \qquad (8.29)$$

The ergodic properties of the SVM are qualitatively different for recurrent and for transient transition kernels. In particular, when $p(\cdot,\cdot)$ is recurrent all equilibria are trivial, i.e. $\mu_\rho = (1-\rho)\delta_0 + \rho\delta_1$, while when $p(\cdot,\cdot)$ is transient there are also non-trivial equilibria, i.e. ergodic μ_ρ parameterized by the density ρ. When starting from ν_ρ, $\xi(\cdot,t)$ converges in law to μ_ρ as $t \to \infty$.

8.4.2 Main theorems

Theorem 8.7 *Let $d \geq 1$, $\kappa \in [0,\infty)$, $\rho \in (0,1)$, $\gamma \in (0,\infty)$ and $p \in \mathbb{N}$.*
(i) For all $\kappa \in [0,\infty)$, the limit in (8.8) exists and is finite, and is the same for the two choices of starting measure in (8.29).
(ii) On $\kappa \in [0,\infty)$, $\kappa \to \lambda_p(\kappa)$ is continuous.

The following dichotomy holds (see Fig. 8.3):

Theorem 8.8 *Suppose that $p(\cdot,\cdot)$ has finite variance. Fix $\rho \in (0,1)$, $\gamma \in (0,\infty)$ and $p \in \mathbb{N}$.*
(i) If $1 \leq d \leq 4$, then $\lambda_p(\kappa) = \gamma$ for all $\kappa \in [0,\infty)$.
(ii) If $d \geq 5$, then $\rho\gamma < \lambda_p(\kappa) < \gamma$ for all $\kappa \in [0,\infty)$.

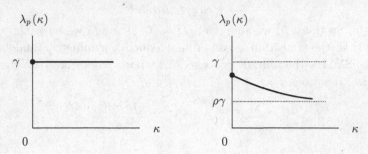

Fig. 8.3. Qualitative picture of $\kappa \mapsto \lambda_p(\kappa)$ for a symmetric random walk with finite variance in $d = 1, 2, 3, 4$, and in $d \geq 5$.

Theorem 8.9 *Suppose that $p(\cdot, \cdot)$ has finite variance. Fix $\rho \in (0, 1)$ and $\gamma \in (0, \infty)$. If $d \geq 5$, then $p \mapsto \lambda_p(0)$ is strictly increasing.*

8.4.3 Discussion

Theorem 8.8 shows that the Lyapunov exponents exhibit a dichotomy similar to those found for ISRW and SEP (see Fig. 8.3). The crossover in dimensions is at $d = 5$ rather than at $d = 3$. Theorem 8.9 shows that the system is intermittent at $\kappa = 0$ when the Lyapunov exponents are nontrivial, which is similar as well.

We conjecture that the following properties hold, whose analogues for ISRW and SEP are known to be true:

Conjecture 8.3 *Let $d \geq 5$. Then on $[0, \infty)$, $\kappa \mapsto \lambda_p(\kappa)$ is strictly decreasing and convex with $\lim_{\kappa \to \infty} \lambda_p(\kappa) = \rho\gamma$.*

We close with a conjecture about the scaling behaviour for $\kappa \to \infty$.

Conjecture 8.4 *Let $d \geq 5$, $\rho \in (0, \infty)$ and $p \in \mathbb{N}$. Assume (8.25). Then*

$$\lim_{\kappa \to \infty} 2d\kappa[\lambda_p(\kappa) - \rho\gamma] = \rho(1-\rho)\gamma^2 \frac{G_d^*}{G_d} + 1_{\{d=5\}}(2d)^3 \left[\rho(1-\rho)\gamma^2 \frac{1}{G_d}p\right]^2 \mathcal{P}_5 \tag{8.30}$$

with

$$G_d = \int_0^\infty p_t(0,0)\, dt,$$
$$G_d^* = \int_0^\infty t\, p_t(0,0)\, dt, \tag{8.31}$$

and

$$\mathcal{P}_5 = \sup_{\substack{f \in H^1(\mathbb{R}^5) \\ \|f\|_2 = 1}} \left[\int_{\mathbb{R}^5} dx\, |f(x)|^2 \int_{\mathbb{R}^5} dy\, |f(y)|^2 \frac{1}{16\pi^2 \|x - y\|} \right.$$
$$\left. - \int_{\mathbb{R}^5} dx\, |\nabla f(x)|^2 \right]. \tag{8.32}$$

8.5 Concluding remarks

The theorems listed in Sections 8.2–8.4 show that the intermittent behaviour of the reactant for the three types of catalyst exhibits interesting similarities and differences. ISRW, SEP and SVM each show a dichotomy of strongly catalytic versus weakly catalytic behaviour, for ISRW between divergence and convergence of the Lyapunov exponents, for SEP and SVM between maximality and non-maximality. Each also shows an interesting dichotomy in the dimension for the scaling behaviour at large diffusion constants, with $d = 3$ being critical for ISRW and SEP, and $d = 5$ for SVM. For ISRW and SEP the same polaron term appears in the scaling limit, while for SVM an analogous but different polaron-like term appears. Although the techniques we use for the three models differ substantially, there is a universal principle behind their scaling behaviour. See the heuristic explanation offered in [6] and [7].

Both ISRW and SEP are conservative and reversible dynamics. The reversibility allows for the use of spectral techniques, which play a key role in the analysis. The SVM, on the other hand, is a non-conservative and irreversible dynamics. The non-reversibility precludes the use of spectral techniques, and this dynamics is therefore considerably harder to handle.

Both for SEP and SVM, the graphical representation is a powerful tool. For SEP this graphical representation builds on random walks, for SVM on coalescing random walks (see Liggett [15]).

The reader is invited to look at the original papers for details.

Bibliography

[1] Carmona, R.A., Koralov, L. and Molchanov, S.A. (2001). Asymptotics for the almost-sure Lyapunov exponent for the solution of the parabolic Anderson problem. *Random Oper. Stochastic Equations*, **9**, 77–86.

[2] Carmona, R.A. and Molchanov, S.A. (1994). Parabolic Anderson Problem and Intermittency. *AMS Memoir 518. American Mathematical Society.*

248   Jürgen Gärtner, Frank den Hollander, Grégory Maillard

[3] Carmona, R.A., Molchanov S.A. and Viens, F. (1996). Sharp upper bound on the almost-sure exponential behavior of a stochastic partial differential equation. *Random Oper. Stoch. Equations*, **4**, 43–49.

[4] Cranston, M., Mountford, T.S. and Shiga, T. (2002). Lyapunov exponents for the parabolic Anderson model. *Acta Math. Univ. Comeniane*, **71**, 163–188.

[5] Donsker M.D. and Varadhan S.R.S. (1983). Asymptotics for the polaron. *Comm. Pure Appl. Math.*, **36**, 505–528.

[6] Gärtner, J. and den Hollander, F. (2006). Intermittency in a catalytic random medium. *Ann. Prob.*, **34**, 2219–2287.

[7] Gärtner, J., den Hollander, F. and Maillard, G. (2007). Intermittency on catalysts: symmetric exclusion. *Electr. J. Prob.*, **12**, 516–573.

[8] Gärtner, J., den Hollander, F. and Maillard, G. *Intermittency on catalysts: scaling for three-dimensional simple symmetric exclusion.* Preprint.

[9] Gärtner, J., den Hollander, F. and Maillard, G. *Intermittency on Catalysts: Voter Model.* Work in progress.

[10] Gärtner, J. and König, W. (2005). The parabolic Anderson model. In: *Interacting Stochastic Systems* , ed. J.-D. Deuschel and A. Greven. Berlin: Springer, pp. 153–179.

[11] Gärtner, J., König, W. and Molchanov, S.A. (2007). Geometric characterization of intermittency in the parabolic Anderson model. *Ann. Prob.*, **35**, 439–499.

[12] Kesten, H. and Sidoravicius, V. (2003). Branching random walk with catalysts. *Electr. J. Prob.*, **8**, 1–51.

[13] Kipnis, C. and Landim, C. (1999). *Scaling Limits of Interacting Particle Systems.* Grundlehren der Mathematischen Wissenschaften **320**. Berlin: Springer.

[14] Lieb, E.H. (1977). Existence and uniqueness of the minimizing solution of Choquard's nonlinear equation. *Stud. Appl. Math.*, **57**, 93–105.

[15] Liggett, T.M. (1985). *Interacting Particle Systems.* Grundlehren der Mathematischen Wissenschaften **276**. New York: Springer.

[16] Sznitman, A.-S. (1998). *Brownian Motion, Obstacles and Random Media.* Berlin: Springer.

9

Stochastic dynamical systems in infinite dimensions

Salah-Eldin A. Mohammed[†]

Department of Mathematics
Southern Illinois University
Carbondale, Ill. 62901, USA

Abstract

We study the local behaviour of infinite-dimensional stochastic semiflows near hyperbolic equilibria. The semiflows are generated by stochastic differential systems with finite memory, stochastic evolution equations and semilinear stochastic partial differential equations.

9.1 Introduction

In this paper, we summarize some results on the existence and qualitative behavior of stochastic dynamical systems in infinite dimensions. The three main examples covered are stochastic systems with finite memory (stochastic functional differential equations: sfde's), semilinear stochastic evolution equations (see's) and stochastic partial differential equations (spde's). Owing to limitations of space, our summary is by no means intended to be exhaustive: The emphasis will be mainly on the local behavior of infinite-dimensional stochastic dynamical systems near hyperbolic equilibria (or stationary solutions).

The main highlights of the paper are:

- Infinite-dimensional cocycles
- Ruelle's spectral theory for compact linear cocycles in Hilbert space
- Stationary points (equilibria). Hyperbolicity

[†] The research of the author is supported in part by NSF Grants DMS-9703852, DMS-9975462, DMS-0203368, DMS-0705970 and Alexander von Humboldt-Stiftung (Germany).
Trends in Stochastic Analysis, ed. J. Blath, P. Mörters and M. Scheutzow.
Published by Cambridge University Press. ©Cambridge University Press 2008.

- Existence of stable/unstable manifolds near equilibria
- Cocycles generated by regular sfde's. Singular sfde's
- Cocycles generated by semilinear see's and spdc's
- Solutions of anticipating semilinear sfde's and see's.

9.2 What is a stochastic dynamical system?

We begin by formulating the idea of a *stochastic semiflow* or an infinite-dimensional *cocycle* which is central to the analysis in this work.

First, we establish some notation. Let $(\Omega, \mathcal{F}, P)$ be a probability space. Denote by $\bar{\mathcal{F}}$ the P-completion of $\mathcal{F}$, and let $(\Omega, \bar{\mathcal{F}}, (\mathcal{F}_t)_{t \geq 0}, P)$ be a complete filtered probability space satisfying the usual conditions [32].

If E is a topological space, we denote by $\mathcal{B}(E)$ its Borel σ-algebra. If E is a Banach space, we may give the space $L(E)$ of all bounded linear operators on E the *strong topology*, viz. the smallest topology with respect to which all evaluations $L(E) \ni T \mapsto T(x) \in E$, $x \in E$, are continuous. Denote by $\mathcal{B}_s(L(E))$ the σ-algebra generated by the strong topology on $L(E)$. Let $\mathbf{R}$ denote the set of all reals, and $\mathbf{R}^+ := [0, \infty)$. We say that a process $T : \mathbf{R}^+ \times \Omega \to L(E)$ is *strongly measurable* if it is $(\mathcal{B}(\mathbf{R}^+) \otimes \mathcal{F}, \mathcal{B}_s(L(E)))$-measurable.

Let k be a positive integer and $0 < \epsilon \leq 1$. If E and N are real Banach spaces with norms $|\cdot|$, we will denote by $L^{(k)}(E, N)$ the Banach space of all continuous k-multilinear maps $A : E^k \to N$ with the uniform norm $\|A\| := \sup\{|A(v_1, v_2, \cdots, v_k)| : v_i \in E, |v_i| \leq 1, i = 1, \cdots, k\}$. We let $L^{(k)}(E)$ stand for $L^{(k)}(E, E)$. Suppose $U \subseteq E$ is an open set. A map $f : U \to N$ is said to be *of class* $C^{k,\epsilon}$ if it is C^k and if $D^{(k)} f : U \to L^{(k)}(E, N)$ is ϵ-Hölder continuous on bounded sets in U. A $C^{k,\epsilon}$ map $f : U \to N$ is said to be *of class* $C_b^{k,\epsilon}$ if all its derivatives $D^{(j)} f, 1 \leq j \leq k$, are globally bounded on U, and $D^{(k)} f$ is ϵ-Hölder continuous on U. If $U \subset E$ is open and bounded, denote by $C^{k,\epsilon}(U, N)$ the Banach space of all $C^{k,\epsilon}$ maps $f : U \to N$ given the norm:

$$\|f\|_{k,\epsilon} := \sup_{\substack{x \in U \\ 0 \leq j \leq k}} \|D^j f(x)\| + \sup_{\substack{x_1, x_2 \in U \\ x_1 \neq x_2}} \frac{|D^k f(x_1) - D^k f(x_2)|}{|x_1 - x_2|^\epsilon}.$$

We now define a *cocycle* on Hilbert space.

Definition 9.1 (Cocycle) *Let* $\theta : \mathbf{R} \times \Omega \to \Omega$ *be a* $(\mathcal{B}(\mathbf{R}) \otimes \mathcal{F}, \mathcal{F})$-*measurable group of* P-*preserving transformations on the probability space* $(\Omega, \mathcal{F}, P)$, H *a real separable Hilbert space, k a non-negative integer and*

$\epsilon \in (0,1]$. *A $C^{k,\epsilon}$ perfect cocycle (U,θ) on H is a $(\mathcal{B}(\mathbf{R}^+) \otimes \mathcal{B}(H) \otimes \mathcal{F}, \mathcal{B}(H))$-measurable random field $U : \mathbf{R}^+ \times H \times \Omega \to H$ with the following properties:*

   (i) For each $\omega \in \Omega$, the map $\mathbf{R}^+ \times H \ni (t,x) \mapsto U(t,x,\omega) \in H$ is continuous; and for fixed $(t,\omega) \in \mathbf{R}^+ \times \Omega$, the map $H \ni x \mapsto U(t,x,\omega) \in H$ is $C^{k,\epsilon}$.

   (ii) $U(t+s,\cdot,\omega) = U(t,\cdot,\theta(s,\omega)) \circ U(s,\cdot,\omega)$ for all $s,t \in \mathbf{R}^+$ and all $\omega \in \Omega$.

   (iii) $U(0,x,\omega) = x$ for all $x \in H, \omega \in \Omega$.

Using Definition 9.1, it is easy to check that a cocycle (U,θ) corresponds to a one-parameter semigroup on $H \times \Omega$.

Throughout this paper we will assume that each P-preserving transformation $\theta(t,\cdot) : \Omega \to \Omega$ is ergodic.

9.3 Spectral theory of linear cocycles: hyperbolicity

The question of hyperbolicity is central to many studies of finite, and infinite-dimensional (stochastic) dynamical systems. This question focuses on the characterization of almost-sure "saddle-like behavior" of the nonlinear stochastic dynamical system when linearized at a given statistical equilibrium. Statistical equilibria are viewed as random points in the infinite-dimensional state space called *stationary points* of the nonlinear cocycle. For the underlying stochastic differential equation, the stationary points correspond to *stationary solutions*.

The main results in this section are the *spectral theorem* for a compact *linear* infinite-dimensional cocycle (Theorem 9.4) and the *saddle-point property* in the hyperbolic case (Theorem 9.6). A discrete version of the spectral theorem was established in the fundamental work of D. Ruelle [34], using multiplicative ergodic theory techniques. A continuous version of the spectral theorem is developed in [19] within the context of linear stochastic systems with finite memory. See also work by the author and M. Scheutzow on regular stochastic systems with finite memory [22], and joint work with T.S. Zhang and H. Zhao [27]. The spectral theorem gives a deterministic discrete *Lyapunov spectrum* or *set of exponential growth rates* for the linear cocycle. The proof of the spectral theorem uses infinite-dimensional discrete multiplicative ergodic theory techniques and interpolation arguments in order to control the excursions of the cocycle between discrete times. A linear cocycle is *hyperbolic* if its Lyapunov spectrum does not contain zero.

For a nonlinear cocycle, a stationary point is defined to be *hyperbolic* if the linearized cocycle (at the stationary point) is hyperbolic. Under such a hyperbolicity condition, one may obtain a *local stable manifold theorem* for the non linear cocycle near the stationary point (Theorem 9.8).

Throughout the paper we will use the following convention:

Definition 9.2 (Perfection) *A family of propositions $\{P(\omega) : \omega \in \Omega\}$ is said to hold perfectly in ω if there is a sure event $\Omega^* \in \mathcal{F}$ such that $\theta(t, \cdot)(\Omega^*) = \Omega^*$ for all $t \in \mathbf{R}$ and $P(\omega)$ is true for every $\omega \in \Omega^*$.*

We now define a *stationary point* for a cocycle (U, θ) in Hilbert space H.

Definition 9.3 (Stationary point) *An $\mathcal{F}$-measurable random variable $Y : \Omega \to H$ is said be a stationary random point for the cocycle (U, θ) if it satisfies the following identity:*

$$U(t, Y(\omega), \omega) = Y(\theta(t, \omega)) \tag{9.1}$$

for all $t \in \mathbf{R}^+$, perfectly in $\omega \in \Omega$.

The reader may note that the above definition is an infinite-dimensional analog of a corresponding concept of cocycle-invariance that was used by the author in joint work with M. Scheutzow to give a proof of the stable manifold theorem for stochastic ordinary differential equations (sode's) (Definition 3.1, [21]). Definition 9.3 above essentially gives a useful realization of the idea of an invariant measure for a stochastic dynamical system generated by an sode, a stochastic functional differential equation (sfde), a stochastic evolution equation (see) or an spde. Such a realization allows us to analyze the local *almost-sure* stability properties of the stochastic semiflow in the neighborhood of the stationary point. The existence (and uniqueness/ergodicity) of a stationary random point for various classes of spde's and see's has been studied by many researchers; see for example [7] and the references therein.

The following spectral theorem gives a fixed discrete set of Lyapunov exponents for a compact linear cocycle (T, θ) on H. The discreteness of the Lyapunov spectrum is a consequence of the compactness of the cocycle, while the ergodicity of the shift θ guarantees that the spectrum is deterministic. This fact allows us to define hyperbolicity of the linear cocycle (T, θ) and hence that of the stationary point Y of a nonlinear cocycle (U, θ).

Theorem 9.4 (Oseledec–Ruelle) *Let H be a real separable Hilbert space. Suppose (T, θ) is an $L(H)$-valued strongly measurable cocycle such that there exists $t_0 > 0$ with $T(t, \omega)$ compact for all $t \geq t_0$. Assume that $T : \mathbf{R}^+ \times \Omega \to L(H)$ is strongly measurable and*

$$E \sup_{0 \leq t \leq 1} \log^+ \|T(t, \cdot)\|_{L(H)} + E \sup_{0 \leq t \leq 1} \log^+ \|T(1 - t, \theta(t, \cdot))\|_{L(H)} < \infty.$$

Then there is a sure event $\Omega_0 \in \mathcal{F}$ such that $\theta(t, \cdot)(\Omega_0) \subseteq \Omega_0$ for all $t \in \mathbf{R}^+$ and, for each $\omega \in \Omega_0$, the limit

$$\Lambda(\omega) := \lim_{t \to \infty} [T(t, \omega)^* \circ T(t, \omega)]^{1/(2t)}$$

exists in the uniform operator norm. Each linear operator $\Lambda(\omega)$ is compact, non-negative and self-adjoint with a discrete spectrum

$$e^{\lambda_1} > e^{\lambda_2} > e^{\lambda_3} > \cdots$$

where the Lyapunov exponents λ_i's are distinct and non-random. Each eigenvalue $e^{\lambda_i} > 0$ has a fixed finite non-random multiplicity m_i and a corresponding eigenspace $F_i(\omega)$, with $m_i := \dim F_i(\omega)$. Set $i = \infty$ when $\lambda_i = -\infty$. Define

$$E_1(\omega) := H, \quad E_i(\omega) := \left[\oplus_{j=1}^{i-1} F_j(\omega)\right]^{\perp}, \ i > 1, \ E_\infty := \ker \Lambda(\omega).$$

Then

$$E_\infty \subset \cdots \subset \cdots \subset E_{i+1}(\omega) \subset E_i(\omega) \cdots \subset E_2(\omega) \subset E_1(\omega) = H,$$

$$\lim_{t \to \infty} \frac{1}{t} \log |T(t, \omega)x| = \begin{cases} \lambda_i & \text{if } x \in E_i(\omega) \backslash E_{i+1}(\omega), \\ -\infty & \text{if } x \in E_\infty(\omega), \end{cases}$$

and $T(t, \omega)(E_i(\omega)) \subseteq E_i(\theta(t, \omega))$ for all $t \geq 0$, $i \geq 1$.

The figure below illustrates the Oseledec–Ruelle theorem.

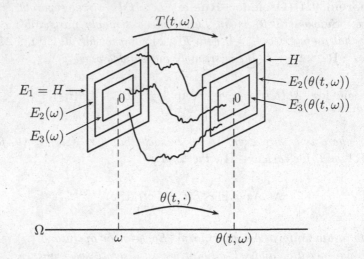

Proof The proof is based on a discrete version of Oseledec's multiplicative ergodic theorem and the perfect ergodic theorem ([33], pp. 303–304; cf. [31], [19], Lemma 5). Details of the extension to continuous time are given in [19] within the context of linear stochastic functional differential equations with finite memory. The arguments in [19] extend directly to general linear cocycles in a separable Hilbert space. Cf. [10].

Definition 9.5 *Let (T, θ) be a linear cocycle on a Hilbert space H satisfying all the conditions of Theorem 9.4. The cocycle (T, θ) is said to be hyperbolic if its Lyapunov spectrum $\{\cdots < \lambda_{i+1} < \lambda_i < \cdots < \lambda_2 < \lambda_1\}$ does not vanish, in the sense that $\lambda_i \neq 0$ for all $i \geq 1$.*

The following result is a "random saddle point property" for hyperbolic linear cocycles. A proof is given in [19], Theorem 4, Corollary 2 and [24], Theorem 5.3 within the context of stochastic differential systems with finite memory; but the arguments therein extend immediately to linear cocycles in a separable Hilbert space.

Theorem 9.6 (The saddle point property) *Let (T, θ) be a hyperbolic linear cocycle on a Hilbert space H. Assume that*

$$E \log^+ \sup_{0 \leq t_1, t_2 \leq 1} \|T(t_2, \theta(t_1, \cdot))\|_{L(H)} < \infty;$$

and denote by $\{\cdots < \lambda_{i+1} < \lambda_i < \cdots < \lambda_2 < \lambda_1\}$ the non-vanishing Lyapunov spectrum of (T, θ). Pick $i_0 > 1$ such that $\lambda_{i_0} < 0 < \lambda_{i_0-1}$.

Then the following assertions hold perfectly in $\omega \in \Omega$: There exist stable and unstable subspaces $\{S(\omega), U(\omega)\}$, $\mathcal{F}$-measurable (into the Grassmannian), such that

(i) $H = \mathcal{U}(\omega) \oplus \mathcal{S}(\omega)$. *The unstable subspace $\mathcal{U}(\omega)$ is finite-dimensional with a fixed non-random dimension, and the stable subspace $\mathcal{S}(\omega)$ is closed with a finite non-random codimension. In fact, $\mathcal{S}(\omega) := E_{i_0}$.*

(ii) *(Invariance)*

$$T(t,\omega)(\mathcal{U}(\omega)) = \mathcal{U}(\theta(t,\omega)), \; T(t,\omega)(\mathcal{S}(\omega)) \subseteq \mathcal{S}(\theta(t,\omega)),$$

for all $t \geq 0$,

(iii) *(Exponential dichotomies)*

$$|T(t,\omega)(x)| \geq |x| e^{\delta_1 t} \quad \text{for all} \quad t \geq \tau_1^*, x \in \mathcal{U}(\omega),$$

$$|T(t,\omega)(x)| \leq |x| e^{-\delta_2 t} \quad \text{for all} \quad t \geq \tau_2^*, x \in \mathcal{S}(\omega),$$

where $\tau_i^* = \tau_i^*(x,\omega) > 0, i = 1, 2$, are random times and $\delta_i > 0, i = 1, 2$, are fixed.

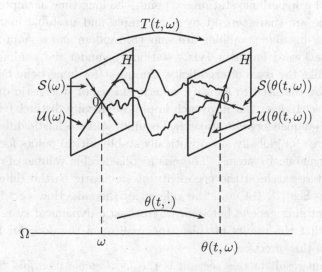

We are now in a position to define the concept of *hyperbolicity* for a stationary point Y of the nonlinear cocycle (U, θ):

Definition 9.7 *Let (U, θ) be a $C^{k,\epsilon}$ ($k \geq 1, \epsilon \in (0,1]$) perfect cocycle on a separable Hilbert space H and there exists $t_0 > 0$ such that $U(t, \cdot, \omega)$:*

$H \to H$ *takes bounded sets into relatively compact sets for each* $(t, \omega) \in$ $(t_0, \infty) \times \Omega$. *A stationary point* $Y : \Omega \to H$ *of the cocycle* (U, θ) *is said to be hyperbolic if*

(a) For any $a \in (t_0, \infty)$,

$$\int_\Omega \log^+ \sup_{0 \le t_1, t_2 \le a} \|DU(t_2, Y(\theta(t_1, \omega)), \theta(t_1, \omega))\|_{L(H)} \, dP(\omega) < \infty.$$

(b) The linearized cocycle $(DU(t, Y(\omega), \omega), \theta(t, \omega))$ *has a non-vanishing Lyapunov spectrum* $\{\cdots < \lambda_{i+1} < \lambda_i < \cdots < \lambda_2 < \lambda_1\}$, *viz.* $\lambda_i \ne 0$ *for all* $i \ge 1$.

Note that, in Definition 9.7, the linearized cocycle $(DU(t, Y(\omega), \omega), \theta(t, \omega))$ has a discrete non-random Lyapunov spectrum because of the compactness hypothesis on (U, θ) and the integrability condition (a). This follows immediately from the Oseledec–Ruelle spectral theorem (Theorem 9.4).

9.4 The local stable manifold theorem

In this section, we will show that, within a stationary random neighborhood of a hyperbolic stationary point, the long-time asymptotics of the cocycle are characterized by local stable and unstable manifolds. The stable/unstable manifolds are smooth, random and asymptotically forward/backward invariant (viz. stationary) under the nonlinear cocycle. Unlike the issue of ergodicity, the quest for hyperbolic behavior is driven by the need to identify generic classes of stochastic dynamical systems. Indeed, our approach is philosophically distinct from the search for uniquely ergodic statistical equilibria in stochastic differential equations, or for globally asymptotically stable critical points for deterministic dynamical systems. There is a considerable volume of current and recent research on the ergodicity of stochastic partial differential equations. See [7], [14] and the references therein. However, little is known regarding generic behavior of stochastic dynamical systems. It is hoped that the results in this paper will open the door for further research in this direction.

The main result in this section is the *local stable manifold theorem* (Theorem 9.8 below). This result characterizes the asymptotic behavior of the cocycle (U, θ) in a random neighborhood of a hyperbolic stationary point. The local stable manifold theorem is the main tool that we use to analyze the almost-sure stability of cocycles generated by stochastic systems with memory, semilinear see's and spde's. The proof of the theorem is a non-trivial refinement and extension to the continuous-time setting of discrete-time results due to D. Ruelle [33, 34]. An outline of the main ideas in the proof of Theorem 9.8 is given after the statement of the theorem. For further details the reader may consult [21], [22] and [27].

In what follows, we denote by $B(x, \rho)$ the open ball, radius ρ and center $x \in H$, and by $\bar{B}(x, \rho)$ the corresponding closed ball.

Theorem 9.8 (The local stable manifold theorem) *Let (U, θ) be a $C^{k,\epsilon}$ ($k \geq 1, \epsilon \in (0, 1]$) perfect cocycle on a separable Hilbert space H such that, for each $(t, \omega) \in (0, \infty) \times \Omega$, $U(t, \cdot, \omega) : H \to H$ takes bounded sets into relatively compact sets. For any $\rho \in (0, \infty)$, denote by $\| \cdot \|_{k,\epsilon}$ the $C^{k,\epsilon}$-norm on the Banach space $C^{k,\epsilon}(\bar{B}(0, \rho), H)$. Let Y be a hyperbolic stationary point of the cocycle (U, θ) satisfying the following integrability property:*

$$\int_\Omega \log^+ \sup_{0 \leq t_1, t_2 \leq a} \|U(t_2, Y(\theta(t_1, \omega)) + (\cdot), \theta(t_1, \omega))\|_{k,\epsilon} \, dP(\omega) < \infty \quad (*)$$

for any fixed $0 < \rho, a < \infty$ and $\epsilon \in (0, 1]$. Denote by $\{\cdots < \lambda_{i+1} < \lambda_i < \cdots < \lambda_2 < \lambda_1\}$ the Lyapunov spectrum of the linearized cocycle $(DU(t, Y(\omega), \omega), \theta(t, \omega), t \geq 0)$. Define $\lambda_{i_0} := \max\{\lambda_i : \lambda_i < 0\}$ if at least one $\lambda_i < 0$. If all finite λ_i are positive, set $\lambda_{i_0} := -\infty$. (Thus λ_{i_0-1} is the smallest positive Lyapunov exponent of the linearized cocycle, if at least one $\lambda_i > 0$; when all the λ_i's are negative, set $\lambda_{i_0-1} := \infty$.) Fix $\epsilon_1 \in (0, -\lambda_{i_0})$ and $\epsilon_2 \in (0, \lambda_{i_0-1})$. Then there exist

(i) *a sure event $\Omega^* \in \mathcal{F}$ with $\theta(t, \cdot)(\Omega^*) = \Omega^*$ for all $t \in \mathbf{R}$,*

(ii) *$\bar{\mathcal{F}}$-measurable random variables $\rho_i, \beta_i : \Omega^* \to (0, 1)$, $\beta_i > \rho_i > 0$, $i = 1, 2$, such that, for each $\omega \in \Omega^*$, the following is true: There are $C^{k,\epsilon}$ ($\epsilon \in (0, 1]$) submanifolds $\tilde{S}(\omega), \tilde{U}(\omega)$ of $\bar{B}(Y(\omega), \rho_1(\omega))$ and $\bar{B}(Y(\omega), \rho_2(\omega))$ (resp.) with the following properties:*

(a) *For $\lambda_{i_0} > -\infty$, $\tilde{S}(\omega)$ is the set of all $x \in \bar{B}(Y(\omega), \rho_1(\omega))$ such that*

$$|U(n, x, \omega) - Y(\theta(n, \omega))| \leq \beta_1(\omega) \, e^{(\lambda_{i_0}+\epsilon_1)n}$$

for all integers $n \geq 0$. If $\lambda_{i_0} = -\infty$, then $\tilde{S}(\omega)$ is the set of all $x \in \bar{B}(Y(\omega), \rho_1(\omega))$ such that

$$|U(n, x, \omega) - Y(\theta(n, \omega))| \leq \beta_1(\omega) \, e^{\lambda n}$$

for all integers $n \geq 0$ and any $\lambda \in (-\infty, 0)$. Furthermore,

$$\limsup_{t \to \infty} \frac{1}{t} \log |U(t, x, \omega) - Y(\theta(t, \omega))| \leq \lambda_{i_0} \qquad (9.2)$$

for all $x \in \tilde{S}(\omega)$. Each stable subspace $S(\omega)$ of the linearized co-cycle $(DU(t, Y(\cdot), \cdot), \theta(t, \cdot))$ is tangent at $Y(\omega)$ to the submani-fold $\tilde{S}(\omega)$, viz. $T_{Y(\omega)}\tilde{S}(\omega) = S(\omega)$. In particular, codim $\tilde{S}(\omega) =$ codim $S(\omega)$, is fixed and finite.

(b)

$$\limsup_{t \to \infty} \frac{1}{t} \log \left[\sup \left\{ \frac{|U(t, x_1, \omega) - U(t, x_2, \omega)|}{|x_1 - x_2|} : \right. \right.$$

$$\left. \left. x_1 \neq x_2, \, x_1, x_2 \in \tilde{S}(\omega) \right\} \right] \leq \lambda_{i_0}.$$

(c) *(Cocycle-invariance of the stable manifolds):*
There exists $\tau_1(\omega) \geq 0$ such that

$$U(t, \cdot, \omega)(\tilde{S}(\omega)) \subseteq \tilde{S}(\theta(t, \omega)) \qquad (9.3)$$

for all $t \geq \tau_1(\omega)$. Also

$$DU(t, Y(\omega), \omega)(S(\omega)) \subseteq S(\theta(t, \omega)), \quad t \geq 0. \qquad (9.4)$$

(d) *For $\lambda_{i_0-1} < \infty$, $\tilde{\mathcal{U}}(\omega)$ is the set of all $x \in \bar{B}(Y(\omega), \rho_2(\omega))$ with the property that there is a discrete-time "history" process $y(\cdot, \omega) : \{-n : n \geq 0\} \to H$ such that $y(0, \omega) = x$ and, for each integer $n \geq 1$, one has $U(1, y(-n, \omega), \theta(-n, \omega)) = y(-(n-1), \omega)$ and*

$$|y(-n, \omega) - Y(\theta(-n, \omega))| \leq \beta_2(\omega) e^{-(\lambda_{i_0-1} - \epsilon_2)n}.$$

If $\lambda_{i_0-1} = \infty$, $\tilde{\mathcal{U}}(\omega)$ is the set of all $x \in \bar{B}(Y(\omega), \rho_2(\omega))$ with the property that there is a discrete-time "history" process $y(\cdot, \omega) : \{-n : n \geq 0\} \to H$ such that $y(0, \omega) = x$ and, for each integer $n \geq 1$,

$$|y(-n, \omega) - Y(\theta(-n, \omega))| \leq \beta_2(\omega) e^{-\lambda n},$$

for any $\lambda \in (0, \infty)$. Furthermore, for each $x \in \tilde{\mathcal{U}}(\omega)$, there is a unique continuous-time "history" process also denoted by $y(\cdot, \omega) : (-\infty, 0] \to H$ such that $y(0, \omega) = x$, $U(t, y(s, \omega), \theta(s, \omega)) = y(t + s, \omega)$ for all $s \leq 0, 0 \leq t \leq -s$, and

$$\limsup_{t \to \infty} \frac{1}{t} \log |y(-t, \omega) - Y(\theta(-t, \omega))| \leq -\lambda_{i_0-1}.$$

Each unstable subspace $\mathcal{U}(\omega)$ of the linearized cocycle

$$(DU(t, Y(\cdot), \cdot), \theta(t, \cdot))$$

is tangent at $Y(\omega)$ to $\tilde{\mathcal{U}}(\omega)$, viz. $T_{Y(\omega)}\tilde{\mathcal{U}}(\omega) = \mathcal{U}(\omega)$. In particular, $\dim \tilde{\mathcal{U}}(\omega)$ is finite and non-random.

(e) Let $y(\cdot, x_i, \omega), i = 1, 2,$ be the history processes associated with $x_i = y(0, x_i, \omega) \in \tilde{\mathcal{U}}(\omega),\ i = 1, 2.$ Then

$$\limsup_{t \to \infty} \frac{1}{t} \log \left[\sup \left\{ \frac{|y(-t, x_1, \omega) - y(-t, x_2, \omega)|}{|x_1 - x_2|} : \right.\right.$$
$$\left.\left. x_1 \neq x_2,\ x_i \in \tilde{\mathcal{U}}(\omega), i = 1, 2 \right\} \right] \leq -\lambda_{i_0-1}.$$

(f) *(Cocycle-invariance of the unstable manifolds):*

There exists $\tau_2(\omega) \geq 0$ such that

$$\tilde{\mathcal{U}}(\omega) \subseteq U(t, \cdot, \theta(-t, \omega))(\tilde{\mathcal{U}}(\theta(-t, \omega))) \tag{9.5}$$

for all $t \geq \tau_2(\omega)$. Also

$$DU(t, \cdot, \theta(-t, \omega))(\mathcal{U}(\theta(-t, \omega))) = \mathcal{U}(\omega), \quad t \geq 0;$$

and the restriction

$$DU(t, \cdot, \theta(-t, \omega))|\mathcal{U}(\theta(-t, \omega)) : \mathcal{U}(\theta(-t, \omega)) \to \mathcal{U}(\omega), \quad t \geq 0,$$

is a linear homeomorphism onto.

(g) The submanifolds $\tilde{\mathcal{U}}(\omega)$ and $\tilde{\mathcal{S}}(\omega)$ are transversal, viz.

$$H = T_{Y(\omega)}\tilde{\mathcal{U}}(\omega) \oplus T_{Y(\omega)}\tilde{\mathcal{S}}(\omega).$$

Assume, in addition, that the cocycle (U, θ) is C^∞. Then the local stable and unstable manifolds $\tilde{\mathcal{S}}(\omega), \tilde{\mathcal{U}}(\omega)$ are also C^∞.

The following figure illustrates the local stable manifold theorem.

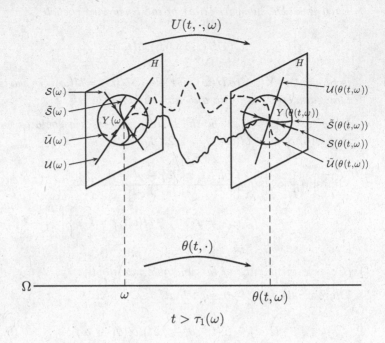

Owing to limitations of space, it is not possible to give a complete proof of the local stable manifold theorem (Theorem 9.8). However, we will outline below its main ingredients. For further details, the reader may consult [21], [22], [27].

Proof [An outline of the proof of Theorem 9.8:]

- Develop perfect continuous-time versions of Kingman's subadditive ergodic theorem as well as the ergodic theorem (see [27], Lemma 2.3.1 (ii), (iii)). The linearized cocycle $(DU(t,Y),\theta(t))$ at the hyperbolic stationary point Y can be shown to satisfy the hypotheses of these perfect ergodic theorems. As a consequence of the perfect ergodic theorems one obtains stable/unstable subspaces for the linearized cocycle, which will constitute tangent spaces to the local stable and unstable manifolds of the nonlinear cocycle (U,θ).

- The nonlinear cocycle (U,θ) may be "centered" around the hyperbolic equilibrium $Y(\theta(t))$ by using the auxiliary perfect cocycle (Z,θ):

$$Z(t,\cdot,\omega) := U(t,(\cdot)+Y(\omega),\omega) - Y(\theta(t,\omega)), \ t \in \mathbf{R}^+, \omega \in \Omega.$$

Hence $0 \in H$ becomes a fixed hyperbolic equilibrium for the auxiliary cocycle (Z, θ). We then use hyperbolicity of Y, the continuous-time integrability condition $(*)$ on the cocycle and perfect versions of the ergodic and subadditive ergodic theorems to show the existence of local stable/unstable manifolds for the discrete auxiliary cocycle $(Z(n, \cdot, \omega), \theta(n, \omega))$ near 0 (cf. [34], Theorems 5.1 and 6.1). These manifolds are random objects and are perfectly defined for $\omega \in \Omega$. Local stable/unstable manifolds for the discrete cocycle $U(n, \cdot, \omega)$ near the equilibrium Y are then obtained via translating the corresponding local manifolds for Z by the stationary point $Y(\omega)$. Using interpolation between discrete times and the (continuous-time) integrability condition $(*)$, it can be shown that the above manifolds for the discrete-time cocycle $(U(n, \cdot, \omega), \theta(n, \omega)), n \geq 1$, also serve as perfectly defined local stable/unstable manifolds for the *continuous-time* cocycle (U, θ) near Y (see [21], [22], [27], [34]).

• Using the integrability condition $(*)$ on the nonlinear cocycle and its Fréchet derivatives, it is possible to control the excursions of the continuous-time cocycle (U, θ) between discrete times. In view of the perfect subadditive ergodic theorem, these estimates show that the local stable manifolds are asymptotically invariant under the nonlinear cocycle. The asymptotic invariance of the unstable manifolds is obtained via the concept of a *stochastic history process* for the cocycle. The existence of a stochastic history process is needed because the cocycle is not invertible.

This completes the outline of the proof of Theorem 9.8.

9.5 Stochastic systems with finite memory

In order to formulate the stochastic dynamics of systems with finite memory (sfde's), we will first describe the class of *regular* sfde's which admit locally compact smooth cocycles.

It is important to note that not all sfde's are regular: indeed, consider the simple one-dimensional linear stochastic delay differential equation (sdde):

$$\left. \begin{array}{r} \mathrm{d}x(t) = x(t-1)\, \mathrm{d}W(t), \quad t > 0, \\ (x(0), x_0) = (v, \eta) \in \mathbf{R} \times L^2([-1, 0], \mathbf{R}), \end{array} \right\} \tag{9.6}$$

with initial condition $(v, \eta) \in \mathbf{R} \times L^2([-1, 0], \mathbf{R})$. In (9.6), we use the symbol $^{(v,\eta)}x_t \in L^2([-1,0], \mathbf{R})$ to represent the *segment* (or *slice*) of the solution path $^{(v,\eta)}x : [-1,\infty) \times \Omega \to \mathbf{R}$ at time $t \geq 0$, viz.: $^{(v,\eta)}x_t(s) := {}^{(v,\eta)} x(t+s)$, $s \in [-1,0]$, $t \geq 0$. The trajectory $\{^{(v,\eta)}x_t : t \geq 0, v \in \mathbf{R}, \eta \in L^2([-1,0], \mathbf{R})\}$ of (9.6) does not admit a measurable version $\mathbf{R}^+ \times \mathbf{R} \times L^2([-1,0], \mathbf{R}) \times \Omega \to L^2([-1,0], \mathbf{R})$ that is pathwise continuous (or even linear) in $\eta \in L^2([-1,0], \mathbf{R})$, (see [17], pp. 144–149; [18]). Sfde's such as (9.6) above, which do not admit continuous stochastic semiflows, are called *singular*.

At this point, we should note that in spite of the easy estimate

$$E\|^{(0,\eta_1)}x_1 - {}^{(0,\eta_2)} x_1\|_2^{2p} \leq C\|\eta_1 - \eta_2\|_2^{2p}, \quad \eta_1, \eta_2 \in L^2([-1,0], \mathbf{R}), \, p \geq 1,$$

Kolmogorov's continuity theorem fails to yield a pathwise continuous version of the random field $\{^{(0,\eta)}x_1 : \eta \in L^2([-1,0], \mathbf{R})\}$.

Owing to the pathological behavior of infinite-dimensional stochastic dynamical systems such as (9.6), it is imperative that one should address perfection issues for such systems with due care.

The construction of the cocycle for regular sfde's is based on the theory of stochastic flows for stochastic *ordinary* differential equations (sode's) in finite dimensions. Once the cocycle is established, we then identify sufficient regularity and growth conditions on the coefficients of the sfde that will allow us to apply the local stable manifold theorem (Theorem 9.8). This yields the existence of local stable/unstable manifolds near hyperbolic stationary solutions of the regular sfde.

9.5.1 Existence of cocycles for regular sfde's

Let $(\Omega, \mathcal{F}, P)$ be the Wiener space where $\Omega := C(\mathbf{R}, \mathbf{R}^p; 0)$ is the space of all continuous paths $\omega : \mathbf{R} \to \mathbf{R}^p$ with $\omega(0) = 0$, $\mathcal{F}$ is the Borel σ-field generated by the topology of uniform convergence on compacta, and P is the Wiener measure on $C(\mathbf{R}^+, \mathbf{R}^p; 0)$. Denote by $\bar{\mathcal{F}}$ the P-completion of $\mathcal{F}$, and by $\mathcal{F}_t$ the P-completion of the sub-σ-algebra of $\mathcal{F}$ generated by all evaluations $\Omega \ni \omega \to \omega(u) \in \mathbf{R}^p$, $u \leq t$. Thus $(\Omega, \bar{\mathcal{F}}, (\mathcal{F}_t)_{t\geq 0}, P)$ is a complete filtered probability space satisfying the usual conditions [32]. Fix an arbitrary delay $r > 0$ and a positive integer dimension d.

Consider the stochastic functional differential equation (sfde):

$$\left.\begin{aligned} dx(t) &= H(x(t), x_t)\, dt + G(x(t))\, dW(t), \quad t \geq 0 \\ x(0) &= v \in \mathbf{R}^d, \quad x_0 = \eta \in L^2([-r, 0], \mathbf{R}^d). \end{aligned}\right\} \tag{9.7}$$

A solution of (9.7) is a process $x : [-r, \infty) \times \Omega \to \mathbf{R}^d$ whereby x_t denotes the segment

$$x_t(\cdot, \omega)(s) := x(t + s, \omega), \quad s \in [-r, 0], \ \omega \in \Omega, \ t \geq 0,$$

and (9.7) holds a.s. The state space for (9.7) is the Hilbert space $M_2 := \mathbf{R}^d \times L^2([-r, 0], \mathbf{R}^d)$ endowed with the norm

$$\|(v, \eta)\|_{M_2} := (|v|^2 + \|\eta\|_{L^2}^2)^{1/2}, \quad v \in \mathbf{R}^d, \eta \in L^2([-r, 0], \mathbf{R}^d).$$

The drift is a globally bounded $C^{k,\delta}$ functional $H : M_2 \to \mathbf{R}^d$, the noise coefficient is a $C_b^{k+1,\delta}$ mapping $G : \mathbf{R}^d \to \mathbf{R}^{d \times p}$, and W is p-dimensional Brownian motion on $(\Omega, \bar{\mathcal{F}}, (\mathcal{F}_t)_{t \geq 0}, P)$:

$$W(t, \omega) := \omega(t), \quad t \in \mathbf{R}^+, \omega \in \Omega.$$

Denote by $\theta : \mathbf{R}^+ \times \Omega \to \Omega$ the P-preserving ergodic Brownian shift

$$\theta(t, \omega)(s) := \omega(t + s) - \omega(t), \quad t, s \in \mathbf{R}, \omega \in \Omega.$$

It is known that the sfde (9.7) admits a unique family of trajectories $\{(^{(v,\eta)}x(t), ^{(v,\eta)}x_t) : t \geq 0, (v, \eta) \in M_2\}$ (see [17], [20]). In our next result, we will show that the ensemble of all these trajectories can be viewed as a $C^{k,\epsilon}$ $(0 < \epsilon < \delta)$ locally compact cocycle (U, θ) on M_2 satisfying $U(t, (v, \eta), \cdot) = (^{(v,\eta)}x(t), ^{(v,\eta)}x_t)$ for all $(v, \eta) \in M_2$ and $t \geq 0$, a.s. (Definition 9.1). The cocycle property is still maintained if H and G are allowed to be stationary, or if the diffusion coefficient $G(x(t))$ is replaced by a smooth memory-dependent term of the form $G(x(t), g(x_t))$ where the path $\mathbf{R}^+ \ni t \mapsto g(x_t) \in \mathbf{R}^d$ is locally of bounded variation. More general noise terms such as Kunita-type spatial semimartingales may also be allowed [16]. The construction of the cocycle uses the finite-dimensional stochastic flow for the diffusion term coupled with a nonlinear variational technique. The nonlinear variational approach reduces the sfde (9.7) to a *random* (pathwise) neutral functional integral equation.

Stability issues for linear versions of the sfde (9.7) are studied in [26], [19], [20], [23]–[25].

For general white noise, an invariant measure on M_2 for the one-point motion of the sfde (9.7) gives a stationary point of the cocycle (U, θ) by enlarging the probability space. On the other hand, if $Y : \Omega \to M_2$ is a stationary random point for (U, θ) independent of the Brownian motion $W(t)$, $t \geq 0$, then the distribution $\rho := P \circ Y^{-1}$ of Y is an invariant measure for the one-point motion of (9.7). This is because Y and W are independent [22], Part II.

Theorem 9.9 *Under the given assumptions on the coefficients H and G, the trajectories of the sfde (9.7) induce a locally compact $C^{k,\epsilon}$ $(0 < \epsilon < \delta)$ perfect cocycle (U, θ) on M_2, where $U : \mathbf{R}^+ \times M_2 \times \Omega \to M_2$ satisfies the following conditions:*

(i) *For each $\omega \in \Omega$ and $t \geq r$ the map $U(t, \cdot, \omega) : M_2 \to M_2$ carries bounded sets into relatively compact sets. In particular, each Fréchet derivative, $DU(t, (v, \eta), \omega) : M_2 \to M_2$, of $U(t, \cdot, \omega)$ with respect to $(v, \eta) \in M_2$, is a compact linear map for $t \geq r, \omega \in \Omega$.*

(ii) *The map $DU : \mathbf{R}^+ \times M_2 \times \Omega \to L(M_2)$ is $(\mathcal{B}(\mathbf{R}^+) \otimes \mathcal{B}(M_2) \otimes \mathcal{F}, \mathcal{B}_s(L(M_2)))$-measurable. Furthermore, the function*

$$\mathbf{R}^+ \times M_2 \times \Omega \ni (t, (v, \eta), \omega) \mapsto \|DU(t, (v, \eta), \omega)\|_{L(M_2)} \in \mathbf{R}^+$$

is $(\mathcal{B}(\mathbf{R}^+) \otimes \mathcal{B}(M_2) \otimes \mathcal{F}, \mathcal{B}(\mathbf{R}^+))$-measurable.

(iii) *If $Y : \Omega \to M_2$ is a stationary point of (U, θ) such that*

$$E(\|Y\|_{M_2}^{\epsilon_0}) < \infty$$

for some $\epsilon_0 > 0$, then the following integrability condition

$$\int_\Omega \log^+ \sup_{0 \leq t_1, t_2 \leq a} \|U(t_2, Y(\theta(t_1, \omega)) + (\cdot), \theta(t_1, \omega))\|_{k,\epsilon} \, dP(\omega) < \infty$$
(9.8)

holds for any fixed $0 < \rho, a < \infty$ and $\epsilon \in (0, \delta)$.

Idea of proof. The construction and regularity of the cocycle (U, θ) is based on the following observation:

The sfde (9.7) is equivalent to the random neutral integral equation:

$$\zeta(t, x(t, \omega), \omega) = v + \int_0^t F(u, \zeta(u, x(u, \omega), \omega), x(u, \omega), x_u(\cdot, \omega), \omega) \, du,$$
(9.9)

$t \geq 0, (v, \eta) \in M_2$. In the above integral equation, $F : [0, \infty) \times \mathbf{R}^d \times M_2 \times \Omega \to \mathbf{R}^d$ is given by

$$F(t, z, v, \eta, \omega) := \{D\psi(t, z, \omega)\}^{-1} H(v, \eta),$$

$t \geq 0, z, (v, \eta) \in M_2, \omega \in \Omega$; and $\zeta : [0, \infty) \times \mathbf{R}^d \times \Omega \to \mathbf{R}^d$ is the inverse flow defined by

$$\zeta(t, x, \omega) := \psi(t, \cdot, \omega)^{-1}(x), \quad t \geq 0, x \in \mathbf{R}^d, \omega \in \Omega,$$

where ψ is the $C^{k+1,\epsilon}$ $(0 < \epsilon < \delta)$ perfect cocycle of the sode

$$\left.\begin{array}{l} d\psi(t) = G(\psi(t)) \, dW(t), \quad t \geq 0, \\ \psi(0) = x \in \mathbf{R}^d. \end{array}\right\}$$

[16, 15]. The existence, perfection and regularity properties of the cocycle (U, θ) may be read from the integral equation (9.9). The integrability property (9.8) also follows from (9.9) coupled with spatial estimates on the finite-dimensional flows ψ and ζ [22].

Example 9.10 Consider the affine linear sfde

$$
\left.
\begin{array}{c}
dx(t) = H(x(t), x_t)\, dt + G\, dW(t), \quad t > 0 \\
x(0) = v \in \mathbf{R}^d, \quad x_0 = \eta \in L^2([-r, 0], \mathbf{R}^d)
\end{array}
\right\} \tag{9.10}
$$

where $H : M_2 \to \mathbf{R}^d$ is a continuous linear map, G is a fixed $(d \times p)$-matrix, and W is p-dimensional Brownian motion. Assume that the linear deterministic $(d \times d)$-matrix-valued fde

$$
dy(t) = H \circ (y(t), y_t)\, dt \tag{9.11}
$$

has a semiflow

$$
T_t : L(\mathbf{R}^d) \times L^2([-r, 0], L(\mathbf{R}^d)) \to L(\mathbf{R}^d) \times L^2([-r, 0], L(\mathbf{R}^d)), \, t \geq 0,
$$

which is uniformly asymptotically stable [13]. Set

$$
Y := \int_{-\infty}^{0} T_{-u}(I, 0) G\, dW(u) \tag{9.12}
$$

where I is the identity $(d \times d)$-matrix. It is easy to see that the trajectories of the affine sfde (9.10) admit an affine linear cocycle $U : \mathbf{R}^+ \times M_2 \times \Omega \to M_2$. Integration by parts and the *helix property*

$$
W(t_2, \theta(t_1, \omega)) = W(t_2 + t_1, \omega) - W(t_1, \omega), \quad t_1, t_2 \in \mathbf{R}, \, \omega \in \Omega, \tag{9.13}
$$

imply that Y has a measurable version satisfying the perfect identity

$$
U(t, Y(\omega), \omega) = Y(\theta(t, \omega)), \quad t \in \mathbf{R}^+, \, \omega \in \Omega.
$$

Note that the stationary point Y (as given by (9.12)) is Gaussian and thus has finite moments of all orders. (See [17], Theorem 4.2, Corollary 4.2.1, pp. 208–217.) More generally, if the semigroup generated by the linear fde (9.11) is hyperbolic, then the sfde (9.10) has a stationary point [17, 26].

Theorem 9.11 *[22] (The stable manifold theorem for sfde's] Assume the given regularity hypotheses on H and G in the sfde (9.7). Let $Y : \Omega \to M_2$ be a hyperbolic stationary point of the sfde (9.7) such that $E(\|Y(\cdot)\|_{M_2}^{\epsilon_0}) < \infty$ for some $\epsilon_0 > 0$. Then the cocycle (U, θ) of (9.7) satisfies the conclusions of the stable manifold theorem (Theorem 9.8)*

with $H = M_2$. *If, in addition, the coefficients H, G of (9.7) are C_b^∞,
then the local stable and unstable manifolds $\tilde{S}(\omega), \tilde{U}(\omega)$ are C^∞, perfectly
in ω.*

Proof [Outline of proof of Theorem 9.11] In view of the integrability
property (9.8), the local compactness of $U(t, \cdot, \omega)$, $t \geq r$, and the ergodicity of the Brownian shift θ, it is possible to define hyperbolicity for the
stationary point $Y : \Omega \to M_2$. The conditions of the local stable manifold theorem (Theorem 9.8) now apply to the cocycle (U, θ). So Theorem 9.11 follows from Theorem 9.8 with $H = M_2$ and $x = (v, \eta) \in M_2$.

9.6 Semilinear see's

In this section, we will first address the question of the existence of
a regular cocycle for semilinear stochastic evolution equations (see's) in
Hilbert space. Using the cocycle together with suitable integrability estimates, we will establish a local stable manifold theorem near hyperbolic
stationary points for these equations.

The existence of local stable/unstable manifolds for nonlinear stochastic evolution equations (see's) and stochastic partial differential equations (spde's) has been an open problem since the early nineties (see
[10], [3], [2], [8], [9]). The analysis in this section will be carried out in
the spirit of Section 9.5, although the construction of the cocycle will
require entirely different techniques. Further details are made available
in the forthcoming paper by the author with T.S. Zhang and H. Z. Zhao
[27].

In [10], the existence of a random evolution operator and its Lyapunov spectrum is established for a linear stochastic heat equation on
a bounded Euclidean domain, driven by finite-dimensional white noise.
For linear see's with finite-dimensional white noise, a stochastic semiflow (i.e. a random evolution operator) is obtained in [3]. A multiplicative ergodic theorem for hyperbolic spde's is developed in [11]. Subse-
quent work on the dynamics of nonlinear spde's has focused mainly on
the question of existence of *continuous* semiflows and the existence and
uniqueness of invariant measures and/or stationary solutions. Existence
of global invariant, stable/unstable manifolds (through a fixed point)
for semilinear see's is established in [4] and [5], when the global Lipschitz constant is relatively small with respect to the spectral gaps of the
second-order term.

The main objectives in this section are to:

- construct a Fréchet differentiable, locally compact cocycle for mild/weak trajectories of the semilinear see
- derive appropriate estimates on the cocycle of the see so as to guarantee applicability of the local stable manifold theorem;
- show the existence of local stable/unstable manifolds near a hyperbolic stationary point, in the spirit of Theorem 9.8.

9.6.1 Smooth cocycles for semilinear see's and spde's

As was indicated at the beginning of Section 9.5, there are no general techniques which give the existence of infinite-dimensional smooth cocycles. In this section we will use a combination of lifting techniques, chaos-type expansion and variational methods in order to construct smooth cocycles for semilinear see's.

The problem of the existence of semiflows for see's (and spde's) is nontrivial, mainly because of the well-known fact that finite-dimensional methods for constructing (even continuous) stochastic flows break down in the infinite-dimensional setting of spde's and see's. More specifically, for see's in Hilbert space, our construction employs a "chaos-type" representation in the Hilbert–Schmidt operators, using the linear terms, (see [27], Theorems 1.2.1–1.2.4). This technique bypasses the need for Kolmogorov's continuity theorem. A variational technique is then employed in order to handle the nonlinear terms. Applications to specific classes of spde's are given in Section 9.7.

It should be noted that the case of *nonlinear* multiplicative noise is largely open: it is not known to us whether see's driven by nonlinear *multidimensional* white noise admit perfect (smooth, or even continuous) cocycles.

We now formulate the setup for the class of semilinear see's we wish to consider.

Denote by $(\Omega, \mathcal{F}, (\mathcal{F}_t)_{t \geq 0}, P)$ the complete filtered Wiener space of all continuous paths $\omega : \mathbf{R} \to E$, $\omega(0) = 0$, where E is a real separable Hilbert space, $\Omega := C(\mathbf{R}^+, E; 0)$ has the compact open topology, $\mathcal{F}$ is the Borel (completed) σ-field of Ω; $\mathcal{F}_t$ is the sub-σ-field of $\mathcal{F}$ generated by all evaluations $\Omega \ni \omega \mapsto \omega(u) \in E, u \leq t$; and P is Wiener measure on Ω. Define the group of P-preserving ergodic Wiener shifts $\theta : \mathbf{R} \times \Omega \to \Omega$ by $(\Omega, \mathcal{F}, P)$:

$$\theta(t, \omega)(s) := \omega(t + s) - \omega(t), \quad t, s \in \mathbf{R}, \, \omega \in \Omega.$$

Let H be a real (separable) Hilbert space, with norm $|\cdot|_H$. Recall that $\mathcal{B}(H)$ is its Borel σ-algebra, and $L(H)$ the Banach space of all bounded linear operators $H \to H$ given the uniform operator norm $\|\cdot\|_{L(H)}$.

Let W denote E-valued Brownian motion $W : \mathbf{R} \times \Omega \to E$ with separable covariance Hilbert space $K \subset E$, a Hilbert–Schmidt embedding. Write

$$W(t) := \sum_{k=1}^{\infty} W^k(t) f_k, \ t \in \mathbf{R},$$

where $\{f_k : k \geq 1\}$ is a complete orthonormal basis of K; $W^k, k \geq 1$, are standard independent one-dimensional Wiener processes (see [6], Chapter 4). The series $\sum_{k=1}^{\infty} W^k(t) f_k$ converges absolutely in E but not necessarily in K. Note that (W, θ) is a helix:

$$W(t_1 + t_2, \omega) - W(t_1, \omega) = W(t_2, \theta(t_1, \omega)), \quad t_1, t_2 \in \mathbf{R}, \omega \in \Omega.$$

Denote by $L_2(K, H)$ the Hilbert space of all Hilbert–Schmidt operators $S : K \to H$, furnished with the norm

$$\|S\|_2 := \left[\sum_{k=1}^{\infty} |S(f_k)|_H^2 \right]^{1/2}.$$

Consider the semilinear Itô stochastic evolution equation (see):

$$\left. \begin{aligned} du(t,x) &= -Au(t,x)\, dt + F\big(u(t,x)\big)\, dt + Bu(t,x)\, dW(t), \\ u(0,x) &= x \in H \end{aligned} \right\} \tag{9.14}$$

in H.

In the above see, $A : D(A) \subset H \to H$ is a closed linear operator on H. Assume A has a complete orthonormal system of eigenvectors $\{e_n : n \geq 1\}$ with corresponding positive eigenvalues $\{\mu_n : n \geq 1\}$; i.e. $Ae_n = \mu_n e_n$, $n \geq 1$. Suppose $-A$ generates a strongly continuous semigroup of bounded linear operators $T_t : H \to H$, $t \geq 0$. Assume that $F : H \to H$ is (Fréchet) $C_b^{k,\epsilon}$ ($k \geq 1, \epsilon \in (0,1]$); thus F has a continuous and globally bounded Fréchet derivative $F : H \to L(H)$. Suppose $B : H \to L_2(K, H)$ is a bounded linear operator.

The stochastic Itô integral in the see (9.14) is defined in the following sense (see [6], Chapter 4): Let $\psi : [0,a] \times \Omega \to L_2(K, H)$ be jointly measurable, $(\mathcal{F}_t)_{t \geq 0}$-adapted and

$$\int_0^a E\|\psi(t)\|_{L_2(K,H)}^2\, dt < \infty.$$

Define the Itô integral

$$\int_0^a \psi(t)\,dW(t) := \sum_{k=1}^{\infty} \int_0^a \psi(t)(f_k)\,dW^k(t)$$

where the H-valued Itô integrals on the right-hand side are with respect to the one-dimensional Wiener processes W^k, $k \geq 1$. The above series converges in $L^2(\Omega, H)$ because

$$\sum_{k=1}^{\infty} E\left| \int_0^a \psi(t)(f_k)\,dW^k(t) \right|^2 = \int_0^a E\|\psi(t)\|_{L_2(K,H)}^2\,dt < \infty.$$

The following standing hypotheses will be invoked throughout this section.

Hypothesis (A): $\displaystyle\sum_{n=1}^{\infty} \mu_n^{-1}\|B(e_n)\|_{L_2(K,H)}^2 < \infty.$

Hypothesis (B): Assume that $B : H \to L_2(K,H)$ extends to a bounded linear operator $B \in L(H, L(E, H))$, and the series $\displaystyle\sum_{k=1}^{\infty} \|B_k\|^2$ converges, where $B_k \in L(H)$ is defined by $B_k(x) := B(x)(f_k)$, $x \in H$, $k \geq 1$.

Observe that Hypothesis (A) is implied by the following two requirements:

(a) The operator $B : H \to L_2(K,H)$ is Hilbert–Schmidt.
(b) $\displaystyle\liminf_{n \to \infty} \mu_n > 0.$

The requirement (b) above is satisfied if $A = -\Delta$, where Δ is the Laplacian on a compact smooth d-dimensional Riemannian manifold M with boundary, under Dirichlet boundary conditions. Moreover, Hypothesis (A) does not place any restriction on the dimension of M.

A *mild solution* of the semilinear see (9.14) is a family of $(\mathcal{B}(\mathbf{R}^+) \otimes \mathcal{F}, \mathcal{B}(H))$-measurable, $(\mathcal{F}_t)_{t \geq 0}$-adapted processes $u(\cdot, x, \cdot) : \mathbf{R}^+ \times \Omega \to H$, $x \in H$, satisfying the following Itô stochastic integral equation:

$$
\begin{aligned}
u(t, x, \cdot) = T_t x &+ \int_0^t T_{t-s} F(u(s, x, \cdot))\,ds \\
&+ \int_0^t T_{t-s} B u(s, x, \cdot)\,dW(s), \quad t \geq 0,
\end{aligned}
\tag{9.15}
$$

(see [6], [7]).

Theorem 9.12 *Under Hypotheses (A) and (B), the see* (9.14) *admits a perfect* $(\mathcal{B}(\mathbf{R}^+) \otimes \mathcal{B}(H) \otimes \mathcal{F}, \mathcal{B}(H))$*-measurable* $C^{k,\epsilon}$ *cocycle* (U, θ) *with* $U : \mathbf{R}^+ \times H \times \Omega \to H$. *Furthermore, if* $Y : \Omega \to H$ *is a stationary point of* (U, θ) *such that* $E|Y|^{\epsilon_0} < \infty$ *for some* $\epsilon_0 > 0$, *then the integrability estimate* $(*)$ *of Theorem 9.8 holds. Indeed, the following (stronger) spatial estimates hold*

$$E\left\{ \sup_{\substack{0 \le t_1, t_2 \le a \\ x \in H}} \frac{|U(t_2, x, \theta(t_1, \cdot))|^{2p}}{(1 + |x|^{2p})} \right\} < \infty, \quad p \ge 1,$$

and

$$E \sup_{\substack{0 \le t_1, t_2 \le a \\ x \in H, \, 1 \le j \le k}} \left\{ \|D^{(j)} U(t_2, x, \theta(t_1, \cdot))\|_{L^{(j)}(H,H)} \right\} < \infty.$$

Proof. We will only sketch the proof of Theorem 9.12. For more details of the arguments involved the reader may consult [27], Theorem 1.2.6 and [28].

Step 1:

We first construct an $L(H)$-valued linear cocycle for mild solutions of the following associated linear see ($F \equiv 0$ in (9.14)):

$$\left. \begin{aligned} du(t, x, \cdot) &= -Au(t, x, \cdot) \, dt + Bu(t, x, \cdot) \, dW(t), \quad t > 0, \\ u(0, x, \omega) &= x \in H. \end{aligned} \right\} \tag{9.16}$$

A *mild solution* of the above linear see is a family of jointly measurable, $(\mathcal{F}_t)_{t \ge 0}$-adapted processes $u(\cdot, x, \cdot) : \mathbf{R}^+ \times \Omega \to H$, $x \in H$, such that

$$u(t, x, \cdot) = T_t x + \int_0^t T_{t-s} Bu(s, x, \cdot) \, dW(s), \quad t \ge 0.$$

The above integral equation holds x-*almost surely*, for each $x \in H$. The crucial question here is whether $u(t, x, \omega)$ is pathwise continuous linear in x perfectly in ω? In view of the failure of Kolmogorov's continuity theorem in infinite dimensions (as pointed out in Section 9.5), we will use a chaos-type expansion technique to show that $u(t, \cdot, \omega) \in L(H)$ perfectly in $\omega \in \Omega$, for all $t \ge 0$. In order to do this, we first lift the linear see (9.16) to the Hilbert space $L_2(H)$ of all Hilbert–Schmidt operators $H \to H$. This is achieved as follows:

- Lift the semigroup $T_t : H \to H, t \ge 0$, to a strongly continuous semigroup of bounded linear operators $\tilde{T}_t : L_2(K, H) \to L_2(K, H), t \ge 0$, defined by the composition $\tilde{T}_t(C) := T_t \circ C$, $C \in L_2(K, H)$, $t \ge 0$.

- Lift the Itô stochastic integral

$$\int_0^t \tilde{T}_{t-s}(\{[B \circ v(s)](x)\}) \, dW(s), \quad x \in H, t \geq 0,$$

to $L_2(H)$ for adapted square-integrable $v : \mathbf{R}^+ \times \Omega \to L_2(H)$. Denote the lifting by $\int_0^t T_{t-s}Bv(s) \, dW(s) \in L_2(H)$. That is:

$$\left[\int_0^t T_{t-s}Bv(s) \, dW(s)\right](x) = \int_0^t \tilde{T}_{t-s}(\{[B \circ v(s)](x)\}) \, dW(s)$$

for all $t \geq 0$, x-a.s.

Step 2:

Next we solve the "lifted" linear see using the following "chaos-type" series expansion in $L_2(H)$ for its solution $\Phi(t,\omega) \in L_2(H), t > 0, \omega \in \Omega$:

$$\Phi(t, \cdot) = T_t + \sum_{n=1}^{\infty} \int_0^t T_{t-s_1}B \int_0^{s_1} T_{s_1-s_2}B$$

$$\cdots \int_0^{s_{n-1}} T_{s_{n-1}-s_n} BT_{s_n} \, dW(s_n) \cdots dW(s_2) \, dW(s_1). \tag{9.17}$$

In the above expansion, the iterated Itô stochastic integrals are lifted integrals in $L_2(H)$. More specifically, denote by $\Psi^n(t) \in L_2(H)$ the general term in the series (9.17), viz.

$$\Psi^n(t) := \int_0^t T_{t-s_1}B \int_0^{s_1} T_{s_1-s_2}B$$

$$\cdots \int_0^{s_{n-1}} T_{s_{n-1}-s_n} BT_{s_n} \, dW(s_n) \cdots dW(s_2) \, dW(s_1),$$

for $t \geq 0, n \geq 1$. Observe that

$$\left.\begin{array}{l} \Psi^n(t) = \displaystyle\int_0^t T_{t-s_1}B\Psi^{n-1}(s_1) \, dW(s_1), \quad n \geq 2, \\[3mm] \Psi^1(t) = \displaystyle\int_0^t T_{t-s_1}BT_{s_1} \, dW(s_1), \end{array}\right\} \tag{9.18}$$

for $t \geq 0$. Using Hypotheses (A) and (B) and induction on $n \geq 1$, one may obtain the following estimate from (9.18):

$$E \sup_{0 \leq s \leq t} \|\Psi^n(s)\|_{L_2(H)}^2 \leq K_1 \frac{(K_2 t)^{n-1}}{(n-1)!}, \quad t \in [0, a],$$

for fixed $a > 0$ and for all integers $n \geq 1$, where K_1, K_2 are positive constants depending only on a. The above estimate implies that the series on the right-hand-side of (9.17) converges absolutely in $L^2(\Omega, L_2(H))$ for any fixed $t > 0$.

Step 3:

We now approximate the Brownian noise W in (9.16) by a sequence of smooth helices

$$W_n(t, \omega) := n \int_{t-1/n}^{t} W(u, \omega) \, du - n \int_{-1/n}^{0} W(u, \omega) \, du, \quad t \geq 0, \, \omega \in \Omega.$$

Thus we obtain a perfect linear cocycle $\Phi(t, \omega) \in L_2(H)$, $t > 0$, $\omega \in \Omega$, for (9.16):

$$\Phi(t + s, \omega) = \Phi(t, \theta(s, \omega)) \circ \Phi(s, \omega), \quad s, t \geq 0, \, \omega \in \Omega$$

satisfying the estimate

$$\sup_{0 \leq s \leq t \leq a} \|\Phi(t - s, \theta(s, \omega))\|_{L(H)} < \infty$$

for any $\omega \in \Omega$ and any fixed $a \in (0, \infty)$.

Step 4:

Now we consider the semilinear Itô see (9.14). Since the linear cocycle (Φ, θ) is a mild solution of (9.16), it is not hard to see that solutions of the random integral equation

$$U(t, x, \omega) = \Phi(t, \omega)(x) + \int_{0}^{t} \Phi(t - s, \theta(s, \omega))(F(U(s, x, \omega))) \, ds, \quad (9.19)$$

for each $t \geq 0$, $x \in H$, give a version of the mild solution of the see (9.14). Using successive approximations on the above integral equation together with the cocycle property for (Φ, θ), we obtain a $C^{k,\epsilon}$ perfect cocycle (U, θ) for mild solutions of the semilinear see (9.14).

Step 5:

The integrability estimate $(*)$ of Theorem 9.8, as well as the two estimates in Theorem 9.12, follow from the random integral equation (9.19) and a "Gronwall-type" argument using Lemma 2.1 in [28]. Cf. the proof of Theorem 2.2 in [28].

We may now state the stable manifold theorem for the semilinear see (9.14). It is a direct consequence of Theorems 9.8 and 9.12.

Theorem 9.13 *Mohammed et al. [27] (The stable manifold theorem for semilinear see's] In the see (9.14) assume Hypotheses (A) and (B) and let F be $C_b^{k,\epsilon}$. Let $Y : \Omega \to H$ be a hyperbolic stationary point of (9.14) such that $E(\|Y(\cdot)\|_H^{\epsilon_0}) < \infty$ for some $\epsilon_0 > 0$. Then the local stable manifold theorem (4.1) holds for the cocycle (U, θ) of (9.14). If F is C_b^∞, the local stable and unstable manifolds $\tilde{\mathcal{S}}(\omega), \tilde{\mathcal{U}}(\omega)$ of (9.14) are C^∞, perfectly in ω.*

9.7 Examples: semilinear spde's

In this section, we will examine applications of the ideas in Section 9.6 to two classes of semilinear spde's: *Semilinear parabolic spde's with Lipschitz nonlinearity* and *stochastic reaction diffusion equations with dissipative nonlinearity*. In particular, we obtain smooth globally defined stochastic semiflows for semilinear spde's driven by cylindrical Brownian motion. In constructing such semiflows, it turns out that, in addition to smoothness of the nonlinear terms, one requires some level of dissipativity or Lipschitz continuity of the nonlinear terms. A discussion of the stochastic semiflow for Burgers equations with additive infinite-dimensional noise is given in [27].

Consider the semilinear spde

$$\left.\begin{aligned}
du(t) &= \frac{1}{2}\Delta u(t)dt + f(u(t))dt + \sum_{i=1}^\infty \sigma_i u(t)\, dW^i(t), \quad t > 0, \\
u(0) &= \psi \in H_0^k(\mathcal{D}).
\end{aligned}\right\} \quad (9.20)$$

In the above spde, Δ is the Laplacian $\dfrac{1}{2}\displaystyle\sum_{i,j=1}^d \dfrac{\partial^2}{\partial \xi_i^2}$ on a bounded domain $\mathcal{D}$ in $\mathbf{R}^d$, with a smooth boundary $\partial \mathcal{D}$ and Dirichlet boundary conditions. The nonlinearity in (9.20) is given by a C_b^∞ function $f : \mathbf{R} \to \mathbf{R}$. We consider weak solutions of (9.20) with initial conditions ψ in the Sobolev space $H_0^k(\mathcal{D})$, the completion of $C_0^\infty(\mathcal{D}, \mathbf{R})$ under the Sobolev norm

$$\|u\|_{H_0^k(\mathcal{D})}^2 := \sum_{|\alpha| \le k} \int_{\mathcal{D}} |D^\alpha u(\xi)|^2 \, d\xi,$$

with $d\xi$ Lebesgue measure on $\mathbf{R}^d$. The noise in (9.20) is given by a family $W^i, i \ge 1$, of independent one-dimensional standard Brownian motions with $W^i(0) = 0$ defined on the canonical complete filtered Wiener space $(\Omega, \bar{\mathcal{F}}, (\mathcal{F}_t)_{t\in\mathbf{R}}, P)$. The Brownian shift on $\Omega := C(\mathbf{R}, \mathbf{R}^\infty; 0)$ is denoted by θ. Furthermore, we assume that $\sigma_i \in H_0^s(\mathcal{D})$ for all $i \geq 1$, and the series

$$\sum_{i=1}^{\infty} \|\sigma_i\|_{H_0^s}^2$$

converges, where $s > k + \frac{d}{2} > d$. Note also that f induces the C_b^{∞} (Nemytskii) map

$$F : H_0^k(\mathcal{D}) \to H_0^k(\mathcal{D}), \; F(\psi) := f \circ \psi, \; \psi \in H_0^k(\mathcal{D}).$$

Under these conditions and using similar ideas to those in Section 9.6, one can show that the random field of weak solutions of the initial-value problem (9.20) yields a perfect smooth cocycle (U, θ) on the Sobolev space $H_0^k(\mathcal{D})$ which satisfies the integrability estimate $(*)$ of Theorem 9.8 with $H := H_0^k(\mathcal{D})$. Suppose $Y : \Omega \to H_0^k(\mathcal{D})$ is a hyperbolic stationary point of the cocycle (U, θ) of (9.20) such that $E \log^+ \|Y\|_{H_0^k} < \infty$. Then the local stable manifold theorem (Theorem 9.8) applies to the cocycle (U, θ) in a neighborhood of Y. Indeed, we have:

Theorem 9.14 *Assume the above hypotheses on the coefficients of the spde (9.20). Then the weak solutions of (9.20) induce a C^{∞} perfect cocycle $U : \mathbf{R}^+ \times H_0^k(\mathcal{D}) \times \Omega \to H_0^k(\mathcal{D})$. Suppose the cocycle (U, θ) of (9.20) has a hyperbolic stationary point $Y : \Omega \to H_0^k(\mathcal{D})$ such that $E \log^+ \|Y\|_{H_0^k} < \infty$. Then (U, θ) has a perfect family of C^{∞} local stable and unstable manifolds in $H_0^k(\mathcal{D})$ satisfying all the assertions of Theorem 9.8 with $H := H_0^k(\mathcal{D})$.*

For further details on the proof of Theorem 9.14, see [27].

We close this section by discussing the dynamics of the following stochastic reaction diffusion equation with dissipative nonlinearity:

$$\left. \begin{array}{l} du = \nu \Delta u \, dt + u(1 - |u|^{\alpha}) \, dt + \displaystyle\sum_{i=1}^{\infty} \sigma_i u(t) \, dW^i(t), \quad t > 0, \\[3mm] u(0) = \psi \in L^2(\mathcal{D}), \end{array} \right\} \quad (9.21)$$

defined on a bounded domain $\mathcal{D} \subset \mathbf{R}^d$ with a smooth boundary $\partial \mathcal{D}$. In (9.21), $\mathcal{D}$ and the $W^i, i \geq 1$, are as in (9.20), and the series

$$\sum_{i=1}^{\infty} \|\sigma_i\|_{H_0^s}^2$$

converges for $s > 2 + \frac{d}{2}$. For weak solutions of (9.21), one can construct a C^1 cocycle (U, θ) on the Hilbert space $H := L^2(\mathcal{D})$ [27]. Under appropriate choice of the diffusion parameter ν, a unique stationary solution of (9.21) exists [7].

The following local stable manifold theorem holds for (9.21) [27].

Theorem 9.15 *Assume the above hypotheses on the coefficients of the spde (9.21). Let $\alpha < 4/d$. Then the weak solutions of (9.21) generate a C^1 cocycle $U : \mathbf{R}^+ \times L^2(\mathcal{D}) \times \Omega \to L^2(\mathcal{D})$. Suppose $Y : \Omega \to L^2(\mathcal{D})$ is a hyperbolic stationary point of the cocycle (U, θ) such that $E \log^+ \|Y\|_{L^2} < \infty$. Then (U, θ) has a perfect family of C^1 local stable and unstable manifolds in $L^2(\mathcal{D})$ satisfying the assertions of Theorem 9.8 with $H := L^2(\mathcal{D})$.*

A proof of Theorem 9.15 is given in [27].

9.8 Applications: anticipating semilinear systems

In this section we give dynamic representations of infinite-dimensional cocycles on their stable/unstable manifolds at stationary points. This is done via substitution theorems which provide pathwise solutions of semilinear sfde's or see's when the initial conditions are random, anticipating and sufficiently regular in the Malliavin sense. The need for Malliavin regularity of the substituting initial condition is dictated by the infinite-dimensionality of the stochastic dynamics. Indeed, existing substitution theorems (see [12], [1]) do not apply in our present context because the substituting random variable may not take values in a relatively compact or σ-compact space.

9.8.1 Anticipating semilinear sfde's

Consider the following Stratonovich version of the sfde (9.7) of Section 9.5

$$
\left.
\begin{aligned}
dx(t) &= H(x(t), x_t)\, dt \\
&\quad - \frac{1}{2} \sum_{k=1}^{p} G_k^2(x(t))\, dt + G(x(t)) \circ dW(t), \quad t > 0, \\
(x(0), x_0) &= Y,
\end{aligned}
\right\} \quad (9.22)
$$

with anticipating random initial condition $Y : \Omega \to M_2 := \mathbf{R}^d \times L^2([-r, 0], \mathbf{R}^d)$ and with linear noise coefficient $G : \mathbf{R}^d \to \mathbf{R}^{d \times p}$. Using a coordinate basis $\{f_k\}_{k=1}^p$ of $\mathbf{R}^p$, write the p-dimensional Brownian motion W in the form

$$W(t) = \sum_{k=1}^p W^k(t)f_k, \quad t \geq 0,$$

where the W^k, $1 \leq k \leq p$, are independent standard one-dimensional Wiener processes. The linear maps $G_k \in L(\mathbf{R}^d)$, $1 \leq k \leq p$, are defined by $G_k(v) := G(v)(f_k)$, $v \in \mathbf{R}^d$, $1 \leq k \leq p$. Assume the rest of the conditions in Section 9.5.

The following theorem establishes the existence of a solution to (9.22) when Y is $Y : \Omega \to M_2$ which is sufficiently regular in the Malliavin sense; i.e. $Y \in \mathbb{D}^{1,4}(\Omega, M_2)$, the Sobolev space of all $\mathcal{F}$-measurable random variables $Y : \Omega \to H$ which have fourth-order moments together with their Malliavin derivatives $\mathcal{D}Y$ [29], [30]. Throughout this section, we will denote Fréchet derivatives by D and Malliavin derivatives by $\mathcal{D}$.

Theorem 9.16 *In the semilinear sfde* (9.22) *assume that H is C_b^1 and $Y \in \mathbb{D}^{1,4}(\Omega, M_2)$. Then* (9.22) *has a solution $x \in L^\infty([0,a], \mathbb{D}^{1,2}(\Omega, \mathbf{R}^d))$ satisfying*

$$\sup_{t \in [0,a]} |x(t,\omega)| \leq K(\omega)\big[1 + \|Y(\omega)\|_{M_2}\big], \quad a.a. \ \omega \in \Omega,$$

for any $a \in (0, \infty)$, where K is a positive random variable having moments of all orders. When H is C_b^2, a similar substitution result holds for the linearized version of (9.22).

Proof [Sketch of Proof of Theorem 9.16] Denote by $\Psi(t, \cdot, \omega) \in L(\mathbf{R}^d), t \in \mathbf{R}^+$, $\omega \in \Omega$, the linear cocycle for the linear Itô sode

$$\left.\begin{aligned} d\Psi(t) &= G \circ \Psi(t)\, dW(t), \quad t \geq 0, \\ \Psi(0) &= I \in L(\mathbf{R}^d). \end{aligned}\right\}$$

From the construction in Section 9.5, the semilinear sfde (9.22) has a perfect cocycle $U : \mathbf{R}^+ \times M_2 \times \Omega \to M_2$ satisfying the following random functional integral equation:

$$\left.\begin{aligned} p_1(U(t, (v, \eta), \omega)) &= \Psi(t, \omega)(v) \\ &\quad + \int_0^t \Psi(t - u, \theta(u, \omega))(H(U(u, (v, \eta), \omega)))\, du, \\ p_2(U(0, (v, \eta), \omega)) &= \eta \in L^2([-r, 0], \mathbf{R}^d), \end{aligned}\right\}$$

$$(9.23)$$

for each $\omega \in \Omega$, $t \geq 0$, $(v, \eta) \in M_2$. In (9.23), $p_1 : M_2 \to \mathbf{R}^d$, $p_2 : M_2 \to L^2([-r, 0], \mathbf{R}^d)$ denote the projections onto the first and second factors respectively.

We will show that $U(t, Y)$, $t \geq 0$, is a solution of (9.22) satisfying the conclusion of the theorem. To this aim, it is sufficient to show that Y can be substituted in place of the parameter (v, η) in the semilinear Stratonovich integral equation

$$
\left.
\begin{aligned}
p_1 U(t, (v, \eta)) &= v + \int_0^t H(U(u, (v, \eta)))\, du \\
&\quad - \frac{1}{2} \sum_{k=1}^p \int_0^t G_k^2(p_1 U(u, (v, \eta)))\, du \\
&\quad + \int_0^t G(p_1 U(u, (v, \eta))) \circ dW(u), \quad t > 0, \\
U(0, (v, \eta)) &= (v, \eta) \in M_2.
\end{aligned}
\right\} \tag{9.24}
$$

One can easily make the substitution $(v, \eta) = Y$ in the two Lebesgue integrals on the right-hand side of (9.24). So it is sufficient to show that a similar substitution also works for the Stratonovich integral; that is

$$
\int_0^t G(p_1 U(u, (v, \eta))) \circ dW(u)\bigg|_{(v,\eta)=Y} = \int_0^t G(p_1 U(u, Y)) \circ dW(u) \tag{9.25}
$$

a.s. for all $t \geq 0$. We will establish (9.25) in two steps: first, we show that it holds if Y is replaced by its finite-dimensional projections $Y_n : \Omega \to H_n$, $n \geq 1$, where H_n is the linear subspace spanned by $\{e_i : 1 \leq i \leq n\}$ from a complete orthonormal basis $\{e_j\}_{j=1}^\infty$ of M_2; secondly, we pass to the limit as n goes to ∞ (in (9.26) below). Denote $g(t, (v, \eta)) := G(p_1 U(t, (v, \eta)))$, $t \geq 0$, $(v, \eta) \in H_n$. Using martingale estimates for $p_1 U(t, (v, \eta))$ it is easy to see that g satisfies all requirements of Theorem 5.3.4 in [29]. Therefore,

$$
\int_0^t G(p_1 U(u, (v, \eta))) \circ dW(u)\bigg|_{(v,\eta)=Y_n} = \int_0^t G(p_1 U(u, Y_n)) \circ dW(u) \tag{9.26}
$$

a.s. for all $n \geq 1$ and $t \geq 0$. The next step is to establish the a.s. limit

$$
\lim_{n \to \infty} \int_0^t G(p_1 U(s, Y_n)) \circ dW(s) = \int_0^t G(p_1 U(s, Y)) \circ dW(s), \quad t \geq 0. \tag{9.27}
$$

Set $f(s) := G(p_1 U(s, Y))$, $f_n(s) := G(p_1 U(s, Y_n))$, $s \geq 0$, $n \geq 1$. To prove (9.27), we will show first that f and f_n are sufficiently regular to allow for the following representations of the Stratonovich integrals in terms of Skorohod integrals:

$$\int_0^t f(s) \circ dW(s) = \int_0^t f(s)\, dW(s) + \frac{1}{2}\int_0^t \nabla f(s)\, ds \qquad (9.28)$$

and

$$\int_0^t f_n(s) \circ dW(s) = \int_0^t f_n(s)\, dW(s) + \frac{1}{2}\int_0^t \nabla f_n(s)\, ds, \qquad (9.29)$$

where

$$\begin{aligned}
\nabla f(s) &:= (\mathcal{D}^+ f)(s) + (\mathcal{D}^- f)(s),\ (\mathcal{D}^+ f)(s) \\
&:= \lim_{t\to s+} \mathcal{D}_s f(t),\ (\mathcal{D}^- f)(s) := \lim_{t\to s-} \mathcal{D}_s f(t),
\end{aligned} \qquad (9.30)$$

for any $s \geq 0$. In view of the expression

$$\mathcal{D}_s f(t) = Gp_1 \mathcal{D}_s U(t, Y) + Gp_1 DU(t, Y)\mathcal{D}_s Y, \quad s, t \geq 0, \qquad (9.31)$$

the integrability estimates (9.32) below, and the fact that $Y \in \mathbb{D}^{1,4}(\Omega, M_2)$, it can be shown that

$$\int_0^a \int_0^a E|\mathcal{D}_s f(t)|^2\, dt\, ds < \infty.$$

This implies that $f \in \mathbb{L}^{1,2}$, and so the Stratonovich integral in (9.28) is well-defined. Similarly for $f_n \in \mathbb{L}^{1,2}$, $n \geq 1$. The following integrability estimates on the cocycle U of the semilinear sfde are obtained using the integral equation (9.23) and a Gronwall-type lemma in (see [28], Lemma 2.1; cf. proofs of Theorems 2.2, 2.3):

$$\left.\begin{aligned}
& E \sup_{\substack{0 \leq t \leq a \\ (v,\eta)\in M_2}} \frac{|U(t,(v,\eta),\cdot)|^{2p}}{(1 + \|(v,\eta)\|_{M_2}^{2p})} < \infty, \\
& E \sup_{\substack{0 \leq t \leq a \\ (v,\eta)\in M_2}} \|DU(t,x,\cdot)\|^{2p} < \infty, \\
& E \sup_{\substack{0 \leq t \leq a \\ (v,\eta)\in M_2}} \|D^2 U(t,(v,\eta),\cdot)\|^{2p} < \infty, \\
& E\left[\sup_{s,u \leq t \leq a} \|\mathcal{D}_s \Psi(t-u,\theta(u,\cdot))\|_{L(\mathbf{R}^d)}^{2p} \right] < \infty, \\
& E\left[\sup_{\substack{0 \leq t \leq a \\ (v,\eta)\in H}} \frac{|DU(t,(v,\eta),\cdot)|_{M_2}^{2p}}{(1 + \|(v,\eta)\|_{M_2}^{2p})} \right] < \infty,
\end{aligned}\right\} \qquad (9.32)$$

for any $0 < a < \infty$ and $p \geq 1$.

Using the estimates (9.32) again, together with (9.31) and the integral equation (9.23), a lengthy computation shows that

$$\lim_{l \to \infty} \int_0^a \sup_{s < t \le s + (1/l)} E(|\mathcal{D}_s f(t) - (\mathcal{D}^+ f)(s)|) \, ds = 0 \qquad (9.33)$$

and

$$\lim_{l \to \infty} \int_0^a \sup_{0 \vee [s - (1/l)] \le t < s} E(|\mathcal{D}_s f(t) - (\mathcal{D}^- f)(s)|) \, ds = 0. \qquad (9.34)$$

Similar statements also hold for each f_n, $n \ge 1$. This justifies (9.28) and (9.29).

To complete the proof of (9.27), we take limits as $n \to \infty$ in (9.29) and note that

$$\lim_{n \to \infty} \int_0^a \int_0^a E|\mathcal{D}_s f_n(t) - \mathcal{D}_s f(t)|^2 \, dt \, ds = 0 \qquad (9.35)$$

and

$$\lim_{n \to \infty} \int_0^a \nabla f_n(s) \, ds = \int_0^a \nabla f(s) \, ds. \qquad (9.36)$$

Relations (9.35) and (9.36) follow from (9.31), the fact $Y \in \mathbb{D}^{1,4}(\Omega, M_2)$, the estimates (9.32) and the dominated convergence theorem. This completes the proof of the substitution formula (9.25).

In the second part of this section, we describe a similar substitution formula for the semilinear see of Section 9.6.

9.8.2 Anticipating semilinear see's

Here we adopt the setting and hypotheses of Section 9.6. Specifically, we consider the following Stratonovich version of the see (9.14):

$$\left. \begin{aligned} du(t,x) &= -Au(t,x) \, dt + F\big(u(t,x)\big) \, dt \\ &\quad - \frac{1}{2} \sum_{k=1}^{\infty} B_k^2 u(t,x) \, dt + Bu(t,x) \circ dW(t), \ t > 0, \\ u(0,x) &= x \in H. \end{aligned} \right\} \qquad (9.37)$$

The following result is obtained using similar techniques to those used in the proof of Theorem 9.16:

Theorem 9.17 *Assume that the see (9.37) satisfies all the conditions of Section 9.6. Let $Y \in \mathbb{D}^{1,4}(\Omega, H)$ be a random variable, and U :*

$\mathbf{R}^+ \times H \times \Omega \to H$ *be the C^1 cocycle generated by all mild solutions of the Stratonovich see (9.37). Then $U(t, Y)$, $t \geq 0$, is a mild solution of the (anticipating) Stratonovich see*

$$
\left.
\begin{aligned}
\mathrm{d}U(t, Y) = &-AU(t, Y)\,\mathrm{d}t \\
&+ F\big(U(t, Y)\big)\,\mathrm{d}t \\
&- \frac{1}{2} \sum_{k=1}^{\infty} B_k^2 U(t, Y)\,\mathrm{d}t + BU(t, Y) \circ \mathrm{d}W(t),\, t > 0, \\
U(0, Y) = &\, Y.
\end{aligned}
\right\} \quad (9.38)
$$

In particular, if $Y \in \mathbb{D}^{1,4}(\Omega, H)$ is a stationary point of the see (9.37), then $U(t, Y) = Y\big(\theta(t)\big)$, $t \geq 0$, is a stationary solution of the (anticipating) Stratonovich see

$$
\left.
\begin{aligned}
\mathrm{d}Y(\theta(t)) = &-AY(\theta(t))\,\mathrm{d}t + F\big(Y(\theta(t))\big)\,\mathrm{d}t \\
&- \frac{1}{2} \sum_{k=1}^{\infty} B_k^2 Y(\theta(t))\,\mathrm{d}t + BY(\theta(t)) \circ \mathrm{d}W(t), t > 0, \\
Y(\theta(0)) = &\, Y.
\end{aligned}
\right\} \quad (9.39)
$$

Details of the proof of the above result are given in [28].

Bibliography

[1] Arnold, L., and Imkeller, P., Stratonovich calculus with spatial parameters and anticipative problems in multiplicative ergodic theory, *Stoch. Proc. Appl.*, **62** (1996), 19–54.

[2] Bensoussan, A., and Flandoli, F., Stochastic inertial manifold, *Stochastics Stoch. Rep.*, **53** (1995), 1–2, 13–39.

[3] Brzezniak, Z. and Flandoli, F., Regularity of solutions and random evolution operator for stochastic parabolic equations *Stochastic Partial Differential Equations and Applications* (Trento, 1990), Res. Notes Math. Ser. **268**, Longman Sci. Tech., Harlow (1992), 54–71.

[4] Duan, J., Lu, K. and Schmalfuss, B., Invariant manifolds for stochastic partial differential equations, *Ann. Prob.*, **31** (2003), 2109–2135.

[5] Duan, J., Lu, K. and Schmalfuss, B., Stable and unstable manifolds for stochastic partial differential equations, *J. Dynamics Diff. Eqns.*, **16**, no. 4 (2004), 949–972.

[6] Da Prato, G., and Zabczyk, J., *Stochastic Equations in Infinite Dimensions*, Cambridge University Press (1992).

[7] Da Prato, G., and Zabczyk, J., *Ergodicity for Infinite Dimensional Systems*, Cambridge University Press (1996).

[8] Flandoli, F., *Regularity Theory and Stochastic Flows for Parabolic SPDE's*, Stochastics Monographs, **9**, Yverdon: Gordon and Breach, 1995.

[9] Flandoli, F., Stochastic flows for nonlinear second-order parabolic SPDE, *Ann. Prob.* **24** (1996), no. 2, 547–558.

[10] Flandoli, F., and Schaumlöffel, K.-U., Stochastic parabolic equations in bounded domains: Random evolution operator and Lyapunov exponents, *Stochastics Stoch. Rep.*, **29**, no. 4 (1990), 461–485.

[11] Flandoli, F., and Schaumlöffel, K.-U., A multiplicative ergodic theorem with applications to a first order stochastic hyperbolic equation in a bounded domain, *Stochastics Stoch. Rep.*, **34**, no. 3–4 (1991), 241–255.

[12] Grorud, A., Nualart, D., and Sanz-Solé, M., Hilbert-valued anticipating stochastic differential equations, *Ann. Inst. Henri Poincaré (B) Probabilités et Statistiques*, **30**, no. 1 (1994), 133–161.

[13] Hale, J. K., *Theory of Functional Differential Equations*, Berlin: Springer-Verlag, (1977).

[14] Hairer, M., and Mattingly, J. C., Ergodicity of the 2D Navier-Stokes equations with degenerate stochastic forcing, *Ann. Math.* **2**, 164 (2006), no. 3, 993–1032.

[15] Ikeda, N., and Watanabe, S., *Stochastic Differential Equations and Diffusion Processes*, second edition, Amsterdam: North-Holland & Kodansha (1989).

[16] Kunita, H., *Stochastic Flows and Stochastic Differential Equations* , Cambridge University Press (1990).

[17] Mohammed, S.-E.A., *Stochastic Functional Differential Equations*, Research Notes in Mathematics, no. 99, Boston, MA: Pitman Advanced Publishing Program (1984).

[18] Mohammed, S.-E. A., Non-linear flows for linear stochastic delay equations, *Stochastics,* **17**, no.3 (1987), 207–212.

[19] Mohammed, S.-E. A., The Lyapunov spectrum and stable manifolds for stochastic linear delay equations, *Stochastics Stoch. Rep.*, **29** (1990), 89–131.

[20] Mohammed, S.-E. A., Stochastic differential systems with memory: theory, examples and applications. In: *Proceedings of The Sixth Workshop on Stochastic Analysis, Geilo, Norway, July 29–August 4, 1996*, ed. L. Decreusefond, Jon Gjerde, B. Oksendal, A.S. Ustunel, Progress in Probability, Birkhäuser (1998), 1–77.

[21] Mohammed, S.-E. A., and Scheutzow, M. K. R., The stable manifold theorem for stochastic differential equations, *Ann. Prob.*, **27** (1999), no. 2, 615–652.

[22] Mohammed, S.-E. A., and Scheutzow, M. K. R., The stable manifold theorem for nonlinear stochastic systems with memory, Part I: Existence of the semiflow, *Funct. Anal.*, **205**, (2003), 271–305. Part II: The local stable manifold theorem, *Funct. Anal.*, **206** (2004), 253–306.

[23] Mohammed, S.-E.A. and Scheutzow, M.K.R., Lyapunov exponents and stationary solutions for affine stochastic delay equations, *Stochastics Stoch. Rep.* **29** (1990), no. 2, 259–283.

[24] Mohammed, S.-E.A. and Scheutzow, M.K.R., Lyapunov exponents of linear stochastic functional differential equations driven by semimartingales, Part I: The multiplicative ergodic theory, *Ann. Inst. Henri Poincaré, Probabilités et Statistiques,* **32** (1996), 69–105.

[25] Mohammed, S.-E.A. and Scheutzow, M.K.R., Lyapunov exponents of linear stochastic functional differential equations driven by semimartingales, Part II: Examples and case studies, *Ann. Prob.*, **6** (1997), no. 3, 1210–1240.

[26] Mohammed, S.-E.A., Scheutzow, M.K.R. and Weizsäcker, H.v., Hyperbolic

state space decomposition for a linear stochastic delay equation, *SIAM Contr. Optimiz.*, 24-3 (1986), 543-551.

[27] Mohammed, S.-E. A., Zhang, T. S. and Zhao, H. Z., The stable manifold theorem for semilinear stochastic evolution equations and stochastic partial differential equations, Part 1: The Stochastic semiflow, Part 2: Existence of stable and unstable manifolds, pp. 98, *Memoirs of the American Mathematical Society* (to appear).

[28] Mohammed, S.-E. A., and Zhang, T. S., The substitution theorem for semilinear stochastic partial differential equations, *Funct. Anal.*, **253** (2007), 122-157.

[29] Nualart, D., Analysis on Wiener space and anticipating stochastic calculus, In: *Lectures on Probability Theory and Statistics. Ecole d'Eté de probabilités de Saint-Flour XXV-1995.* Lectures given at the summer school in Saint-Flour, France, 10-26 July 1995. Berlin: Springer. Lectures Notes in Mathematics 1690, 123-237 (1998).

[30] Nualart, D., The Malliavin Calculus and Related Topics, *Probability and its Applications*, Berlin: Springer-Verlag (1995).

[31] Oseledec, V. I., A multiplicative ergodic theorem. Lyapunov characteristic numbers for dynamical systems, *Trudy Moskov. Mat. Obšč. 19* (1968), 179-210. English transl. *Trans. Moscow Math. Soc.* **19** (1968), 197-221.

[32] Protter, Ph. E., *Stochastic Integration and Stochastic Differential Equations: A New Approach*, Berlin: Springer (1990).

[33] Ruelle, D., Ergodic theory of differentiable dynamical systems, *Publ. Math. Inst. Hautes Etud. Sci. 33* (1979), 275-306.

[34] Ruelle, D., Characteristic exponents and invariant manifolds in Hilbert space, *Ann. Math.* **115** (1982), 243-290.

10

Feynman formulae for evolutionary equations

Oleg G. Smolyanov

Faculty of Mechanics and Mathematics
Moscow State University
119899 Moscow, Russia

Abstract

The Feynman formula is a representation of the solution of an evolutionary equation by a limit of some multiple integrals over Cartesian products of the classical configuration space, or of the classical phase space, when the multiplicity of integrals tends to infinity. From the Feynman formula one can deduce the Feynman–Kac formula, i.e. a representation of the solution by an integral over trajectories. In this paper we consider representations, by the Feynman formula, of solutions both of some Schrödinger type equations and of the corresponding diffusion equations.

10.1 Introduction

The first of Feynman's papers [7], on what one now calls Feynman path integrals over trajectories in the configuration space, contains three main observations. Firstly, it is shown that the solution of the Cauchy problem for a Schrödinger equation can be represented by a limit of a sequence of integrals over Cartesian products of the classical configuration space when the multiplicity of integrals tends to infinity. Secondly, the limit is interpreted as an integral over trajectories in the configuration space. And finally, it is noticed that the integrand contains the exponent of the classical action. Feynman's definition of the Feynman path integrals over trajectories in the phase space, which is formulated in his second paper on the subject, has a similar structure but, in contrast to the

Trends in Stochastic Analysis, ed. J. Blath, P. Mörters and M. Scheutzow.
Published by Cambridge University Press. ©Cambridge University Press 2008.

preceding definition, the Lagrangian in the classical action is expressed through the Hamiltonian function.

A formalization of the first observation is called the Feynman formula; a formalization of the second observation leads to the Feynman path integrals both over trajectories in the phase and in the configuration space;† actually such a formalization means that the integrand is split into two proper multiplicators one of which is a "generalized density" of a Feynman pseudo-measure and the second one is a new integrand.

The present paper gives a brief review of some Feynman formulae representing solutions of Cauchy problems for Schrödinger type equations and heat equations. Though we give main definitions of Feynman type path integrals we do not discuss in detail developing representations of solutions for evolutionary equations by path integrals. In connection with Feynman's interpretation of the integrand as the exponent of the Lagrangian action we would like to mention only that this interpretation can be formalized in the original form only in the case when the Lagrangian is the sum of two functions, one of which depends only on the momentum and the other only on position. In the general case one needs to put in the action under the exponent $\xi(\cdot)$ in the place of the position coordinate, but $i\xi'(\cdot)$ (not $\xi'(\cdot)$) in the place of momentum, and after that one needs to consider different possible interpretations of the products $\xi(\cdot) \cdot \xi'(\cdot)$.

We consider Feynman formulae related both to integrals over trajectories in the phase space and to integrals over trajectories in configuration spaces which are (some domains of) finite-dimensional or infinite-dimensional manifolds. Actually, the Feynman formulae coincide with finite-dimensional approximations for some Gaussian type measures or Feynman type pseudo-measures on infinite-dimensional spaces of trajectories, and also give some approximations for solutions of the corresponding evolutionary equations.‡ It is worth emphasizing that such approximations are not unique but there exist approximations that use finite-dimensional integrals of integrands which are *elementary functions* of coefficients of the equations (or, equivalently, of parameters characterizing the measures or pseudo-measures). It is worth mentioning also that

---

† Usually the corresponding formulae for solutions of Schrödinger type (and heat) equations containing integrals over infinite dimensional spaces of trajectories are called Feynman–Kac formulae, but actually M.Kac considered only solutions of the heat (but not Schrödinger) equation and, moreover, in the formulae he used only integrals over trajectories in configuration but not in phase space.

‡ They are similar to the classical Euler approximations for ordinary differential equations.

finite-dimensional approximations for Feynman type pseudo-measures can be considered as giving a definition of such pseudo-measures.

The main tool in the approach that is used in the paper is the Chernoff formula [5] which is as related to the Feynman formulae under consideration as the famous Trotter formula (independently proved by Yu. L. Daleckii) is related to Feynman formulae discussed in the paper [13] of E. Nelson.

The paper is organized as follows. At the beginning we give definitions of Feynman formulae, Feynman pseudo-measures and Feynman integrals. After that we consider some representations of solutions both of Schrödinger type equations and of the corresponding diffusion equations. In some places in the text the presentation is formal: some analytical assumptions are omitted and only the main ideas of proofs are formulated.

Most of the results presented in the paper were obtained together with H. v. Weizsäcker and our collaborators N.A. Sidorova and O. Wittich.

10.2 Feynman formulae, Feynman pseudo-measures and Feynman integrals

The Feynman formula is a representation of a solution of a Schrödinger type equation (or of a heat equation) by a limit of some multiple integrals over Cartesian products of the classical configuration space or of the classical phase space, when the multiplicity of integrals tends to infinity. In the latter case we speak about the Feynman formula in the phase space; in the former case about the Feynman formula in the configuration space.

In the case of the heat equation the limit of integrals over Cartesian products of the configuration space coincides with an integral over a Gaussian measure; in the case of Schrödinger equation the above limit coincides with the Feynman path integrals.

A Feynman pseudo-measure, on a (typically infinite-dimensional) vector space or on a manifold, is a linear continuous functional on a locally convex space of (some) functions defined on the vector space or on the manifold; the value which the functional takes on a function from its domain is called the Feynman integral of the function (over the vector space or the manifold), or the integral of the function with respect to the Feynman pseudo-measure. If the vector space or the manifold consists of functions taking values in the classical configuration space (resp.

in the classical phase space) then one calls the corresponding Feynman integral the Feynman path integral or integral over trajectories in the configuration space (in the phase space). Of course in turn the classical configuration space and hence the classical phase space can be (e.g. in quantum field theory) infinite-dimensional. In particular the elements of these classical spaces can be functions taking values in some vector space or manifold. But a function of one variable taking values in a space of functions of another variable (which can be multidimensional) can be identified with a function of two variables; hence in the latter case one actually deals with integrals over functions of several variables. In this review we will not discuss explicitly such integrals.

There exist several definitions of Feynman pseudo-measures (for the sake of brevity they are often called Feynman measures): as a limit of integrals over spaces which are isomorphic to Cartesian powers either of the classical phase space or of the classical phase space (Feynman, 1948 [7] and 1951 [8]); as an analytical continuation of a Gaussian measure (Gelfand-Yaglom, 1956) for integrals over trajectories in the configuration space; a similar result for integrals over trajectories in the phase space was obtained in 1990 (see [15]; this result disproves a conjecture of Berezin [4]) by Fourier transform (Cecile deWitt-Morette, 1974); by Parceval's identity (Maslov, 1976 [12], Albeverio and Hoeg-Krohn, 1976 [2], for integrals over trajectories in the configuration space); in the frame of white noise analysis (Hida-Streit, 1975–80, for integrals over trajectories in configuration space). In addition, both the central limit theorem, for Feynman integrals over trajectories in the configuration space, from [14], and a Cameron–Martin formula developed by Elworthy–Truman in [6] (see also [18]) can be used as definitions. Some references can be found also in [15], [16], [2], [1], [11].

In a conceptual sense the most satisfactory definition is one which uses the Fourier transform; for calculation the most convenient definition is the definition by a limit of integrals over spaces which are isomorphic to Cartesian powers of the configuration space or of the phase space. Both the Feynman integral and the Feynman pseudo-measure in the sense of the latter definition are called the sequential integral and the sequential pseudo-measure. Therefore below we first formulate some definitions which use Fourier transform and afterwards define both the sequential Feynman pseudo-measure and integral.

There exists a wide variety of applications of Feynman pseudo-measures and integrals. Of course the most important ones are representations of solutions of evolutionary (pseudo)differential equations. The so-called Feynman quantization, when one passes directly from a classical (Hamiltonian or Lagrangian) system to a Feynman integral, in fact is closely related to that application. Actually when one writes the latter Feynman integral (for a classical Hamiltonian or Lagrangian system) then one is motivated by a Schrödinger type equation; but in infinite-dimensional cases (in particular in quantum field theory) the Schrödinger type equations are usually not well defined and hence it is much more convenient to consider, roughly speaking, only the representations of solutions ignoring the equations.

The Hamiltonian, or symplectic, Feynman pseudo-measure is also the kernel of the inverse Fourier transform. Hence the Hamiltonian Feynman pseudo-measure can be used already in definitions of infinite-dimensional pseudo-differential operators. The passage from the classical Hamiltonian function to the corresponding pseudo-differential operator, whose symbol is just this function is usually called the Schrödinger quantization; hence in the infinite-dimensional case already the Schrödinger quantization needs the (Hamiltonian) Feynman pseudo-measure. Consequently integrals over Cartesian products of the phase space are (in infinite-dimensional cases) integrals over tensor products of Hamiltonian Feynman pseudo-measures; integrals over Cartesian products of the configuration space are (in infinite-dimensional cases) integrals over tensor products of some Feynman pseudo-measures on the configuration space.

Finally the integrals over trajectories both over the configuration space and over the phase space can be used to get explicit formulae for both generalized and classical eigenfunctions.†

Now we present two definitions of the Feynman measures: by Fourier transform and by a limit of integrals over Cartesian products of proper spaces. In both cases we start with a general definition of Feynman

---

† In fact, using the spectral decomposition of the Hamilton operator $\hat{H}$ one can write

$$e^{t\hat{H}} = \int_a^\infty e^{ts} \psi_s(\cdot) \otimes \psi_s(\cdot)\nu(ds),$$

where $\psi_s(\cdot)$ is a (generalized) eigenfunction of $\hat{H}$ corresponding to the eigenvalue $s \in$ spec $\hat{H}$ and $a = \inf(\operatorname{spec} \hat{H})$. This means that the function $t \mapsto e^{t\hat{H}}$ is the Laplace transform of the the vector valued measure $s \mapsto \psi_s(\cdot) \otimes \psi_s(\cdot)\nu$.

On the other hand, the exponent $e^{t\hat{H}}$ can be expressed by an integral over trajectories; hence if one applies the inverse Laplace transform to that integral one gets a formula for eigenfunctions $\psi_s(\cdot)$.

measures (and corresponding integrals) on a vector space; after that we specify this space in different senses.

We start with the definition of the Fourier transform. For any locally convex space H the symbol H^* denotes the vector space of all continuous linear functionals on H. Let E be a (real) vector space and for any $x \in E$ and for any linear functional g on E let $\phi_g(x) = e^{ig(x)}$. Let F_E be a locally convex space of some complex valued functions on E. The elements of $F(E)^*$ are called $F(E)^*$-distributions on E, or simply distributions on E (if we do not want to specify the space $F(E)^*$).

Let G be a vector space of some linear functionals on E separating elements of E and, for any $g \in G$, let $\phi_g \in F_E$. Then the Fourier $(G\text{-})$transform of any element $\eta \in F_E^*$ is the function on G which is denoted by $\widetilde{\eta}$ or by $\mathcal{F}\eta$ and is defined by

$$\widetilde{\eta}(g)(\equiv \mathcal{F}\eta(g)) = \eta(\phi_g).$$

If the set $\{\phi_g : g \in G\}$ is total in F_E (that means that the linear span of $\{\phi_g : g \in G\}$ is dense in F_E) then the element η is uniquely defined by its Fourier transform. The Feynman pseudo-measure on E, with the correlation functional b and mean value a, is defined as follows.

Definition 10.1 (see [16]) *If b is a quadratic functional on G, $a \in E$ and $\alpha \in \mathbb{C}$ then the Feynman α-pseudo-measure on E, with the correlation functional b and mean value a, is the distribution $\Phi_{b,a,\alpha}$ on E whose Fourier transform is defined by $\mathcal{F}\Phi_{b,a,\alpha}(g) = \exp\{\frac{\alpha b(g)}{2} + ig(a)\}$.*

If $\alpha = -1$ and $b(x) \geqslant 0$ for any $x \in G$ then the Feynman α-pseudo-measure is generated by a Gaussian G-cylindrical measure on E (which may not be σ-additive). If $\alpha = i$ then we get the "true" Feynman pseudo-measure; just the Feynman i-pseudo-measures are used in representations of solutions both of Schrödinger type equations and of infinite-dimensional pseudo-differential operators. Below we substitute the term "pseudo-measure" for "i-pseudo-measure"; we also assume that $a = 0$.

Definition 10.2 (The Hamiltonian Feynman pseudo-measure)
Let $E = Q \times P$, where Q and P are LCS, $Q = P^*$, $P = Q^*$ (as vector spaces); the space $G = P \times Q$ is identified with a space of linear functionals on E (for any $g = (p_g, q_g) \in G$ and $x = (q,p) \in E$, $g(x) = p_g(q) + p(q_g)$). Then the Hamiltonian (or symplectic) Feynman pseudo-measure on E is any Feynman pseudo-measure on E with the correlational functional b defined by $b(p_g, q_g) = 2p_g(q_g)$.

Let there exist a linear injective map $B \colon G \to E$ such that $b(g, g) = g(B(g))$ for any $g \in G$ (B is called the correlation operator of the Feynman measure). Let dom B^{-1} be the domain of B^{-1}; the function

$$\mathrm{dom}\, B^{-1} \ni x \to e^{\frac{\alpha^{-1}B^{-1}(x)(x)}{2}}$$

is called a generalized density of the Feynman α-pseudo-measure [19]. Below we use some standard regularizations of finite-dimensional oscillatory integrals.

Example 10.3 If $E = \mathbf{R}^n = G$ then the Feynman pseudo-measure Φ on E, with the correlation operator B, can be identified with the complex-valued measure (having unbounded variation) on the δ-ring of bounded Borel subsets of $\mathbf{R}^n$ defined by the density

$$f(x) = e^{-\frac{i}{2}(B^{-1}x, x)},$$

with respect to a Lebesgue measure. So in this case the generalized density is just the density in the usual sense.

Definition 10.4 (The sequential Feynman pseudo-measure) *Let $\{E_n, n \in \mathbb{N}\}$ be an increasing sequence of finite-dimensional vector subspaces of dom B^{-1}. Then the value of the sequential Feynman α-pseudo-measure $\Phi_{B,\alpha}^{\{E_n\}}$, associated with the sequence $\{E_n, n \in \mathbb{N}\}$, on a function $f : E \to \mathbb{C}$ (this value is called a sequential Feynman integral of f) is defined by*

$$\Phi_{B,\alpha}^{\{E_n\}}(f) = \lim_{n \to \infty} \left(\int_{E_n} e^{\frac{\alpha^{-1}B^{-1}(x)(x)}{2}} \, \mathrm{d}x \right)^{-1} \int_{E_n} f(q, p) e^{\frac{\alpha^{-1}B^{-1}(x)(x)}{2}} \, \mathrm{d}x$$

(where one integrates w.r.t. any Lebesgue measure) if this limit exists.

So the fact that a function belongs to the domain of $\Phi^{\{E_n\}}$ depends only on its restrictions to E_n. The next definition is a special case of the preceding one; but we prefer to formulate it independently.

Definition 10.5 (The sequential Hamiltonian Feynman pseudo-measure) *Let $\{E_n, n \in \mathbb{N}\}$ be an increasing sequence of finite-dimensional vector subspaces of $E(= Q \times P)$ such that, for any $n \in \mathbb{N}$, $E_n = Q_n \times P_n$, where Q_n and P_n are vector subspaces of Q and P respectively. Then the value of the sequential Hamiltonian Feynman pseudo-measure $\Phi^{\{E_n\}}$, associated with the sequence $\{E_n, n \in \mathbb{N}\}$, on a function $f : E \to \mathbb{C}$ (this* *value is called a sequential Hamiltonian Feynman integral of f) is defined by*

$$\Phi^{\{E_n\}}(f) = \lim_{n \to \infty} \left(\int_{E_n} e^{\langle p,q \rangle} \, \mathrm{d}q \mathrm{d}p \right)^{-1} \int_{E_n} f(q,p) e^{\langle p,q \rangle} \, \mathrm{d}q \mathrm{d}p$$

($\langle p,q \rangle = p(q)$) (where one integrates w.r.t. any Lebesgue measure) if this limit exists. Again the fact that a function belongs to the domain of $\Phi^{\{E_n\}}$ depends only on its restrictions to E_n.

If the space E is a space of functions of a real variable then the Feynman integrals are called Feynman path integrals, or integrals over trajectories. They will be briefly discussed in Section 10.3.

10.3 Feynman formulae via the Chernoff theorem

Let X be a Banach space, $\mathcal{L}(X)$ be the space of all continuous linear operators in X, $\|\cdot\|$ be the operator norm on $\mathcal{L}(X)$, and I be the identity operator in X. For any linear operator A in X let $D(A)$ be the domain of A and for any function $F \colon [0,\infty) \to \mathcal{L}(X)$ let $F'(0)$ be the right strong derivative of F at 0.

Theorem 10.6 (Chernoff) *Let $F : [0,\infty) \to \mathcal{L}(X)$ be a strongly continuous function such that $F(0) = I$, $\|F(t)\| \le \exp(at)$ for some $a \in \mathbb{R}$, D be a linear subspace in $D(F'(0))$ and the restriction of $F'(0)$ to D be a closable operator whose closure we denote by C. If C is the generator of a strongly continuous semigroup $\exp(tC)$, then $F(t/n)^n$ converges to $\exp(tC)$ as $n \to \infty$ in the strong operator topology uniformly with respect to $t \in [0,T]$ for each $T > 0$.*

Example 10.7 (Trotter formula) *Let $F(t) = e^{tA} e^{tB}$; then $F'(0) = A + B$, $F(0) = I$, and*

$$e^{t(A+B)} = \lim (e^{\frac{t}{n}A} e^{\frac{t}{n}B})^n.$$

Let now $H \colon \mathbb{R}^N \times \mathbb{R}^N \to \mathbb{C}$ be a locally integrable function and $\tau \in [0,1]$. We define the operator $\widehat{H}_\tau \colon D(\widehat{H}) \to L^2(\mathbb{R}^N)$ by

$$\left(\widehat{H}_\tau \phi \right)(q) = (2\pi)^{-N} \int_{\mathbb{R}^N} \int_{\mathbb{R}^N} H((1-\tau)q + \tau q', p) e^{ip(q-q')} \phi(q') \mathrm{d}q' \mathrm{d}p,$$

$$(10.1)$$

the domain $D(\widehat{H}_\tau)$ of $\widehat{H}_\tau$ is defined to be the set of all $\phi \in L^2(\mathbb{R}^N)$ such that $(\widehat{H}\phi)(\cdot)$ exists. We say that the function $H(\cdot,\cdot)$ is the τ-symbol of the pseudo-differential operator $\widehat{H}_\tau$ and the mapping $\widehat{\ }_\tau \colon H \mapsto \widehat{H}_\tau$ is the (Schrödinger) τ-quantization. If $\tau = 0$ then the τ-quantization is called the qp-quantization; if $\tau = 1$ then the τ-quantization is called pq-quantization.

Definition 10.8 *We say that the Feynman formula is valid for the operator* $\widehat{H}_\tau$ *if*

$$D(\widehat{e^{-i\frac{t}{n}H}})_\tau = L^2(\mathbb{R}^N) \qquad \text{for any } n \in \mathbb{N} \text{ and } t \in [0, \infty),$$

$-i\widehat{H}_\tau$ *is the generator of a strongly continuous semigroup* $\exp(-it\widehat{H}_\tau)$, $t \in [0, \infty)$, *and*

$$e^{-it\widehat{H}_\tau} \phi = \lim_{n \to \infty} \left(\widehat{e^{-i\frac{t}{n}H}} \right)_\tau^n \phi,$$

for all $\phi \in L^2(\mathbb{R}^N)$ *and* $t \in [0, \infty)$.

So to prove that the Feynman formula is valid for the operator $\widehat{H}_\tau$ it is sufficient to define the function $F \colon [0, \infty) \to \mathcal{L}(L^2(\mathbb{R}^N))$ by

$$F(t) = \left(\widehat{e^{-i\frac{t}{n}H}} \right)_\tau$$

and to check that the assumptions of Theorem 10.6 are satisfied.

Let now $f_0, g_0 \colon \mathbb{R}^N \to \mathbb{C}$ be measurable functions such that $\operatorname{Im} f_0 \leq c$, $\operatorname{Im} g_0 \leq c$ for some $c \in \mathbb{R}$ and let $f, g \colon \mathbb{R}^N \to \mathbb{C}$ be defined by

$$f(q, p) = f_0(p), \qquad g(q, p) = g_0(q)$$

and let there exist the closure A of the operator $\hat{f} + \hat{g}$ defined on a subspace $L \subset D(\hat{f}) \cap D(\hat{g})$, such that $-iA$ is the generator of a strongly continuous semigroup (for example the functions $f_0(p) = (p, p)$, $p \in \mathbb{R}^3$ and $g \in L^2(\mathbb{R}^3) + L^\infty(\mathbb{R}^3)$ satisfy the conditions).

Theorem 10.9 *[17] Let* $\widehat{\ }$ *be the* qp- *or* pq-*symbols,* μ *be a Borel measure on* $\mathbb{R}^N \times \mathbb{R}^N$, $h = \widetilde{\mu}$, $k \in L^2(\mathbb{R}^N \times \mathbb{R}^N)$ *be a real-valued function,* $H(q, p) = f(p) + g(q) + h(q, p) + k(q, p)$. *Then the Feynman formula is valid for* $\widehat{H}$.

10.4 Representation of solutions to the Schrödinger equation by the Feynman integral over trajectories in phase space

For any Banach space T and any $a > 0$ let $C_g([0, a], T)$ be the vector space of all functions on $[0, a]$ taking values in T whose distributional derivatives (= generalized derivative, which explains the subscript g) are measures with finite supports and let $C_g^0([0,a],T)$ (resp. $C_g^1([0,a],T)$) denote the vector space of all right continuous (resp. left continuous) functions from $C_g([0,a],T)$. For any $\tau \in (0,1)$ let $C_g^\tau([0,a],T)$ be the collection of functions f having the form

$$f = (1-\tau)g_f^0 + \tau g_f^1,$$

where $g_f^0 \in C_g^0([0,a],T)$, $g_f^1 \in C_g^1([0,a],T)$ and distributional derivatives of g_f^0 and g_f^1 coincide.

Let $\mathcal{Q}$, $\mathcal{P}$ be finite-dimensional Euclidean spaces and, for any $t > 0$, $\tau \in [0,1]$, let Q_t be the image of $C_g^0([0,t],\mathcal{Q})$ in $L_2([0,t])$,

$$P_t^\tau = C_g^\tau([0,t],\mathcal{P}), \qquad E_t^\tau = Q_t \times P_t^\tau.$$

Q_t and P_t^τ are taken in duality by the form $\int_0^t p(s)q'(s)\mathrm{d}s$ where $q'(s)\mathrm{d}s$ denotes the measure which is the distributional derivative of $q(\cdot)$. Elements of E_t^τ are functions with values in $\mathcal{Q} \times \mathcal{P}$.

A sequence $\{E_n\}$ of subspaces of E_t^τ to which we associate the sequential Feynman measure is defined as follows. Let $t_0 = 0$ and, for any $n \in \mathbb{N}$ and any $k \in \mathbb{N}, k \leq 2^n$, let $t_k = k2^{-n}t$. We define E_n to be equal to the collection of functions from E_t^τ the restrictions of which to any interval $((k-1)2^{-n}, k2^{-n})(k \in \mathbb{N})$ are constant functions.

If f is a function on E_t^τ then the sequential Hamiltonian Feynman integral over this space (Hamiltonian Feynman path integral), denoted by

$$\int_{E_t^\tau} f(q(\cdot),p(\cdot))e^{i\int_0^t p(s)q'(s)\mathrm{d}s} \prod_0^t \mathrm{d}q(s)\mathrm{d}p(s),$$

is defined to be equal to $\Phi^{\{E_n\}}(f)$. For any $\tau \in [0,1]$ the Hamiltonian Feynman path integral defines a Hamiltonian Feynman pseudo-measure Φ^τ on E_t^τ:

$$\Phi^\tau(f) = \int_{E_t^\tau} f(q(\cdot),p(\cdot))e^{i\int_0^t p(s)q'(s)\mathrm{d}s} \prod_0^t \mathrm{d}q(s)\mathrm{d}p(s).$$

If $f(q,p) = e^{-i\int_0^t H(q(\tau),p(\tau))d\tau}\phi_0(q(t))$ then the finite-dimensional integrals which were used to define the sequential Hamiltonian Feynman measure coincide with

$$\left[\left(\widehat{e^{-i\frac{t}{n}H}}\right)^n_\tau \phi_0\right](0);$$

hence the Feynman formula, which holds according to the Chernoff theorem, implies the following proposition.

Proposition 10.10 *If the assumptions of Theorem 10.9 are satisfied then*

$$(\exp(-it\widehat{H}_\tau)\phi_0)(q) = \int_{E_t^\tau} e^{-i\int_0^t H(q(\tau)+q.p(\tau))\mathrm{d}\tau} \phi_0(q(t) + q)$$

$$\times e^{i\int_0^t p(s)q'(s)\mathrm{d}s} \prod_0^t \mathrm{d}q(s)\mathrm{d}p(s).$$

There were two complementary lines of investigations of Feynman path integrals. The first of them is related to integrals over trajectories in the configuration space, and the second to integrals over trajectories in the phase space. Those lines are summarized in the following table.

Configuration space	Phase space
Feynman, 1942–48	Feynman, 1951
Trotter-Daleckii 1960–61	Chernoff, 1968
Nelson 1964	Smolyanov–Tokarev–Truman 2002

10.5 Feynman formulae for the Cauchy-Dirichlet problem [20]

For any Riemannian manifold G let ρ_G be the Riemannian metric in G, $\mathrm{scal}_G(q)$ be the scalar curvature of G at q ($\mathrm{scal}_G(q) = \mathrm{tr}\,\mathrm{Ricci}(q)$); if G is a Riemannian submanifold of a Euclidian space then let $a_G(q)$ be the dimension of G times its mean curvature at q. Let K be a compact Riemannian manifold, ν^K be the Borel measure on K generated by the Riemannian volume, $\mathcal{G}$ be a domain in a Riemannian manifold K having the smooth boundary Γ, and let D be the differential operator, in a space of functions on $\mathcal{G}$, defined by $Df = \frac{1}{2}\Delta f + Vf$. One considers the Cauchy–Dirichlet problems for the Schrödinger equation $if'(t) = D(f(t))$ and for the heat equation $f'(t) = D(f(t))$; here $f : [0,a) \to E = L^2(\mathcal{G},\nu), a > 0$. Let $\mathcal{D}_D$ be the self-adjoint operator in $\mathcal{G}$ corresponding to D. If $\psi(=\psi(\cdot))$ is a function on $\mathcal{G}$ then the mapping $\varphi \mapsto [\mathcal{G} \ni x \mapsto \psi(\cdot)\varphi(\cdot)]$ is also denoted by ψ.

We want to find some Feynman formulae for $e^{it\mathcal{D}_D f}$ and $e^{t\mathcal{D}_D f}$, $t \in [0,a)$ for any $f \in C_0$ where C_0 is a collection of continuous functions on

$\mathcal{G}$ vanishing on Γ. Let

$$g_R(t, q_1, q_2) = e^{-\frac{\rho_K^2(q_1, q_2)}{2t}}, \quad g(t, q_1, q_2) = e^{-\frac{\|q_2 - q_1\|^2}{2t}},$$

$$\Phi(t, q_1, q_2) = e^{-\frac{i\|q_2 - q_1\|^2}{2t}},$$

where $\| \cdot \|$ is the norm in $\mathbf{R}^n$, $q_j \in K$.

For $F \colon (0, \infty) \times K \times K \to \mathbb{C}$ and $t \in (0, \infty)$ let $A_F(t)$ be the operator in the space of complex functions on K, defined by: $(A_F(t)f)(q) = \int_K F(t, q, q_1) f(q_1) \nu(dq)$. A similar notation is used for operators in the space of functions on $\mathcal{G}$.

Theorem 10.11 *If $K = \mathbb{R}^n$ and $f \in C_0$ then for any $t > 0$*

$$(e^{t\mathcal{D}_D} f)(q) = \lim_{n \to \infty} (((\sqrt{2\pi t/n}) A_{g_R}(t/n) e^{\frac{t}{n} V(\cdot)})^n f)(q);$$

$$(e^{it\mathcal{D}_D} f)(q) = \lim_{n \to \infty} (((c_1(t/n)(\cdot) A_\Phi(t/n) e^{\frac{t}{n} i V(\cdot)})^n f)(q),$$

$c_1^{-1}(t) = \int_{\mathbb{R}^n} \Phi(t, q_1, q_2) dq_2$ *(r.h.s. does not depend on q_1).*

Theorem 10.12 *If $f \in C_0$, $t > 0$ then:*

$$(e^{t\mathcal{D}_D} f)(q) = \lim_{n \to \infty} c_2(t, n, q)((A_{g_R}(t/n) e^{\frac{t}{n} V(\cdot)}) e^{\frac{t}{6n} scal_K(\cdot)})^n f)(q),$$

where $(c_1(t, n, q))^{-1} = (A_{g_R}(t/n))^n \mathbf{1})(q)$ and $\mathbf{1}$ is the function on K whose values at each point are equal to 1,

$$(e^{t\mathcal{D}_D} f)(q) = \lim_{n \to \infty} c_3(t, n, q)((A_g(t/n) e^{\frac{t}{n} V(\cdot)})$$
$$\times e^{\frac{t}{4n} scal_K(\cdot)} e^{-\frac{t}{8n} \|a_K(\cdot)\|})^n f)(q),$$

where $(c_3(t, n, q))^{-1} = (A_g(t/n))^n \mathbf{1})(q)$.

Similar statements are valid for $e^{it\mathcal{D}_D}$. Additional details can be found in [20] and [24]; the origin of geometrical characteristics in the exponents is explained in [22] and [23].

10.6 Stochastic Schrödinger–Ito equation (Belavkin equation) with two-dimensional white noise [9]

This equation describes the evolution of a quantum system subjected to continuous observation (=measurements) of both momentum and coordinate; as the measurements are not sharp there is no contradiction with

Heisenberg's uncertainty relations. By definition, continuous measurement means taking a limit of repeated instantaneous measuments when both the frequency of measurements tends to infinity and precision of measurements tends to zero. The equation has the following form.

$$d\varphi(t) = \left[\left(-i\hat{\mathcal{H}} - \frac{\mu_1}{2}k(\hat{q})^2 - \frac{\mu_2}{2}h(\hat{p})^2 \right)(\varphi(t)) \right] dt$$
$$- \sqrt{\mu_1}k(\hat{q})(\varphi(t))dW_1(t) - \sqrt{\mu_2}h(\hat{p})(\varphi(t))dW_2(t),$$

where $\hat{\mathcal{H}}$ is the Hamiltonian which is the result of a quantization of a classical Hamiltonian $\mathcal{H}$, $k(\hat{q})$ and $h(\hat{p})$ are (noncommuting) differential operators with symbols $(q, p) \mapsto k(q)$, $(q, p) \mapsto h(p)$, W_1, W_2 are independent standard Wiener processes and $\varphi(t) \in L^2(\mathbb{R})$ is a random wave function describing mixed states. To get a representation of the solution one can use the stochastic analog of the Chernoff theorem; then one can get a randomized Feynman path integral. The details can be found in [9].

10.7 Equations on manifolds of mappings ([21])

In this section we consider representations of solutions for both some Schrödinger type equations and heat equations on infinite-dimensional manifolds of mappings of a closed interval into a compact Riemannian manifold. Equations of this type arise in string theory and also in M-theory; but in the latter case one needs to consider equations on manifolds of mappings of any finite-dimensional manifolds (in general with boundary), into a Riemannian manifold; some similar results for the Lévy Laplacians are obtained in [3].

Let G be a Riemannian submanifold of an Euclidian space and $\| \cdot \|$ be the norm in that space. If $q_j \in G, j = 1, 2, 3, 4$, then the number $\rho(q_1, q_2; q_3, q_4)$ is defined as follows. For any $x, z \in G$, the symbol $\gamma(x, z)$ denotes a shortest geodesic between x, z, and the symbol γ_0 — the geodesic which passes through q_1 in the direction of the vector a', which is obtained by the parallel transport, along the geodesic $\gamma(q_3, q_1)$, of the vector a that is tangent at q_3 to the geodesic $\gamma(q_3, q_4)$ and is directed from q_3 to q_4. Let $q_4' \in \gamma_0$ be such that the distance of q_4' to q_1, along the geodesic γ_0 (in the direction of a'), is equal to the distance between q_3 and q_4 along $\gamma(q_3, q_4)$. Then $\rho_1(q_1, q_2; q_3, q_4)$ is the length of $\gamma(q_2, q_4')$ and $\rho_2(q_1, q_2; q_3, q_4) = \|q_2 - q_4'\|$; if the distances between q_j are sufficiently small then $\rho_j(q_1, q_2; q_3, q_4)$ depend on q_i continuously $(j = 1, 2)$.

The nonnegative functions p_R, Q_R, p_E, Q_E on $[0, \infty) \times G \times G \times G \times G$ are defined by:

$$p_R(t, q_1, q_2, q_3, q_4) = e^{-\frac{(\rho_1(q_1, q_2; q_3, q_4))^2}{2t}};$$

$$Q_R(t, q_1, q_2, q_3, q_4) = \frac{p_R(t, q_1, q_2, q_3, q_4)}{\int_G p_R(t, q_1, q_2, q_3, q_4) \mathrm{d}q_4};$$

$$p_E(t, q_1, q_2, q_3, q_4) = e^{-\frac{(\rho_2(q_1, q_2; q_3, q_4))^2}{2t}};$$

$$Q_E(t, q_1, q_2, q_3, q_4) = \frac{p_E(t, q_1, q_2, q_3, q_4)}{\int_G p_E(t, q_1, q_2, q_3, q_4) \mathrm{d}q_4}.$$

In the following theorems we consider Feynman type formulae for semigroups which give solutions for the Cauchy Problem for the heat equation and for the Schrödinger equation in spaces of functions on $W_2^1([0, a], G)$.

Let Φ be the function on $C([0, a], G)$ defined by

$$\Phi(g) = \int_0^{a_1} V_\Phi(g(t_1)) \mathrm{d}t_1,$$

and let ψ be the function on $C([0, a], G)$ defined by

$$\psi(g) = \int_0^{a_1} V_\psi(g(t_1)) \mathrm{d}t_1,$$

where V_Φ, V_ψ are continuous functions on G, and let

$$\hat{H}f = -\frac{1}{2}\Delta f + \Phi f$$

for any function $f \colon W_{2,z}^1([0, a], G) \to \mathbb{R}^1$. In Theorems 10.13, 10.14, and 10.16 one assumes that the Cauchy Problem for the equation

$$\frac{\partial f}{\partial t} = -\hat{H}f$$

on $[0, \infty) \times W_{2,z}^1([0, a], G)$ has a solution for initial data $(0, \psi)$, and the value that the solution takes at $(t, q) \in (0, \infty) \times W_{2,z}^1([0, a], G)$ is denoted by $e^{t\hat{H}}\psi(q)$. In Theorem 10.15 one assumes that the Cauchy Problem for the equation $i\frac{\partial f}{\partial t} = \hat{H}f$ on $[0, \infty) \times W_{2,z}^1([0, a], G)$ has a solution for initial data $(0, \psi)$ and the value that the solution takes at $(t, q) \in (0, \infty) \times W_{2,z}^1([0, a], G)$ is denoted by $e^{it\hat{H}}\psi(q)$.

Theorem 10.13 *If* $(t, q) \in (0, \infty) \times W^1_{2,z}([0, a], G)$, *then*

$$
e^{t\hat{H}}\psi(q) = \lim_{p \to \infty} \int_{G \times \cdots \times G} e^{\frac{ta}{p^2} \sum_{n=1, k=1}^{p, p} V_\Phi(q_{n,k})} \left| \frac{a}{p} \sum_{n=1}^{p} V_\psi(q_{n,p}) \right|
$$

$$
\times \prod_{n=1}^{p} \prod_{k=1}^{p} Q_R \left| \frac{ta}{p^2}, q_{n-1,k-1}, q_{n,k-1}, q_{n-1,k}, q_{n,k} \right| dq_{n,k}
$$

$$
= \lim_{p \to \infty} \int_{G \times \cdots \times G} e^{\frac{ta}{p^2} \sum_{n=1, k=1}^{p, p} V_\Phi(q_{n,k})} \left| \frac{a}{p} \sum_{n=1}^{p} V_\psi(q_{n,p}) \right|
$$

$$
\times \prod_{n=1}^{p} \prod_{k=1}^{p} Q_E \left| \frac{ta}{p^2}, q_{n-1,k-1}, q_{n,k-1}, q_{n-1,k}, q_{n,k} \right| dq_{n,k},
$$

where $q_{n,0} = q(\frac{n}{p}a), q_{0,k} = q(0)$.

Theorem 10.14 *For all* $(t, q) \in (0, \infty) \times W^1_{2,z}([0, a], G)$ *the following identities hold (in which* $q_{n,0} = q(\frac{n}{p}a), q_{0,k} = q(0)$):

$$
e^{t\hat{H}}\psi(q) = \lim_{p \to \infty} c^1_p \cdot \int_{G \times \cdots \times G} e^{\frac{ta}{p^2} \sum_{n=1, k=1}^{p, p} V_\Phi(q_{n,k})} \left| \frac{a}{p} \sum_{n=1}^{p} V_\psi(q_{n,0}) \right|
$$

$$
\times \prod_{n=1}^{p} \prod_{k=1}^{p} e^{\frac{1}{6} \frac{ta}{p^2} \mathrm{scal}(q_{n,k})} p_R \left| \frac{ta}{p^2}, q_{n-1,k-1}, q_{n,k-1}, q_{n-1,k}, q_{n,k} \right| dq_{n,k},
$$

where

$$
(c^1_p)^{-1} = \int_{G \times \cdots \times G} \prod_{n=1}^{p} \prod_{k=1}^{p} p_R \left| \frac{ta}{p^2}, q_{n-1,k-1}, q_{n,k-1}, q_{n-1,k}, q_{n,k} \right| dq_{n,k};
$$

$$
e^{t\hat{H}}\psi(q) = \lim_{p \to \infty} c^2_p \int_{G \times \cdots \times G} e^{\frac{ta}{p^2} \sum_{n=1, k=1}^{p, p} V_\Phi(q_{n,k})} \left| \frac{a}{p} \sum_{n=1}^{p} V_\psi(q_{n,0}) \right|
$$

$$
\times \prod_{n=1}^{p} \prod_{k=1}^{p} e^{\frac{ta}{4p^2}(\mathrm{scal}(q_{n,k}) - \frac{1}{2}\|m(q_{n,k})\|^2)}
$$

$$
\times p_R \left| \frac{ta}{p^2}, q_{n-1,k-1}, q_{n,k-1}, q_{n-1,k}, q_{n,k} \right| dq_{n,k},
$$

where

$$
(c^2_p)^{-1} = \int_{G \times \cdots \times G} \prod_{n=1}^{p} \prod_{k=1}^{p} p_R \left| \frac{ta}{p^2}, q_{n-1,k-1}, q_{n,k-1}, q_{n-1,k}, q_{n,k} \right| dq_{n,k};
$$

Below we let $\sqrt{i} = e^{i\frac{\pi}{4}}$.

Theorem 10.15 *Let V_Φ and V_ψ be restrictions to G of some functions (denoted by the same symbols) defined and analytical in a domain C^n which contains G and $\cup_{z \in G}\{z + \sqrt{i}(G - z)\}$ and let the tensor of Riemannian curvature have an analytical extension to the same domain. Then for all $(t,q) \in (0,\infty) \times W^1_{2,z}([0,a], G)$ the following identity holds (where $q_{n,0} = q(\frac{n}{p}a), q_{0,k} = q(0)$):*

$$e^{it\hat{H}}\psi(q) = \lim_{p \to \infty} c_p^3 \int_{G \times \cdots \times G} e^{\frac{ita}{p^2}\sum_{n=1,k=1}^{p,p} V_\Phi(q_{n,k})} \left| \frac{a}{p}\sum_{n=1}^{p} V_\psi(q_{n,0}) \right|$$

$$\times \prod_{n=1}^{p}\prod_{k=1}^{p} e^{\frac{ta}{4p^2}(\mathrm{scal}(q_{n,0}+\frac{1}{\sqrt{i}}(q_{n,k}-q_{n0})-\frac{1}{2}\|m(q_{n0}+\frac{1}{\sqrt{i}}(q_{n,k}-q_{n0}))\|^2}$$

$$\times p_E\left| \frac{ta}{p^2}, \sqrt{i}q_{n-1,k-1}, \sqrt{i}q_{n,k-1}, \sqrt{i}q_{n-1,k}, \sqrt{i}q_{n,k} \right| dq_{n,k},$$

where

$$(c_p^3)^{-1} = \int_{G \times \cdots \times G} \prod_{n=1}^{p}\prod_{k=1}^{p} p_E\left| \frac{ta}{p^2}, \sqrt{i}q_{n-1,k-1}, \right.$$

$$\left. \sqrt{i}q_{n,k-1}, \sqrt{i}q_{n-1,k}, \sqrt{i}q_{n,k} \right| dq_{n,k}.$$

Theorem 10.16 *If the assumptions of Theorem 10.13 are satisfied then for all $(t,q) \in (0,\infty) \times W^1_{2,z}([0,a], G)$ the following identity holds*

$$e^{t\hat{H}}\psi(q) = \int_{C([0,t],C([0,a],G))} e^{\int_0^t \int_0^a V_\Phi(q(\tau_1)(\tau_2))d\tau_1 d\tau_2}$$

$$\times \left(\int_0^a V_\psi(q(t)(\tau))\, d\tau \right) \mathcal{W}_{q,t}(dq),$$

where $\mathcal{W}_{q,t}(dq)$ is the Wiener measure on $C([0,t],C([0,a],G))$ concentrated on the set of functions taking value q at 0.

Remark 10.17 The identity

$$e^{it\hat{H}}\psi(q) = \int_{C([0,t],C([0,a],G))} e^{i\int_0^t \int_0^a V_\Phi(q(\tau_1)(\tau_2))d\tau_1 d\tau_2}$$

$$\times \left(\int_0^a V_\psi(q(t)(\tau))\, d\tau \right) \Phi_{q,t}(dq)$$

(whose left-hand side is defined in Theorem 10.15) can be considered as a definition of the Feynman measure $\Phi_{q,t}$ on $C([0,t],C([0,a],G))$.

Again the origin of geometrical characteristics in the exponents can be found in [23].

10.8 Feynman formulae for the diffusion equation with coordinate dependent diffusion coefficient and for the Schrödinger equation with coordinate dependent mass

In this section we consider some Feynman formulae for solutions of Schrödinger type equations with a coordinate dependent mass, which are related both to some integrals over trajectories in the configuration and in the phase space. The equations of the latter type describe quantum evolution of the so-called quasiparticles (for example, in semiconductors); they correspond to diffusion equations with position dependent diffusion coefficients, which are the Kolmogorov equations for some stochastic Ito equations.

Let, for each $x \in \mathbb{R}^n$, $\alpha(x)$ be a positive linear operator in $\mathbb{R}^n$ and let Δ_α be the differential operator in the space of smooth functions on $\mathbb{R}^n$ defined by $(\Delta_\alpha \varphi)(x) := \operatorname{tr} \alpha(x) \varphi''(x)$ where $\varphi''(x)$ is the second derivative of φ at $x \in \mathbb{R}^n$.

Let $\sqrt{\alpha}(x) = \sqrt{\alpha(x)}$ for any $x \in \mathbb{R}^n$. Then the equation

$$f'(t) = \Delta_\alpha f(t) + (f(t))'(\cdot) a(\cdot)$$

is the second Kolmogorov equation for the following stochastic Ito equation

$$d\xi(t) = \sqrt{\alpha(\xi(t))}\, dw(t) + a(\xi(t))\, dt.$$

Here f is a mapping of $I \subset \mathbb{R}$ in a space of (real) functions on $\mathbb{R}^n$ (f can be identified with a function on $I \times \mathbb{R}^n$). By $f'(t)$ one denotes the derivative of f at t and by $(f(t))'(\cdot)$ one denotes the derivative of $f(t)(\cdot)$ $((f(t))'(\cdot)(a(\cdot))$ is the derivative of $f(t)(\cdot)$ along the vector field $a(\cdot)$).

Let us notice that $\Delta_\alpha = D_\alpha + D_\alpha^1$ where D_α is the Laplace operator with respect to a metric on $\mathbb{R}^n$ [10] and D_α^1 is a differential operator of the first order. Let also D_α^s be the Weyl symmetrization of Δ_α; then $\Delta_\alpha = D_\alpha^s + D_\alpha^2$ where D_α^1 is another differential operator of the first order.

Hence to get Feynman formulae for the semigroups e^{tD_α}, $e^{tD_\alpha^s}$ and for the groups e^{itD_α}, $e^{itD_\alpha^s}$ it is sufficient to get some Feynman formulae for the semigroups $e^{tG_{\alpha,d,V}}$ and $e^{itG_{\alpha,d,V}}$, where

$$(G_{\alpha,d,V}\varphi)(x) = \frac{1}{2}\Delta_\alpha\varphi(x) + \varphi'(x)d(x) + V(x)\varphi(x)$$

and the functions $V : \mathbb{R}^n \to \mathbb{R}$ and $d : \mathbb{R}^n \to \mathbb{R}^n$ are smooth; we assume that all semigroups which we investigate exist.

Let $X := L^2(\mathbb{R}^n)$, let F_j, $j = 1, 2$ be the functions defined by $F_j(0) =$ Id and, for any $t > 0$,

$$(F_1(t)\varphi)(x) = \frac{1}{(2\pi t)^{n/2} (\det \alpha(x))^{1/2}}$$
$$\times \int_{\mathbb{R}^n} \exp\left\{ -\frac{(\alpha^{-1}(x)(z - x), (z - x))}{2t} \right\} \varphi(z)\, dz,$$

$$(F_2(t)\varphi)(x) = [c_2(t)]^{-1} \int_{\mathbb{R}^n} \exp\left\{ -\frac{(\alpha^{-1}(z)(z - x), (z - x))}{2t} \right\} \varphi(z)\, dz,$$

where

$$c_2(t) = \int_{\mathbb{R}^n} \exp\left\{ -\frac{(\alpha^{-1}(z)(z - x), (z - x))}{2t} \right\} dz.$$

Below instead of $(\alpha^{-1}(z)v, v)$ ($v \in \mathbb{R}^n$) we write $\alpha^{-1}(z)v^2$.

Theorem 10.18 *For any* $\varphi \in L^2(\mathbb{R}^n)$, $t > 0$, $j = 1, 2$ *one has*

$$\left(e^{t\Delta_\alpha/2} \right) \varphi = \lim_{n \mapsto \infty} \left(F_j\left(\frac{t}{n}\right) \right)^n \varphi.$$

In the proof one uses the Chernoff theorem.

Let now $(F_3(t)\varphi)(x) := \varphi(x + td(x))$, $(F_4(t)\varphi)(x) := e^{tV(x)}\varphi(x)$, and $F_5(t) := F_1(t)F_3(t)F_4(t)$. Hence

$$F_5(t)(\varphi)(x) = c_1(t) \int_{\mathbb{R}^n} e^{tV(x)} \exp\left\{ -\frac{\alpha^{-1}(x)(z - x)^2}{2t} \right\} \varphi(z + td(z))\, dz$$

$$= c_1(t) \int_{\mathbb{R}^n} e^{tV(x)} \exp\left\{ -\frac{\alpha^{-1}(x)(z - td(x) - x)^2}{2t} \right\}$$
$$\times \varphi(z + t(d(z) - d(x)))\, dz$$

$$\approx c_1(t) \int_{\mathbb{R}^n} e^{tV(x)} \exp\left\{ -\frac{\alpha^{-1}(x)(z - x)^2}{2t} \right\}$$
$$\times \exp\left\{ (\alpha^{-1}(x)d(x), (z - x)) - \frac{t\alpha^{-1}(x)(d(x))^2}{2} \right\} \varphi(z)\, dz.$$

Theorem 10.19 *For any* $\varphi \in L^2(\mathbb{R}^n)$ *and* $t > 0$ *one has*

$$e^{tG_{\alpha,d,V}} \varphi = \lim_{n \mapsto \infty} \left(F_5\left(\frac{t}{n}\right) \right)^n \varphi.$$

To prove this one needs to use the identity $(F_5'(0)\varphi)(x) = \frac{1}{2}\Delta_\alpha\varphi(x) + \varphi'(x)d(x) + V(x)\varphi(x)$ and Chernoff's theorem.

Remark 10.20 If $\alpha(x) = constant$ and $V(x) \equiv 0$ then this theorem implies Girsanov's theorem. Hence one can say that Theorem 10.19 implies a generalization of the Girsanov theorem.

Some similar results are also valid for groups $e^{it\Delta_\alpha}$ and e^{itG_α}. We formulate only one of them.

Let, for any $t > 0$, $\Phi(0) =$ Id and

$$(\Phi(t)\varphi)(x) = (c_3(t))^{-1} \int_{\mathbb{R}^n} \exp\left\{-i\frac{\alpha^{-1}(x)(z-x)^2}{2t}\right\} \varphi(z)\,\mathrm{d}z,$$

where

$$c_3(t) = \int_{\mathbb{R}^n} \exp\left\{-i\frac{\alpha^{-1}(x)(z-x)^2}{2t}\right\}\,\mathrm{d}z.$$

We assume that in both integrals one uses some natural regularization.

Theorem 10.21 *For any $\varphi \in L^2(\mathbb{R}^n)$ and $t > 0$ one has*

$$\left(e^{i\Delta_\alpha t/2}\right)\varphi = \lim_{n\mapsto\infty}\left(\Phi\left(\frac{t}{n}\right)\right)^n\varphi.$$

Acknowledgement: I acknowledge the support of Russian Foundation for Basic Research (Grant 06-01-00761-a), Deutsche Forschungsgemeinschaft (Germany) and Ministry of Education of Spain (Grant SAB2005-0200). I also thank two referees due to whom so many misprints have been corrected.

Bibliography

[1] S. Albeverio, Z. Brzeźniak, Oscillatory integrals on Hilbert spaces and Schrödinger equation with magnetic fields, *J. Math. Phys.* **36**, 5 (1995), 2135–2156.

[2] S. Albeverio, R. Hoegh-Krohn, Mathematical Theory of Feynman Path Integrals, *Lecture notes in math.* 523 (Berlin: Springer, 1976).

[3] L. Accardi, O.G. Smolyanov, Feynman formulae for evolution equations with Levy laplacians on infinite-dimensional manifolds, *Dok. Math.*, **73**, 2 (2006), 252–257.

[4] F. A. Berezin, Non-Wiener path integrals, *Theor. Math. Phys.* **6**, 2 (1971), 194–212.

[5] P.R. Chernoff, Note on Product Formulas for Operator Semigroups, *J. Funct. Ana.* **84** (1968), 238–242.

[6] D. Elworthy, A. Truman. Feynman maps, Cameron-Martin formulae and anharmonic oscillators, *Ann. Inst. Henri Poincaré, Physique théorique* **41**, 2 (1984), 115–142.

[7] R.P. Feynman, Space-time approach to nonrelativistic quantum mechanics, *Rev. Mod. Phys.* **20** (1948), 367–387.

[8] R.P. Feynman, An operation calculus having applications in quantum electro-dynamics, *Phys. Rev.* **84** (1951), 108–128.

[9] J. Gough, O.O. Obrezkov, O.G. Smolyanov, Randomized Hamiltonian Feyn-man integrals and stochastic Schrödinger-Ito equations, *Izvest. Math.* **69**, 6 (2005), 3–20.

[10] K. Ito and H.P. McKean, Jr. Diffusion Processes and their Sample Paths (Berlin: Springer, 1965).

[11] G.W. Johnson, M.L. Lapidus, The Feynman integral and Feynman's op-erational calculus, *Oxford Mathematical Monographs* (New York: Oxford University Press, 2000).

[12] V. P. Maslov (Russian) Complex Markov Chains and the Feynman Path Integral for Nonlinear Equations (Moscow: Nauka, 1976).

[13] E. Nelson, Feynman integrals and the Schrödinger equation, *J. Math. Phys.* **5**, 3 (1964), 332–343.

[14] O.G. Smolyanov, A.Yu. Khrennikov, Central limit theorem for generalized measures on rigged Hilbert spaces, *Soviet Math. Dokl.* **31** (1985), 301–304.

[15] O.G. Smolyanov, E.T. Shavgulidze, Functional Integrals (Moscow State Uni-versity Press, 1990, in Russian).

[16] O.G. Smolyanov, M.O. Smolyanova, Transformations of the Feynman inte-gral under nonlinear transformations of the phase space, *Theor. Math. Phys.* **100**, 1 (1995), 803–810.

[17] O.G. Smolyanov, A.G. Tokarev, A. Truman, Hamiltonian Feynman path integrals via the Chernoff formula, *J.Math.Phys.* **43**, 10 (2002) 5161–5171.

[18] O.G.Smolyanov, A.Truman, Change of variable formulas for Feynman pseu-domeasures, *Theor. Math. Phys.* **119**, 3 (1999) 677–686.

[19] O.G.Smolyanov, H. von Weizsäcker, O.Wittich, Smooth probability mea-sures and associated differential operators, *Infin. Dimens. Anal. Quantum Prob. Relat. Topics* **2**, 1 (1999), 51–78.

[20] O.G.Smolyanov, H. von Weizsäcker, O.Wittich, The Feynman Formula for the Cauchy Problems in Domains with Boundary, *Dokl. Math.* **69**, 2 (2004), 257–261.

[21] O.G.Smolyanov, H. von Weizsäcker, O.Wittich, Construction of diffusions on sets of Mappings from an Interval to Compact Riemannian Manifolds, *Dokl. Math.* **71**, 3 (2005), 390–395.

[22] O.G.Smolyanov, H. von Weizsäcker, O.Wittich, Surface integrals in Rieman-nian Spaces and Feynman Formulae, *Dokl. Math.* **73**, 3 (2006), 432–437.

[23] O.G.Smolyanov, H. von Weizsäcker, O.Wittich, Chernoff's theorem and dis-crete time approximations of Brownian motion on manifolds, *Potential Anal.* **26** (2007), 1–29.

[24] O.G.Smolyanov, H. von Weizsäcker, O.Wittich, Surface measures and initial boundary value problem generated by diffusion with drift, *Dokl. Math.*, **76**, 1 (2007), 606–610.

11

Deformation quantization in infinite dimensional analysis

Rémi Léandre

Institut de Mathématiques
Université de Bourgogne
21000 Dijon, France

Abstract

We present a survey on various aspects of deformation quantization in infinite-dimensional analysis.

11.1 Introduction

Let M be a manifold. On it we consider the algebra of smooth function $C^\infty(M)$ with values in $\mathbb{C}$. We say that M is endowed with a Poisson structure if there exists a bilinear map $\{,\}$ from $C^\infty(M) \times C^\infty(M)$ into $C^\infty(M)$ such that for all f, g, h belonging to this algebra:

$$\{1, f\} = 0 \tag{11.1}$$

$$\{f, g\} = -\{g, f\} \tag{11.2}$$

$$\{fg, h\} = f\{g, h\} + g\{f, h\} \tag{11.3}$$

$$\{f, \{g, h\}\} + \{g, \{h, f\}\} + \{h, \{f, g\}\} = 0. \tag{11.4}$$

Let ω be a symplectic structure on M, i.e. a smooth closed 2-form ω which is non-degenerated. There is a canonical Poisson structure which is associated to ω. Let f be an element of the algebra of smooth functions on M. Let df be the derivative of f. This realizes a section of the bundle of 1-forms on M. Let X be a vector field on M. The Hamiltonian vector field X_f is defined as follows:

$$< df, X >= \omega(X_f, X).$$

Trends in Stochastic Analysis, ed. J. Blath, P. Mörters and M. Scheutzow.
Published by Cambridge University Press. ©Cambridge University Press 2008.

The Poisson structure is given by:

$$\{f, g\} = \; < \mathrm{d}f, X_g > .$$

The simplest example is the case where we consider $\mathbb{R}^n$ endowed with a constant 2-form $(\omega_{i,j})$. Then

$$\{f, g\} = \sum \omega^{i,j} \frac{\partial f}{\partial x_i} \frac{\partial g}{\partial x_j}, \tag{11.5}$$

where $(\omega^{i,j})$ is the inverse matrix of the matrix $(\omega_{i,j})$. If we consider the phase space $\mathbb{R}^n \oplus \mathbb{R}^{n*}$ of $\mathbb{R}^n$, there is the canonical 1-form which to (q, p) over (q_0, p_0) on the phase space associates $< p_0, q >$: p_0 is a 1-form on $\mathbb{R}^n$ and q is a vector on $\mathbb{R}^n$. Since this 1-form is canonically written, we can still define this 1-form σ on the phase space (the cotangent bundle) $T^*(M)$ of a manifold M. Then $\mathrm{d}\sigma$ defines a symplectic structure on $\mathbb{R}^n \oplus \mathbb{R}^{n*}$ or $T^*(M)$.

If we consider the algebra $A = C^\infty(M)$ associated to the Poisson manifold M, the program for its quantization in quantum mechanics consists of the following request:

- We would like to construct a complex Hilbert space Ξ.
- To each $f \in A$, we would like to associate a densely define operator ξ_f on Ξ which satisfies the following property:

$$[\xi_f, \xi_g] = \xi_{\{f,g\}}.$$

Therefore the Poisson bracket is transformed through this construction into a classical Lie bracket. Generally, this program is not fullfilled.

Bayen et al. ([5,6]) produce a way to replace the Poisson bracket by an ordinary Lie bracket.

They consider $A[[h]]$ the space of formal series in the algebra A. By a star product $*$, we mean a bilinear operation from $A[[h]] \times A[[h]]$ into $A[[h]]$ satisfying the following requirements:

i) The space $A[[h]]$ is an algebra for $*$.

ii) The star-product $*$ is $\mathbb{C}[[h]]$ linear, where $\mathbb{C}[[h]]$ is the algebra of formal series with component in $\mathbb{C}$.

iii) If f, g belong to A,

$$f * g = \sum h^n P_n[f, g],$$

where $P_n[f, g]$ are bidifferential operators vanishing on constants. Moreover $P_0[f, g] = fg$.

iv) The equality $P_1[f, g] - P_1[g, f] = 2\{f, g\}$ holds.

Two star-products $*_1$ and $*_2$ are equivalent if there exists a sequence Q_n of differential operators such that, by $\sum Q_n h^n$, $*_1$ is transformed to $*_2$. We suppose moreover $Q_0 = I$.

The simplest case of a star-product is the Moyal product [41] for a constant symplectic structure $(\omega_{i,j})$ on $\mathbb{R}^n$ where

$$f * g = \sum \frac{i^n}{2^n n!} h^n P_n[f,g]$$

$$P_n[f,g] = \sum \omega^{i_1,j_1} \cdots \omega^{i_n,j_n} \frac{\partial^n}{\partial x_{i_1} \cdots \partial x_{i_n}} f \frac{\partial^n}{\partial x_{j_1} \cdots \partial x_{j_n}} g$$

if f, g belong to A.

This star-product was extended to the case of a symplectic manifold endowed with a symplectic flat connection ([5, 6]). A symplectic connection is a connection ∇ such that for all vector fields X and Y:

$$d.\omega(X,Y) = \omega(\nabla.X,Y) + \omega(X,\nabla.Y),$$

$$\nabla_X Y - \nabla_Y X = [X,Y].$$

This last property means that the connection has no torsion.

Several works were done in order to define the deformation quantization of a general symplectic manifold (see [10], [45], [17]): the construction in [10] was done step by step. Fedosov [17] glues together all the Moyal products on $T_x(M)$ via a suitable flat connection by using the Weyl bundle of [45]. (For readers interested in an introduction to Fedosov's work, we refer to [51]). This flat connection operates on the bundle of formal series on $T.(M)$: on the tangent bundle only, this flat connection does not exist generally!

For review papers on deformation quantization, we refer the reader to [15], [34], [51], and [54] for the physicist side.

Our first aim in this review paper is to replace the algebra A (without any topology) by a general Frechet unital commutative algebra $A_{\infty-}$.

The Frechet algebra $A_{\infty-}$ is the intersection of A_l where A_l is an Hilbert space. If F, G belong to all A_l, FG belongs to all A_l and

$$\|FG\|_l \leq C \|F\|_{l_1} \|G\|_{l_1}.$$

for some big l_1 depending only on l.

A Poisson structure on $A_{\infty-}$ is given by a continuous map $\{,\}$ from $A_{\infty-} \times A_{\infty-}$ into $A_{\infty-}$ such that for all F, G belonging to $A_{\infty-}$

$$\|\{F,G\}\|_l \leq C \|F\|_{l_1} \|G\|_{l_1}$$

for some large l_1 depending only on l. Moreover we suppose that $\{\}$ still satisfies (11.1), (11.2), (11.3), (11.4).

A star-product $*$ on $A_{\infty-}$ is given by an operation on $A_{\infty-}[[h]]$, the algebra of formal series with components in the Frechet algebra $A_{\infty-}$ satisfying the same requirements as before. Moreover if we suppose that

$$F * G = \sum h^n P_n[F, G]$$

then

$$\|P_n[F, G]\|_l \leq C\|F\|_{l_1}\|G\|_{l_1}$$

for some large l_1 depending only on l and n.

The notion of equivalences of two star-products can be extended in this situation: Q_n is supposed only continuous on $A_{\infty-}!$.

In the specific example, if we consider specific examples of function spaces, we can ask if the deformation operators P_n are constituted of bidifferential operators. This leads to two types of deformations:

- continuous deformations
- differential deformations.

If we consider the algebra of functions on smooth compact manifolds, these two notions are equivalent (see [9], [42], [46]), but the problem of the equivalence of these two theories is open in infinite dimensions.

We are motivated by an algebra of functionals related to some infinite-dimensional manifolds. This follows the work of Dito [11, 12, 13], Witten [52], and Duetsch-Fredenhagen [16] who performed some deformation quantization in field theory. Dito [13] has defined deformation quantization on a Hilbert space. We consider the following algebras of functionals which are useful in infinite-dimensional analysis:

(i) The test algebra of functionals which belong to all the Sobolev spaces of the Malliavin Calculus; we will call the Malliavin test algebra. We consider the classical Wiener product on it.

(ii) The test algebra for the Wick product of white noise analysis (or Hida Calculus).

We are also concerned with the quantization of the free path space of a manifold: we consider for that the Taubes limit model [49], that is, a family of Fock spaces parametrized by the finite-dimensional manifold M.

This paper is divided into the following sections:

- Wiener space and Fock space
- Quantization and the Malliavin Calculus [14]
- Quantization and white noise analysis [29]
- Fedosov quantization for the Taubes limit model [31]

We thank G. Dito for helpful comments.

11.2 Wiener space and Fock space

For details in this part, we refer the reader to [38], [39] and [40]. To clarify the exposition, we begin by recalling the relation between analysis on one-dimensional Gaussian measure and Bosonic Fock space, we then recall this relation in finite dimension where there is no problem of measure theory, and we finish by recalling this relation on an infinite-dimensional Hilbert space where there are some problems in constructing the Gaussian measure.

Let us consider the set of formal series $F(x) = \sum \lambda_n x^n$ on $\mathbb{R}$ where λ_n are complex. We put on this set the Hilbert structure

$$\|F\|^2 = \sum |\lambda_n|^2 < \infty.$$

We get a Hilbert space E. This Hilbert space is isomorphic to $L^2(\mathrm{d}\mu)$ where $\mathrm{d}\mu$ is the Gaussian measure $Z^{-1}\exp[-|x|^2/2]\mathrm{d}x$ on $\mathbb{R}$: to x^n we associate the Hermite polynomial $h_n(x)$.

We get three products on the system of h_n:

- The classical product which to h_n and h_m associates the polynomial $h_n h_m$, which is not a Hermite polynomial.
- The classical Wick product $: . :^{cl}$. We write

$$h_m h_n = c(n, m) h_{n+m} + \sum_{k < n+m} c(n, m, k) h_k$$

and we put

$$: h_n . h_m :^{cl} = c(n, m) h_{n+m}.$$

- The normalized Wick product which corresponds to the ordinary product on the algebraic side of this correspondence:

$$: h_n . h_m := h_{n+m}.$$

There are several differential operations: $\frac{\partial}{\partial x}$ which corresponds on the algebraic side to an annihilation operator $a_{e_1}^{cl}$. Namely x^n can be assimilated to $e_1^{\hat{\otimes}n}$, the nth normalized symmetric product of the orthonormal element e_1 of $\mathbb{R}$ in the symmetric tensor algebra of $\mathbb{R}$ and

$$a_{e_1}^{cl} e_1^{\hat{\otimes}n} = \sqrt{n} e_1^{\hat{\otimes}n-1} \tag{11.6}$$

if $n \neq 0$ and equals 0 if $n = 0$ (see [38] for this normalization). Then $a_{e_1}^{cl}$ is a derivation for the ordinary product and $: . . :^{cl}$.

We can choose another normalization to define another annihilation operator. We put

$$a_{e_1} e_1^{\hat{\otimes}n} = n e_1^{\hat{\otimes}n-1}$$

if $n \neq 0$ and equals 0 if $n = 0$. This new annihilation operator is clearly a derivation for $: . . :$.

There are two sides to the one-dimensional Gaussian analysis (see [38], [39], [40]):

- an algebraic side, which uses algebraic computations on the one-dimensional Fock space
- an analytical side, which uses measure theory.

In particular, the adjoint of an annihilation operator on the Fock space, called a creation operator, which can be computed algebraically, corresponds through this isomorphism to an integration by parts formula on the Gaussian space.

Since there is no problem in defining Gaussian measures on a finite-dimensional Euclidean space, we can extend without difficulty these considerations in finite dimension.

Let us consider the symmetric algebra of $\mathbb{R}^d$ with orthonormal basis e_i. Let

$$B = ((1, n_1), \dots, (d, n_d)).$$

We can associate to B the symmetric normalized tensor product:

$$F_B = e_1^{\hat{\otimes}n_1} \hat{\otimes} \cdots \hat{\otimes} e_d^{\hat{\otimes}n_d}.$$

We consider the symmetric tensor algebra of $\mathbb{R}^d$, $S(\mathbb{R}^d)$ of finite sums $F = \sum \lambda_B F_B$ where λ_B belongs to $\mathbb{C}$. We do the completion of it for the Hilbert structure

$$\|F\|^2 = \sum |\lambda_B|^2$$

and we get an object E called the Bosonic Fock space associated to $\mathbb{R}^d$. Classical annihilation operators a_i^{cl} act on $S(\mathbb{R}^d)$ only on $e_i^{\hat{\otimes}n_i}$ and can be extended algebraically to infinite sums, because their adjoint can be computed algebraically. We can do the same considerations for the normalized annihilation operators.

To B, we can associate the polynomial on $\mathbb{R}^d$:

$$F_B(x) = <x, e_1>^{n_1} \cdots <x, e_d>^{n_d}.$$

We consider the Gaussian measure on $\mathbb{R}^d$ $d\mu = Z^{-1} \exp[-|x|^2/2]dx$ and to F_B we associate

$$H_B(x) = h_{n_1}(<x, e_1>) \cdots h_{n_d}(<x, e_d>).$$

The Bosonic Fock space corresponds to $L^2(d\mu)$ by the linear map which to F_B associates H_B.

Through this correspondence, $\frac{\partial}{\partial x_i}$ on the analytical side corresponds to $a_{e_i}^{cl}$ which acts through (11.6) only on $e_i^{\otimes n_i}$ in F_B and not on the other components. We will omit the normalization details. We can define through this correspondence a Wick product $: . :^{cl}$ on $L^2(d\mu)$ and the annihilation classical annihilation operators are a derivation for the classical Wick product. We can define the annihilation operator a_{e_i} which corresponds to $\frac{\partial}{\partial x_i}$ on the polynomial function $F_B(x)$. This is clearly a derivation for $: . :$, the normalized Wick product on $L^2(d\mu)$, which corresponds to multiplication between polynomial functions $F_B(x)$.

Let us summarize:

- If $h = \sum \lambda_i e_i$, $\sum \lambda_i \frac{\partial}{\partial x_i}$ on the Gaussian side corresponds to $\sum \lambda_i a_{e_i}^{cl}$ on the side of the Bosonic Fock space.

- $a_h = \sum \lambda_i a_{e_i}$ differs from a_h^{cl} on the Gaussian side by normalizing constants and is equal to $\sum \lambda_i \frac{\partial}{\partial x_i}$ on the polynomial side. It is therefore a derivation for the normalized Wick product $: . :$.

These differentials are defined for finite sums of F_B and H_B, but we can perform the closure because we can define their adjoint: the adjoint of an annihilation operator is a creation operator, which can be algebraically computed.

We would like to pass to the infinite dimension. We consider a separable Hilbert space H with an orthonormal basis e_i. We consider now the system of

$$B = ((i_1, n_1), \ldots, (i_{|B|}, n_{|B|}))$$

with $i_1 < i_2 < \cdots < i_{|B|}$ and where all the n_i are different from 0. We can still define F_B in the symmetric tensor algebra of H and the polynomial $F_B(x)$. The Bosonic Fock space E is the space of series $F = \sum \lambda_B F_B$ such that

$$\|F\|^2 = \sum |\lambda_B|^2 < \infty.$$

We can still define a_h^{cl} on E if $h \in H$ and a_h on E. On finite sum $F(x) = \sum \lambda_B F_B(x)$, a_h corresponds to the directional derivative in the direction h. We can extend the notion of classical Wick product : . :^{cl} or normalized Wick product : . : for finite sum in E: a_h^{cl} is a derivation for the classical Wick product and a_h a derivation for the normalized Wick product. These differential operations can be closed because they have adjoints. Moreover a_h^{cl} differs by normalizing constants from a_h.

The problem is to find the Gaussian counterpart of these considerations. In order to clarify the exposition, we will choose the Hilbert space H of maps from $[0, 1]$ into $\mathbb{R}$ such that $h(0) = 0$ and such that

$$\|h\|^2 = \int_0^1 |\frac{\mathrm{d}}{\mathrm{d}s} h(s)|^2 \mathrm{d}s < \infty.$$

The formal measure

$$\mathrm{d}\mu = Z^{-1} \exp[-\|h\|^2/2] \, \mathrm{d}D(h),$$

where $\mathrm{d}D(h)$ is the formal Lebesgue measure on H, is the measure of the Brownian motion with values in $\mathbb{R}$ and lives in fact on the continuous paths B_t. We consider an orthonormal basis h^i of H. $< h^i, B. >$ can be interpreted rigorously as the Itô integral $\int_0^1 \mathrm{d}/\mathrm{d}s \, h^i(s) \delta B_s$. Then the Hermite polynomial $h_n(< h^i, B. >)$ can be interpreted with the help of the theory of Wiener chaos. If $g(s_1, \ldots, s_n)$ is a symmetric function from $[0, 1]^n$ into $\mathbb{R}$ such that

$$\int_{[0,1]^n} |g(s_1, \ldots, s_n)|^2 \mathrm{d}s_1 \ldots \mathrm{d}s_n < \infty,$$

we introduce the Wiener chaos

$$
\begin{aligned}
I_n(g) &= \int_{0 < s_1 < .. < s_n < 1} g(s_1, \ldots, s_n) \delta B_{s_1} \cdots \delta B_{s_n} \\
&= \frac{1}{n!} \int_{[0,1]^n} g(s_1, \ldots, s_n) \delta B_{s_1} \ldots \delta B_{s_n}.
\end{aligned}
$$

To h^i we associate $g^i(s_1, \ldots, s_n) = \prod \mathrm{d}/\mathrm{d}s_j \, h(s_j)$. We get the relation:

$$h_n(< h^i, B. >) = \sqrt{n!} I_n(g^i). \tag{11.7}$$

(Let us remark that in this case the right-hand side of (11.7) for g^i is not equal to $< h^i, B. >^n$ because of the Itô formula.) We can repeat in this set up the considerations above word for word, but with the Hermite polynomials H_B being replaced by Wiener chaos.

On the Gaussian side, a_h^{cl} corresponds to the stochastic derivative ∇_h on the chaos in the direction h: we do the the transformation $B. \rightarrow B. + \epsilon h$, $I_n(g)$ is replaced by by $I_n(g)(\epsilon)$ and we take its derivative when $\epsilon \rightarrow 0$. For the finite sum of Wiener chaos, the adjoint ∇_h^* can be computed and corresponds to the algebraic adjoint a_h^{cl*} on the side of the Bosonic Fock space. Therefore these differential operations can be closed. It is the same for a_h, which differs from a_h^{cl} by normalizing constants.

We can choose three directions:

- We consider the ordinary product on the Wiener space (called the Wiener product on the Bosonic Fock space side) and we consider ∇_h (or a_h^{cl}): this defines a derivation for the Wiener product.
- We consider the classical Wick product on the Wiener space, which corresponds to the component of length $n + m$ in the chaos decomposition of $I_n(g)I_m(g)$, and we consider ∇_h or its algebraic counterpart a_h^{cl}: they act as a derivation for this product.
- We consider the normalized Wick product : . : and the annihilation operators a_h on the Bosonic Fock space: they act as a derivation for this product.

There are two classical sides of the Wiener analysis:

- Either we consider the map $h \rightarrow a_h^{cl} F$ or the map $h \rightarrow \nabla_h G$ if G is a Wiener chaos, conveniently completed in all the $L^p(d\mu)$. This is the way chosen in the Malliavin Calculus (see [21], [35], [36], [37], [43], [50]).
- Or we consider the map $F \rightarrow a_h^{cl} F$ or the map $F \rightarrow a_h F$. They are completed algebraically for infinite sums by considering the Hida weighted Fock space. This is the way chosen by the Hida Calculus (see [8], [19], [20], [44]).

Let us remark that these considerations can be generalized to any separable Hilbert space H. There is a dense continuous imbedding of H into a separable Banach space B such that the formal Gaussian probability measure

$$d\mu = Z^{-1} \exp[-\|h\|^2/2] \, dD(h)$$

exists in fact as a measure on B endowed with its Borelian σ-algebra.

11.3 Quantization and the Malliavin Calculus

This section is concerned with the probabilistic side of the previous section. Let us remark that, if the Malliavin Calculus has a lot of precursors (see Albeverio and Hoegh-Krohn [1], Berezanskii [7], Fomin [4], Gross [18]), one of the main originalities of the Malliavin Calculus was to complete the known differential operations on the Wiener space in all the L^p, such that the space of test functionals of the Malliavin Calculus is an algebra. Moreover the test functionals are generally almost surely defined, since there is no Sobolev embedding in infinite dimension.

This part is relevant to the so-called *Malliavin transfer principle*: a formula which is true in the deterministic context remains true in the stochastic context, but almost surely. In order that the reader will be convinced by this statement, we begin by recalling Dito's theory.

Let H be the Hilbert space of functions from $[0,1]$ into $\mathbb{R}^d$ such that:

$$\int_0^1 |\mathrm{d}/\mathrm{d}s\, h(s)|^2 \mathrm{d}s = \|h\|^2 < \infty\,;\, h(0) = 0$$

On $H \oplus H^* = H_t$ we consider the classical pairing 1-form and the associated symplectic structure. The matrix of this constant symplectic structure is bounded as well as its inverse. We consider the algebra $A = C_{H.S}^\infty(H_t)$ of Hilbert–Schmidt Fréchet smooth functionals on H_t with values in $\mathbb{C}$. A is constituted of Frechet smooth functionals and moreover we suppose that the derivative of order r of F is Hilbert–Schmidt. This means that if $F \in A$:

$$< \nabla^r F, h^1, \ldots, h^r >$$
$$= \int_{[0,1]^r} < \nabla^r F(s_1, \ldots, s_r), \mathrm{d}/\mathrm{d}s\, h_{s_1}^1, \ldots, \mathrm{d}/\mathrm{d}s\, h_{s_r}^r > \mathrm{d}s_1 \cdots \mathrm{d}s_r$$

$$(11.8)$$

where

$$\|\nabla^r F\|^2 = \int_{[0,1]^r} |\nabla^r F(s_1, \ldots, s_r)|^2 \mathrm{d}s_1 \ldots \mathrm{d}s_r < \infty. \qquad (11.9)$$

We remark that (11.9) is not in general satisfied if we consider any general Frechet smooth functional. We have two partial derivatives:

- the derivative $\nabla_1 F$ in the direction of H. This gives an element of H^*
- the derivative $\nabla_2 F$ in the direction of H^*. This gives an element of H.

The Poisson bracket is then defined by

$$\{F, G\} = < \nabla_1 F, \nabla_2 G > - < \nabla_1 G, \nabla_2 F > . \tag{11.10}$$

We remark that, if we consider an orthormal basis of H_t, the matrix of the Poisson structure is *bounded*.

Let $\Lambda^{\alpha,\beta} = 1$ if $\alpha = 1$, $\beta = -1$ and equal to 0 on the diagonal which defines the canonical 2×2 antisymmetric matrix. Following Dito [12], we put

$$P_n[F, G] = \sum \Lambda^{\alpha_1,\beta_1} \ldots \Lambda^{\alpha_n,\beta_n} < \nabla^n_{\alpha_1,\ldots,\alpha_n} F, \nabla^n_{\beta_1,\ldots,\beta_n} G > . \tag{11.11}$$

If F, G belong to A, $P_n[F, G]$ belongs to A owing to the integrability condition (11.9).

Theorem 11.1 (Dito [13]) *Let F, G be elements of A. Then*

$$F * G = \sum \frac{i^n}{2^n n!} h^n P_n[F, G] \tag{11.12}$$

defines a differential star-product of A.

This means that:

- The space $A[[h]]$ is an algebra for $*$.
- The infinite-dimensional star-product $*$ is $\mathbb{C}[[h]]$ linear, where $\mathbb{C}[[h]]$ is the algebra of formal series with components in $\mathbb{C}$.
- If F, G belong to A,

$$F * G = \sum \frac{i^n}{2^n n!} h^n P_n[F, G]$$

  where $P_n[F, G]$ are bidifferential operators vanishing on constants. Moreover $P_0[F, G] = FG$.
- The equality $P_1[F, G] - P_1[G, F] = 2\{F, G\}$ holds.

We would like to put some integrability conditions on A: we would like to consider Sobolev spaces in infinite dimension. Unfortunately, the Lebesgue measure in infinite dimension does not exist as a measure. (see [28], [30] in order to define the Haar measure in infinite dimension as a distribution in the Hida–Streit approach). Gaussian measures in infinite dimension are very well understood. For instance, the probability measure on H_t written in the physicist's style

$$d\mu(h) = Z^{-1} \exp[-\|h\|^2/2] \, dD(h),$$

where $dD(h)$ is the formal Lebesgue measure on H_t, can be interpreted rigorously as a measure.

As a measure, $d\mu$ exists on $W \oplus W^* = W_t$ where W and W^* are the space of continuous functions from $[0,1]$ into $\mathbb{R}^d$ (W^* is not the dual of the Banach space W, but we keep this notation in respect of the previous considerations). The reason for choosing the Wiener spaces W and W^* is that we have the notion of L^p spaces. In particular, through Hölder's inequality, the intersection of all the L^p ($p \in [1, \infty[$) is a Frechet *algebra*.

We take derivatives of functionals on W_t in the directions of H_t, which can be split in partial derivatives in the direction of H or H^*. They can be completed as usual using the relevant integration by parts formulae in order to define Sobolev spaces: this allows us to define Sobolev spaces as in finite dimensions. According to the lines of the Malliavin Calculus, a functional is said to be smooth in the Malliavin sense (see [21], [35], [36], [37], [43], [50]) if for all integers r, all $p \in [1, \infty]$,

$$\|F\|_{r,p} = E[\|\nabla^r F\|^p]^{1/p} < \infty \qquad (11.13)$$

where $\nabla^r F$ is given by:

$$< \nabla^r F, h^1, \ldots, h^r >$$
$$= \int_{[0,1]^r} < \nabla^r F(s_1, \ldots, s_r), d/ds\, h^1_{s_1}, \ldots, d/ds\, h^r_{s_r} > ds_1 \cdots ds_r$$

$$(11.14)$$

and the random variable $\|\nabla^r F\|$ is given by (11.9).

The main difference between (11.8) and (11.14) is that (11.14) is only true *almost surely*, while (11.8) is valid path by path. We consider the space $A_{\infty-}$ constituted of functionals smooth in the Malliavin sense: for all r, p, (11.13) is valid. Let us recall one of the main theorems of the Malliavin Calculus: $A_{\infty-}$ is a Frechet *algebra*.

We can define the stochastic Poisson bracket as follows:

$$\{F, G\} = < \nabla_1 F, \nabla_2 G > - < \nabla_1 G, \nabla_2 F > . \qquad (11.15)$$

It is the same formula as (11.10) but valid only *almost surely*. Thanks to (11.13), we have:

Theorem 11.2 (Dito–Léandre [14]) *The Poisson bracket $\{,\}$ defines a Poisson structure on the Frechet algebra $A_{\infty-}$ in the sense of the introduction.*

If F, G belongs to the Frechet algebra $A_{\infty-}$, we put

$$P_n[F, G] = \sum \Lambda^{\alpha_1, \beta_1} \ldots \Lambda^{\alpha_n, \beta_n} < \nabla^n_{\alpha_1, \ldots \alpha_n} F, \nabla^n_{\beta_1, \ldots \beta_n} G > \qquad (11.16)$$

and we put

$$F * G = \sum \frac{i^n}{2^n n!} h^n P_n[F, G]. \tag{11.17}$$

These are the same formulae as (11.11) and (11.12), but they hold only *almost surely*. The integrability conditions are provided by (11.13). This shows us [14]

Theorem 11.3 (Dito–Léandre [14]) *The stochastic star-product given by (11.17) defines a quantization by deformation of the Frechet algebra $A_{\infty-}$ in the sense of the introduction.*

Let us stress the two main differences between this section and the next section:

- The matrix of the Poisson structure is bounded, such that the stochastic Poisson bracket acts continuously on $A_{\infty-} \times A_{\infty-}$.
- We consider the Wiener product on $L^2(d\mu)$ identified with the Bosonic Fock space associated to H_t. In particular, the Wiener product does not keep the natural filtration on the Bosonic Fock space given by the length of the chaos, due to the Itô formula, unlike the normalized Wick product: this important property of the normalized Wick product will play a big role in the final section of this work. Let us recall that this filtration plays an essential role in Fedosov's work [17].

11.4 Quantization and white noise analysis

We are concerned in this part with an example where the Poisson bracket is not bounded.

Let us consider the free loop space H of $\mathbb{R}^d$ of maps h from S^1 into $\mathbb{R}^d$ endowed with the Hilbert structure:

$$\|h\|^2 = \int_{S^1} |h(s)|^2 ds + \int_{S^1} |d/ds h(s)|^2 ds < \infty \tag{11.18}$$

An orthonormal basis of H is given by

$$h_I(s) = \sqrt{Ck^2 + 1}^{-1} e_i f_k(s)$$

where $I = (i, k)$ are ordered by lexicographic order and the system of e_i constitutes an orthonormal basis of $\mathbb{R}^d$. Also $f_k(s) = \cos[2\pi ks]$ if $k \geq 0$ and $f_k(s) = \sin[2\pi ks]$ if $k < 0$.

On $\mathbb{R}^d$ we consider a non-degenerate constant 2-form $\omega = (\omega_{i,j})$ and we introduce the non-degenerate constant 2-form on H:

$$\Omega(h^1, h^2) = \int_{S^1} \omega(h^1(s), h^2(s)) \mathrm{d}s \qquad (11.19)$$

The matrix of this constant symplectic form in the basis h_i is denoted by $\Omega_{I,J}$ and its inverse by $\Omega^{I,J}$. To H is associated the complex Hilbert space $L^2(\mathrm{d}\mu)$ where $\mathrm{d}\mu$ is a Gaussian measure on the Banach space B of *continuous* maps from S^1 into $\mathbb{R}^d$. We can consider the derivative ∇_{h_I} in the direction of h_I (see second part). The generalization of (11.5) and (11.15) leads to the following stochastic Poisson bracket:

$$\{F, G\} = \sum \Omega^{I,J} \nabla_{h_I} F \nabla_{h_J} G. \qquad (11.20)$$

This leads to some divergences. Namely

$$\Omega_{I,J} = (Ck_I^2 + 1)^{-1} \omega_{i_J, i_J} \delta_{k_I, k_J}$$

if $I = (i_I, k_I)$ and $J = (i_J, k_J)$ and $\delta_{k,k'}$ is the Kronecker symbol.

If we consider the Malliavin test algebra associated to these data, $\{,\}$ does not act on this space. For instance, we can consider $F = \sum \lambda_I h_I$ where F belongs to the first chaos and $G = \sum \mu_J h_J$. Then F and G belong to the Malliavin test algebra if $\sum |\lambda_I|^2 < \infty$ and if $\sum |\mu_I|^{2\cdot} < \infty$ and λ_I and μ_J are constants. But the formula (11.19) gives

$$\{F, G\} = \sum \lambda_I \mu_J \Omega^{I,J},$$

a quantity which diverges generally since the matrix $(\Omega^{I,J})$ inverse of the matrix $(\Omega_{I,J})$ is not bounded.

In order to understand this singular Poisson bracket, we will use white noise analysis (or the Hida Calculus), which incorporates very singular objects as for instance the speed of the Brownian motion (see [8], [19], [20], [44]). A basic tool of the Hida Calculus is the Fock space. With respect to the previous part, we will make the following changes:

- Instead of using the Malliavin test algebra on the Wiener space associated to the Wiener product, we will use the Hida test algebra in the Fock space for the normalized Wick product.
- Instead of using the H-derivative of the Malliavin Calculus on the Wiener space, we will use the algebra of creation and annihilation operators on the Wiener space, but not the classical one to take account of the normalized Wick product.

Let $B = (I_1, \ldots, I_{|B|})$ where the I_j are ordered by lexicographic order. We consider the set of formal series

$$F = \sum \lambda_B F_B \tag{11.21}$$

where F_B is the normalized symmetric tensor product of the h_{I_j}, $I_j \in B$, $\lambda_B \in \mathbb{C}$. We consider the Hida weight on B

$$\|B\|_r = \prod_{I \in B} (C_1 k_I^2 + 1)^r$$

and we consider the weighted Fock space for $r > 0, C > 1$, of series (11.21) such that

$$\|F\|_{r,C}^2 = \sum |\lambda_B|^2 \|B\|_r C^{|B|} < \infty.$$

We get an Hilbert space $A_{r,C}$.

We consider the normalized Wick product: $: F_B F_C :$ is the normalized symmetric tensor product of h_{I^B} and h_{I^C} for $I^B \in B$ and $I^C \in C$. We put $A_{\infty-} = \cap A_{r,C}$. It is called the Hida weighted Fock space. We get:

Theorem 11.4 (Léandre–Rogers [33]) *$A_{\infty-}$ is a Fréchet algebra for the normalized Wick product. $A_{\infty-}$ is called the Hida test algebra.*

The main difference between the Hida test algebra and the Malliavin test algebra is the following: if we consider the concretization of the Hida test algebra as functionals on the Banach space B, it is constituted of continuous functionals on B. This explains why we get many more allowed differential operators on the Hida test function space than in the Malliavin Calculus. For instance let us consider a sequence λ_I with polynomial growth: $\sum \lambda_I a_I$ is a continuous differential operator on the Hida test algebra and not on the Malliavin test algebra. For developments of these considerations, we refer to [28] where two star-products on the Hida test algebra were equivalent while they were not equivalent in Dito's theory [13]. Let us recall that there is a big difference between the deformation theory in $\mathbb{R}^d$ where all differential deformations are equivalent and the infinite-dimensional case where a lot of deformations on a Hilbert space are inequivalent [13].

∇_{h_I} corresponds, as we mentioned in Section 11.2, to an annihilation operator a_I on the Bosonic Fock space modulo some normalization. If we replace the Wiener product in (11.20) by the normalized Wick product, we get a notion of the Poisson bracket which fits with the classical tools of the Hida Calculus:

$$\{F, G\} = \sum \Omega^{I,J} : a_I F.a_J G :$$ (11.22)

if F and G belong to the Hida test algebra.

We get

Theorem 11.5 (Léandre [29]). *The Poisson bracket* $\{,\}$ *given by (11.22) defines a Poisson bracket on the Hida test algebra* $A_{\infty-}$ *endowed with the normalized Wick product in the sense of the introduction.*

Proof. The algebraic properties come from the fact that a_I is a derivation for the Wick product and from the fact that the family of annihilation operators commute. The analytical properties are due to the fact that $\Omega^{I,J}$ has 'polynomial growth' and that the components of an element of $A_{\infty-}$ in the basis F_B have a 'quick decay'. □

We can use the same formulae as in the previous section in order to define a Hida–Moyal product by making the following changes compared to (11.16):

- We use the normalized Wick product.
- We replace the bounded matrix of the Poisson structure on H_t in the previous part by the unbounded matrix $\Omega^{I,J}$.
- We replace the stochastic derivative $\nabla.F$ by the annihilation operators a_I.

Therefore if F, G belong to $A_{\infty-}$, we put:

$$P_n[F, G] = \sum \Omega^{I_1,J_1} \cdots \Omega^{I_n,J_n} : a_{I_1} \cdots a_{I_n} F.a_{J_1} \cdots a_{J_n} G :$$

This series converges because $\Omega^{I,J}$ has 'polynomial growth' and because the components of F and G in the basis F_B have a 'quick decay'. We put

$$F * G = \sum \frac{i^n}{2^n n!} h^n P_n[F, G].$$ (11.23)

Theorem 11.6 (Léandre [29]) *The star-product given by (11.23) defines a quantization by deformation of the Poisson bracket (11.22) on* $A_{\infty-}$ *in the sense of the introduction.*

11.5 Fedosov quantization for Taubes limit model

Sections 11.3 and 11.4 dealt with the quantization of infinite-dimensional Hilbert spaces. We are motivated in this part by the quantization of the free loop space of a manifold. Free loop spaces of a manifold play a big role in conformal field theory and in string theory [53]. Taubes [49] considers a limit model of the free loop space of a compact Riemannian manifold M. He considers the family of maps from S^1 into $T_x(M)$, $x \in M$. On the set of loops in $T_x(M)$, he introduces the Hilbert structure

$$\|h\|_{1/2,2} = \int_{S^1} |h(s)|_x^2 ds + \int_{S^1} < \Delta^{1/2} h(s), h(s) >_x ds$$

where $\Delta^{1/2}$ is the square root of the Laplacian on the circle. He gets a Hilbert Sobolev space bundle $H_{1/2,2}(x)$. Taubes considered the Fock space associated to it and obtained a bundle of Bosonic Fock spaces over M. The Gaussian measure on this Hilbert Sobolev space lives on distributions. Taubes was motivated by two-dimensional field theory in order to choose this Hilbert structure. In [22], [24], [25], [26], [27], [32] this Hilbert structure is replaced by the Hilbert structure (11.18): the limit model of Taubes can be curved therefore, by using the theory of stochastic differential equations.

In this part, we consider T_∞, the set of finite energy paths from $[0,1]$ into $T(M)$, the tangent bundle of M: $s \to h(s)$ where for all s, s'

$$p(h(s)) = p(h(s'))$$

where p is the natural projection from $T(M)$ onto M. Moreover, $h(0) = 0$ in $T_{p(h(s))}(M)$. In what follows, we will call x the point of M over which lies the tangent space where a path lives. We put $p_\infty(h) = x$. On the set of paths starting from 0 in $T_x(M)$, we choose the Hilbert structure

$$\|h\|_x^2 = \int_0^1 |d/ds\, h(s)|_x^2 ds < \infty \tag{11.24}$$

If $e_i(x)$ is a local orthormal basis of $T_x(M)$ (M, let us recall, is supposed Riemannian), we put

$$h_I(x)(s) = \begin{cases} \frac{\cos[2\pi k s] - 1}{Ck} e_i(x) & \text{if } k > 0, I = (i,k), \\ \frac{\sin[2\pi k s]}{Ck} e_i(x) & \text{if } k < 0, \\ s e_i(x) & \text{if } k = 0. \end{cases}$$

We can choose C in order to get an orthonormal basis of this Hilbert space. The set of indices I is ordered in lexicographic order.

If $B = (I_1, \ldots, I_{|B|})$, we denote by $F_B(x)$ the normalized symmetric tensor products of the $h_I(x)$, $I \in B$. We introduce the Hida weight

$$\|B\|_r = \prod_{I \in B} (|k_I|^2 + 1)^r$$

and we consider the bundle $A_{r,C}$ on M of weighted Fock space $A_{r,C}(x)$ of series $F(x) = \sum \lambda_B(x) F_B(x)$ where λ_B belong to $\mathbb{C}$ such that

$$\|F(x)\|_{r,C}^2 = \sum |\lambda_B(x)|^2 C^{|B|} \|B\|_r < \infty.$$

This is an infinite-dimensional complex vector bundle associated to the orthonormal frame bundle of M: namely we remark that the space $A_{r,C}$ does not depend of the local orthonormal basis $e_i(x)$ of $T_x(M)$. We consider the Levi-Civita connection ∇ on M. ∇ lifts to a unitary connection on the complex Hermitian bundle $A_{r,C}$.

$A_{r,C,k,p}$ denotes the set of sections F of the bundle $A_{r,C}$ such that

$$\sum_{k' \leq k} \int_M \|\nabla^{k'} F(x)\|_{r,C,x}^p \mathrm{d}x = \|F\|_{r,C,k,p}^p < \infty.$$

Our test function space on the Taubes limit model is the intersection of all $A_{r,C,k,p}$ for $r > 1, C > 1, k > 1, p > 1$. It is called the Hida–Taubes space and is denoted by $A_{\infty-}$.

Theorem 11.7 (Léandre [31]). $A_{\infty-}$ *is a topological Fréchet algebra for the fibrewise normalized Wick product* $: . :$.

We have the natural projection p_∞ from T_∞ onto M. We consider on M the bundle H_1 with fibres that are maps γ from $[0,1]$ into $T_x(M)$ endowed with the Hilbert structure:

$$|\gamma(0)|_x^2 + \int_0^1 |\mathrm{d}/\mathrm{d}s\,\gamma(s)|_x^2 \mathrm{d}s = \|\gamma\|_{1,x}^2$$

(compare with (11.24), where the only difference is that $\gamma(0) \neq 0$!). The tangent bundle of T_∞ is $p_\infty^* H_1$.

Our aim is to quantize the following symplectic structure on T_∞ inherited from the symplectic structure ω on M:

$$\Omega(\gamma^1, \gamma^2) = \omega(\gamma^1(0), \gamma^2(0)) + \int_0^1 \omega(\mathrm{d}/\mathrm{d}s\,\gamma^1(s), \mathrm{d}/\mathrm{d}s\,\gamma^2(s))\mathrm{d}s$$

For that, we will follow the strategy of Fedosov to quantize a symplectic manifold M (see [17]). Let us recall first of all this strategy.

Fedosov considered the Weyl algebra bundle on M. It is the bundle on M with fibres $S(T_x(M))[[h]]$, the formal series with values in the symmetric tensor algebra of $T_x(M)$. We can extend the Moyal product on it $\circ$ by the same formula as (11.5): in infinite dimension, we have done the same in the previous part by replacing the symmetric tensor algebra of the considered Hilbert space by a Hida Fock space! Fedosov introduced a symplectic connection Γ (it is possible to do that, but the choice is not unique!). Starting from Γ, Fedosov constructed a connection on the Weyl bundle ∇^Γ which is flat. ∇^Γ satisfies moreover for all sections ψ_i, $i = 1, 2$, of the Weyl bundle

$$\nabla^{.\Gamma}(\psi_1 \circ \psi_2) = (\nabla^{.\Gamma}\psi_1) \circ \psi_2 + \psi_1 \circ \nabla^{.\Gamma}\psi_2. \tag{11.25}$$

Such flat symplectic connection does not exist generally on the bundle $T(M)$. But, on the Weyl bundle, this connection exists. The main ingredient of Fedosov is to construct step by step this connection, by considering the natural filtration which appears in $S(T_x(M))[[h]]$ by considering twice the degree of the formal variable h and once the length of the considered tensor product. More precisely, in order to construct this connection, Fedosov considers the bundle on M, $S(T(M))[[h]] \otimes \Lambda(T(M))$, where $\Lambda(T(M))$ is the exterior bundle associated to $T(M)$ and introduces a family of Shigekawa complexes in finite dimension on the fibre $S(T_x(M))[[h]] \otimes \Lambda(T_x(M))$ (see [2], [3], [47], [48]). Let us recall that the Shigekawa complex d_x is the classical de Rham complex on $T_x(M)$, but instead of computing its adjoint as is done classically for the Lebesgue measure on $T_x(M)$, the adjoint is performed for the normalized Gaussian measure on $T_x(M)$. These analytical considerations have their counterparts on $S(T_x(M)) \otimes \Lambda(T_x(M)))$, which are used by Fedosov in terms of Bosonic number operator and Fermionic number operator.

Fedosov remarks that there is a bijection Q between elements f of $C^\infty(M)[[h]]$ and flat sections of the Weyl bundle: moreover $Q(f)$ projects naturally on f.

Theorem 11.8 (Fedosov[17]) $f_1 * f_2 = Q^{-1}[Q(f_1) \circ Q(f_2)]$ *defines a star-product on the algebra* $C^\infty(M)$ *endowed with the Poisson structure inherited from the symplectic structure* ω.

Our strategy to produce the Fedosov quantization on the limit model T_∞ is the following: we consider a finite-dimensional approximation T_n

of T_∞ by looking at finite sums

$$\sum_{k=-n,i}^{k=n} \lambda_{i,k} h_{i,k}(x).$$

We have the natural inclusion

$$T_n \subseteq T_{n+1} \subseteq T_\infty.$$

For T_n, we repeat the Fedosov construction, and the main remark is that the Fedosov construction for T_n is compatible with the Fedosov construction for T_{n+1}, because different Fourier modes are orthogonal for the considered symplectic structure. In particular the tangent bundle of T_n, $T(T_n)$, is included in the tangent bundle of T_∞, $p_\infty^* H_1$, and the space $S(T(T_n))[[h]]$ is naturally included in $S(p_\infty^* H_1)[[h]]$. All the constructions in finite dimension are compatible when n increases and can be extended continuously on the Hida–Taubes spaces. So we can define a connection $\nabla^{\Gamma,\infty}$ which acts continuously on the Hida–Taubes sections of $S(p_\infty^*)[[h]]$, which is flat and is still satisfied in this infinite-dimensional setting (11.25).

There is a continuous isomorphism Q_∞ between $A_{\infty-}[[h]]$ and the set of flat sections in Hida–Taubes sense of the bundle $S(p_\infty^* H_1)[[h]]$ (We refer to [31] for details about the analytic difficulties of this statement.)

We can summarize the result of [31] now.

If $I = (i,k)$ we put $\nabla_I = \frac{\partial}{\partial x_i}$ if $k = 0$ and if $k \neq 0$ we denote by ∇_I the annihilation operator associated to h_I on the Bosonic Fock space of the second part (this notation should not confused with the Levi-Civita connection ∇ on M introduced at the beginning of this section). We remark that the matrix of the symplectic structure in the basis considered in the beginning of this section is

$$\Omega_{I,J} = \delta_{k_I,k_J} \omega_{i_I,i_J}(x)$$

such that the matrix of the Poisson bracket is

$$\Omega^{I,J} = \delta_{k_I,k_J} \omega^{i_I,i_J}(x).$$

If F_1 and F_2 belong to $A_{\infty-}$, we put in a system of local coordinates in $p_\infty h$:

$$\{F_1, F_2\} = \sum \Omega^{I,J} : \nabla_I F_1 . \nabla_J F_2 :$$

Theorem 11.9 (Léandre [31]) $\{,\}$ *defines a Poisson bracket on $A_{\infty-}$ in the sense of the introduction.*

Theorem 11.10 (Léandre [31]) *We can define a quantization by deformation of the Poisson Frechet algebra* $(A_{\infty-}, \{,\})$ *by putting*

$$F_1 * F_2 = Q_\infty^{-1}(Q_\infty F_1 \circ Q_\infty F_2).$$

Remark 11.11 *The main ingredient in this construction is that the constructions on the Weyl algebra, which in finite dimension are purely algebraic and without analysis, are very similar to the constructions on the Fock space of the previous section, with the same algebra but now with some analysis in addition.*

Remark 11.12 *Instead of using a family of Hida Fock spaces, it would be tempting to use a family of Malliavin test algebras as it was done for instance in [27] and to consider ordinary products on each Wiener space, that is for their algebraic counterparts Wiener products on each Fock space. But this would lead to some complications because the Wiener product does not keep the filtration which plays an essential role in the step by step construction of the abelian connection of Fedosov.*

Bibliography

[1] S. Albeverio, R. Hoegh-Krohn: Dirichlet forms and diffusion processes on rigged Hilbert spaces. *Z. Wahrsch.* 40, 1977, 1–57.

[2] A. Arai, I. Mitoma: De Rham-Hodge-Kodaira decompositiom in ∞ dimension. *Math. Ann.* 291, 1991, 51–73.

[3] A. Arai, I. Mitoma: Comparison and nuclearity of spaces of differential forms on topological vector spaces. *J. Func. Ana.* 111, 1993, 278–294.

[4] V.I. Averbuh, O.G. Smolyanov, S.V. Fomin: Generalized function and differential equations in linear spaces I: Differential measures. *Tr. Moskov. Ob.* 24, 1971, 133–174.

[5] F. Bayen, M. Flato, C. Fronsdal, A. Lichnerowicz, D. Sternheimer: Deformation theory and quantization. I. *Ann. Phys.* 111, 1978, 61–110.

[6] F. Bayen, M. Flato, C. Fronsdal, A. Lichnerowicz, D. Sternheimer: Deformation theory and quantization. II. *Ann. Phys.* 111, 1978, 111–151.

[7] Y. Berezanskii: The self adjointness of elliptic operators with an infinite number of variables. *Ukrai. Math.* 27, 1975, 729–742.

[8] Y. Berezanskii, Y. Kondratiev: *Spectral Methods in Infinite-dimensional Analysis* Vols I, II. Kluwer, 1995.

[9] A. Connes: Noncommutative differential geometry. *Publi. IHES* 62, 1985, 257–360.

[10] M. De Wilde, P. Lecomte: Existence of star-products and of formal deformations of the Poisson Lie algebra of arbitrary symplectic manifolds. *Lett. Math. Phys* 7, 1983, 487–496.

[11] G. Dito: Star-product approach to quantum field theory: the free scalar field. *Lett. Math. Phys.* 20, 1990, 125–134.

[12] G. Dito: Star-products and nonstandard quantization for Klein-Gordon equation. *J. Math. Phys.* 33 (1992), 791–801.

[13] G. Dito: Deformation quantization on a Hilbert space. In *Noncommutative Geometry and Physics*, ed. U. Carow-watamura, S. Watamura, Y. Maeda. World Scientific, 2005, 139–157.

[14] G. Dito, R. Léandre: Stochastic Moyal product on the Wiener space. *J. Math. Phys.* 48, 2007, 023509.

[15] G. Dito, D. Sternheimer: Deformation quantization: genesis, developments and metamorphoses. In *Deformation Quantization*, ed. G. Halbout. IRMA Lec. Notes. Math. Theor. Phys. Walter de Gruyter. 2002, 9–54.

[16] M. Duetsch, K. Fredenhagen: Perturbative algebraic field theory and deformation quantization. *Fields Ins. Com.* 30, 2001, 151–160.

[17] B. Fedosov: A simple geometric construction of deformation quantization. *J.Diff. Geom.* 40, 1994, 213–238.

[18] L. Gross: Potential theory on Hilbert spaces. *J. Funct. Ana.* 1, 1967, 123–181.

[19] T. Hida: *Analysis of Brownian Functionals* Carleton. Maths. Lect. Notes. 13, 1975.

[20] T. Hida, H.H. Kuo, J. Potthoff, L. Streit: *White Noise: an Infinite Dimensional Calculus*. Kluwer, 1993.

[21] N. Ikeda, S. Watanabe: *Stochastic Differential Equations and Diffusions Processes*, 2nd edn. North-Holland, 1989.

[22] J.D.S. Jones, R. Léandre: A stochastic approach to the Dirac operator over the free loop space. In *Loop Spaces and Groups of Diffeomorphisms. Proc. Steklov Institute* 217, 1997, 253–282.

[23] B. Lascar: Propriétés locales d'espaces de type Sobolev en dimension infinie. *Com. Part.Dif. Equ.*1, 1976, 561–584.

[24] R. Léandre: Brownian motion on a Kähler manifold and elliptic genera of level N. In *Stochastic Analysis and Applications in Physics*, ed. A. Cardoso, M. De Faria, J. Potthoff. NATO series 449, Kluwer, 1994, 193–217.

[25] R. Léandre: A stochastic approach to the Euler-Poincaré characteristic of a quotient of a loop group. *Rev. Math. Phys.* 13, 2001,1307–1315.

[26] R. Léandre: Quotient of a loop group and Witten genus. *J. Math.Phys.* 42, 2001, 1364–1383.

[27] R. Léandre: Wiener Analysis and cyclic cohomology. In *Stochastic Analysis and Mathematical Physics*, ed. R. Rebolledo, J.C. Zambrini. World Scientific, 2004. 115–127.

[28] R. Léandre: Paths integrals in Noncommutative geometry. In *Encyclopedia of Mathematical Physics*, ed. J.P. Françoise. Elsevier, 2006, 8–12.

[29] R. Léandre: Deformation quantization in white noise analysis. In *Geometric Aspects of Integrable Systems*, ed. J.P. Françoise SIGMA. 3, 2007, paper 27 (Electronic).

[30] R. Léandre: Infinite Lebesgue distribution on a current group as an invariant distribution. In *Foundations of Probability and Physics IV*, ed. A. Khrennikov AIP. Proceedings 889, 2007, 332–336.

[31] R. Léandre: Fedosov quantization in white noise analysis. Preprint Universite de Dijon (2006).

[32] R. Léandre, S.S.Roan: A stochastic approach to the Euler-Poincaré number of the loop space of a developable orbifold. *J. Geom. Phys.* 16, 1995,

71–98

[33] R. Léandre, A. Rogers: Equivariant cohomology, Fock space and loop groups. *J.Phy. A: Math. Gen.* 39, 2006, 11929–11946.

[34] Y. Maeda: Deformation quantization and non commutative differential geometry. *Sugaku Expositions*, 16, 2003, 1–2.

[35] P. Malliavin: Stochastic calculus of variations and hypoelliptic operators. In *Pro. ceedings of the International Symposium Stochastic Differential Equations*, ed. K. Itô, Wiley, 1978, 195–263.

[36] P. Malliavin: *Stochastic Analysis*, Springer, 1997.

[37] P.A. Meyer: Le Calcul de Malliavin et un peu de pédagogie. In *R.C.P 34*, ed. J.P. Ramis, Publ. Univ. Strasbourg, 1984, 17–41.

[38] P.A. Meyer: Eléments de probabilités quantiques III. *Séminaire de Probabilité XX. 1984/1985*, ed. J. Azéma M. Yor, Lect.Notes. Math. 1204, Springer, 1986, 229–248.

[39] P.A. Meyer: Eléments de probabilités quantiques IV. *Séminaire de Probabilité XX. 1984/1985*, ed. J. Azéma M. Yor. Lect. Notes Math. 1204, Springer, 1986, 249–285.

[40] P.A. Meyer: *Quantum Probability for Probabilists.* Lect. Notes Math. 1610, Springer, 1995.

[41] J. Moyal: Quantum mechanics as a statistical theory. *Proc. Camb. Phil. Soc.* 45, 1949, 99–124.

[42] F. Nadaud: *Déformations et déformations généralisées.* Thesis, Université de Bourgogne, 2000.

[43] D. Nualart: *Malliavin Calculus and Related Topics*, Springer, 1995.

[44] N. Obata: *White Noise analysis and Fock space.* Lect. Notes Math. 1577, Springer, 1994,

[45] H. Omori, Y. Maeda, Y. Yoshioka: Weyl manifolds and deformation quantization. *Adv. Math.* 85, 1991, 224–255.

[46] G. Pinczon: On the equivalence between continuous and differential deformation theories. *Lett. Math. Phys.* 41, 1997, 143–156.

[47] I. Shigekawa: De Rham-Hodge-Kodaira's decomposition on an abstract Wiener space. *J. Math. Kyoto. Univ.* 26, 191–202.

[48] I. Shigekawa: Vanishing Theorem of the Hodge-Kodaira operator for differential forms on a convex domain of the Wiener space. In *Probability and Geometry*, ed. R. Léandre, *Inf. Dim. Anal. Quant. Probab. Rel. Top.*, 6, Suppl, 2003, 39–53.

[49] C. Taubes: S^1 actions and elliptic genera. *Com. Math. Phys.* 122, 1989, 455-526.

[50] A.S. Ustunel: *An Introduction to Analysis on Wiener Space.* Lect. Notes Math. 1610, Springer, 1995.

[51] A. Weinstein: Deformation quantization. In *Séminaire Bourbaki*, Astérisque 227, 1995, 389–409.

[52] Ed. Witten: Noncommutative geometry and string field theory. *Nucl. Phys. B* 268, 1986, 253–294.

[53] Ed. Witten: The Index of the Dirac operator in loop space. In *Elliptic Curves and Modular Forms in Algebraic Topology*, ed. P.S. Landweber, Lect. Notes Maths. 1326, Springer, 1988, 161–181.

[54] C. Zachos: Deformation quantization: quantum mechanics lives and works in phase-space. *Int. J. Modern. Phys. A* 17, 2002, 297–316.

IV. Stochastic analysis in mathematical biology

12

Measure-valued diffusions, general coalescents and population genetic inference

Matthias Birkner

Weierstraß-Institut für
Angewandte Analysis und Stochastik
Mohrenstraße 39
D-10117 Berlin, Germany

Jochen Blath

Institut für Mathematik
Technische Universität Berlin
Straße des 17. Juni 136
D-10623 Berlin, Germany

Abstract

We review recent progress in the understanding of the interplay between population models, measure-valued diffusions, general coalescent processes and inference methods for evolutionary parameters in population genetics. Along the way, we will discuss the powerful and intuitive (modified) lookdown construction of Donnelly and Kurtz, Pitman's and Sagitov's Λ-coalescents as well as recursions and Monte Carlo schemes for likelihood-based inference of evolutionary parameters based on observed genetic types.

12.1 Introduction

We discuss mathematical models for an effect which in population genetics jargon, somewhat orthogonal to diffusion process nomenclature, is called "genetic drift", namely the phenomenon that the distribution of genetic types in a population changes in the course of time simply due to stochasticity in the individuals' reproductive success and the finiteness of all real populations. We will only consider "neutral" genetic types.

Trends in Stochastic Analysis, ed. J. Blath, P. Mörters and M. Scheutzow.
Published by Cambridge University Press. ©Cambridge University Press 2008.

This contrasts and complements the notion of selection, which refers to scenarios in which one or some of the types confer a direct or indirect reproductive advantage to their bearers. Thus, in the absence of demographic stochasticity, the proportion of a selectively advantageous type would increase in the population, whereas that of neutral types would remain constant. The interplay between small fitness differences among types and the stochasticity due to finiteness of populations leads to many interesting and challenging problems, see e.g. the paper by A. Etheridge, P. Pfaffelhuber and A. Wakolbinger in this volume.

Genetic drift can be studied using two complementary approaches, which are dual to each other, and will be discussed below. Looking "forwards" in time, the evolution of the type distribution can be approximately described by Markov processes taking values in the probability measures on the space of possible types. Looking "backwards", one describes the random genealogy of a sample from the population. Given the genealogical tree, one can then superimpose the mutation process in a second step. The paper by P. Mörters in this volume studies asymptotic properties of these genealogical trees in the limit of a large sample size as an example of the use of the multifractal spectrum.

The classical model for genetic drift is the so-called Wright–Fisher diffusion, which is appropriate when the variability of the reproductive success among individuals is small. Recently, there has been mathematical and biological interest in situations where the variance of the number of offspring per individual is (asymptotically) infinite, and detailed descriptions of the possible limiting objects have been obtained. We review these developments, giving particular emphasis to the interplay between the forwards models, generalized Fleming–Viot processes, and their dual backwards models, Λ-coalescents. We use this opportunity to advertise the lookdown construction of Donnelly and Kurtz (in its [15] "flavour"), which provides a realization-wise coupling for this duality. Furthermore, we show how these approaches can be used to derive recursions for the probabilities of observed types in a sample from a stationary population. These recursions can usually not be solved in closed form and can be difficult to implement exactly, in particular if the space of possible types or the sample size is large. We describe a Monte–Carlo method which allows an approximate solution.

Many important and interesting aspects of mathematical population genetic models are omitted in our review, in particular the possibilities of varying population sizes, selective effects, spatial or other population substructure, multi-locus dynamics and recombination. We also focus on haploid models, meaning that our individuals have only one parent. For an introduction to coalescents with emphasis on biology, see e.g. [32], [42], [33], [51], for background on (classical and generalized) Fleming–Viot processes and variations of Kingman's coalescent, see e.g. [19], [11], [20] and [15].

12.2 Population genetic models with neutral types

Cannings models. In neutral population models, the main (and only) sources of stochasticity are due to random genetic drift and mutation. The first feature is captured in a basic class of population models, namely the so-called *Cannings models* [9, 10]. We will subsequently extend these by adding mutations.

Consider a (haploid) population of constant size (e.g. due to a fixed amount of resources) consisting of, say, N individuals. Suppose the population is undergoing "random mating" with fixed non-overlapping generations and ideally has evolved for a long time, so that it can be considered "in equilibrium". In each generation $t \in \mathbb{Z}$, the distribution of the offspring numbers is given by a non-trivial random vector

$$(\nu_1^{(t)}, \dots, \nu_N^{(t)}) \quad \text{with} \quad \sum_{i=1}^{N} \nu_i^{(t)} = N, \tag{12.1}$$

where $\nu_k^{(t)}$ is the number of children of individual k. The vectors $\nu^{(t)}$, $t \in \mathbb{Z}$ are assumed i.d.d.

Neutrality means that we additionally suppose that the distribution of each such random vector is *exchangeable*, i.e. for each permutation $\sigma \in S_N$, we have that

$$(\nu_{\sigma(1)}, \dots, \nu_{\sigma(N)}) = (\nu_1, \dots, \nu_N) \quad \text{in law.}$$

If these conditions are met, we speak of a Cannings model.

To explain the notion of random genetic drift, imagine that each individual has a certain genetic type. For example, at the genetic locus under consideration, each individual is of one of the types (or alleles) $\{a, A\}$. Each type is passed on unchanged from parent to offspring (we will introduce mutation to this model later).

For each generation t, let X_t denote the number of individuals which carry the "a"-allele. By the symmetries of the model, $\{X_t\}$ is a finite Markov-chain on $\{0, \ldots N\}$ as well as a martingale. In particular, we may represent its dynamics as

$$X_{t+1} = \sum_{i=1}^{X_t} \nu_i^{(t)}. \tag{12.2}$$

Note that, although $\mathbb{E}[X_t] = X_0$ for all t (due to the martingale property), the chain will almost surely be absorbed in either 0 or N in finite time. In fact, the probability that type a will be fixed in the population equals its initial frequency X_0/N. This is a simple example of the power of genetic drift: although in this model there is no evolutionary advantage of one of the types over the other, one type will eventually get fixed (this force will later be balanced by mutation, which introduces new genetic variation).

12.2.1 *"Classical" limit results in the finite variance regime*

Two-type neutral Wright–Fisher model. The classical example from this class is the famous *Wright–Fisher model* [22, 52]. Informally, one can think of the following reproduction mechanism. At generation t, each individual picks one parent uniformly at random from the population alive at time $t - 1$ and copies its genetic type (i.e. either a or A). Denoting by $p_{t-1} = X_{t-1}/N$ the proportion of alleles of type a in generation $t - 1$, the number X_t of a-alleles in generation t is binomial, that is,

$$\mathbb{P}\{X_t = k | X_{t-1}\} = \binom{N}{k} p_{t-1}^k (1 - p_{t-1})^{N-k}.$$

Compliant with (12.1), the offspring vector $(\nu_1, \ldots, \nu_N)$ would be multinomial with N trials and success probabilities $1/N, \ldots, 1/N$.

The Wright–Fisher diffusion as a limit of "many" Cannings models. For large populations, it is often useful to pass to a diffusion limit. To this end, denote by

$$Y^N(t) := \frac{1}{N} X_{\lfloor t/c_N \rfloor}, \quad t \geq 0,$$

where $\lfloor t/c_N \rfloor$ is the integer part of t/c_N, and the time scaling factor is

$$c_N := \frac{\mathbb{E}[\nu_1(\nu_1 - 1)]}{N - 1} = \frac{\mathbb{V}[\nu_1]}{N - 1}, \tag{12.3}$$

the (scaled) "offspring variance". Note that c_N can also be interpreted as the probability that two randomly sampled individuals from the population have the same ancestor one generation ago (this will be important in Section 12.3). The following exact conditions for convergence follow from the conditions given by [41] and a straightforward application of duality, which we will discuss below [see (12.42)]: If

$$c_N \to 0 \quad \text{and} \quad \frac{\mathbb{E}[\nu_1(\nu_1 - 1)(\nu_1 - 2)]}{N^2 c_N} \to 0 \quad \text{as} \quad N \to \infty, \quad (12.4)$$

$\{Y_t^N\}$ converges weakly to a diffusion process $\{Y_t\}$ in $[0,1]$, which is the unique strong solution of

$$dY_t = \sqrt{Y_t(1 - Y_t)}\, dB_t, \quad Y_0 := x \in [0,1],$$

where $\{B_t\}$ is a standard Brownian motion. Equivalently, $\{Y_t\}$ is characterised as a (strong) Markov process with generator

$$Lf(y) = \frac{1}{2}y(1 - y)\frac{d^2}{dy^2}f(y), \quad y \in [0,1], \quad f \in C^2([0,1]). \quad (12.5)$$

To this continuous model, the machinery of one-dimensional diffusion theory may be applied; see e.g. [21] for an introduction. For example, it is easy to compute the mean time to fixation, if $Y_0 = x$, which is

$$m(x) = -2(x \log x + (1 - x) \log(1 - x)).$$

In terms of the original discrete model, if $X_0/N = 1/2$, one obtains

$$(2 \log 2)\frac{N}{\sigma^2} \approx 1.39 \frac{N}{\sigma^2} \text{ generations}$$

(assuming that asymptotically, $\mathbb{V}[\nu_1] \approx \sigma^2$).

Moran's model. An equally famous model for a discrete population, living in *continuous* time, due to P. A. P. Moran, works as follows: Each of the N individuals carries an independent exponential clock (with rate 1). If a bell rings, the corresponding individual (dies and) copies the type of a uniformly at random chosen individual from the current population (including itself). Another way to think about this is to pick the jumps times according to a Poisson–process with rate N and then independently choose a particle which dies and another particle which gives birth.

Note that, even though this model does not literally fit into the Cannings class, its "skeleton chain" *is* a Cannings model with ν uniformly distributed on all the permutations of

$$(2, 0, 1, 1, \ldots, 1).$$

The fraction of type a-individuals in both models, suitably rescaled, converge to the Wright–Fisher diffusion: the continuous-time variant has to be sped up by a factor of N, the skeleton chain by a factor of N^2, as N skeleton steps roughly correspond to one "generation".

Remark 12.1 (several types and higher-dimensional Wright–Fisher diffusions). It is straightforward to extend the discussion above to a situation with finitely-many (say k) genetic types, and obtain analogous limit theorems. Under the same assumptions, the fraction of type i in generation $\lfloor t/c_N \rfloor$ is approximately described by Y_t^i, where

$$Y_t = (Y_t^1, \ldots, Y_t^k) \in \{(y_1, \ldots, y_k) : y_i \geq 0, \sum_i y_i = 1\}$$

is a diffusion with generator $L^{(k)}$, acting on $f \in C^2(\mathbb{R}^k)$ as

$$L^{(k)} f(y) = \frac{1}{2} \sum_{i,j=1}^{k} y_i (\delta_{ij} - y_j) \frac{\partial^2}{\partial y_i \partial y_j} f(y). \tag{12.6}$$

$\square$

Fleming–Viot processes and infinitely many types. To incorporate scenarios with infinitely many possible types, it is most convenient to work with measure-valued processes. For simplicity and definiteness, we choose here $E = [0, 1]$ as the space of possible types, and consider random processes on $\mathcal{M}_1([0, 1])$. For example, let $\tilde{X}(t, i)$ (with values in E) be the type of individual i in generation t in a Cannings model, and let

$$Z_t^N := \frac{1}{N} \sum_{i=1}^{N} \delta_{\tilde{X}(t,i)} \tag{12.7}$$

be the empirical type distribution in generation t. Then, under Assumptions (12.4), if $Z_0^N \Rightarrow \mu \in \mathcal{M}_1([0, 1])$, the rescaled processes $\{Z_{\lfloor t/c_N \rfloor}^N\}$ converge weakly towards a measure-valued diffusion $\{Z_t\}$, which uniquely solves the (well-posed) martingale problem with respect to the generator

$$\mathcal{L}\Phi(\mu) = \sum_{\substack{J \subseteq \{1, \ldots, p\}, \\ |J| = 2}} \int \mu(da_1) \cdots \mu(da_p) \left(\phi(a_1^J, \ldots, a_p^J) - \phi(a_1, \ldots, a_p) \right)$$

$$\tag{12.8}$$

for $\mu \in \mathcal{M}_1([0,1])$ and test functions

$$\Phi(\mu) = \int \phi(a_1, \ldots, a_p)\, \mu(\mathrm{d}a_1) \cdots \mu(\mathrm{d}a_p), \qquad (12.9)$$

where $p \in \mathbb{N}$ and $\phi : [0,1]^p \to \mathbb{R}$ is measurable and bounded, and for $a = (a_1, \ldots, a_p) \in [0,1]^p$ and $J \subseteq \{1, \ldots, p\}$, we put

$$a_i^J = a_{\min J} \text{ if } i \in J, \text{ and } a_i^J = a_i \text{ if } i \notin J, i = 1, \ldots, p, \qquad (12.10)$$

see e.g. [19], Ch. 10, Thm 4.1. Thinking of a as the types of a sample of size p drawn from μ, passage from a to a^J means a coalescence of a_i, $i \in J$.

In particular, if $\mu = \sum_{i=1}^{k} y_i \delta_{a_i}$ for k different points $a_i \in [0,1]$, then

$$Z_t = \sum_{i=1}^{k} Y_t^i \delta_{a_i},$$

where $\{Y_t^i : i = 1, \ldots, k, t \geq 0\}$ is the k-dimensional Wright–Fisher diffusion with generator (12.6).

12.2.2 *Beyond finite variance: occasional extreme reproduction events*

Since the end of the 1990s, more general reproduction mechanisms and their infinite population limits have been studied in the mathematical community (see [45], [43], [15], [41], [46]).

Although the motivation for this came from considerations about the genealogy of population resp. coalescent processes, we describe the corresponding population models forward in time first. Many of the technical assumptions here will become clearer with a reading of the next section.

Implicit in (12.4) is the assumption that each family size ν_i is small compared to the total population size N. A natural generalization, motivated by considering species with potentially very many offspring, is to consider scenarios where occasionally a single family is of appreciable size when compared to N. In this spirit, Eldon and Wakeley [17] proposed a family of Cannings models, where, in a population of size N, ν is a (uniform) permutation of

$$(2, 0, 1, \ldots, 1) \quad \text{or of} \quad (\lfloor \psi N \rfloor, \underbrace{0, 0, \ldots, 0}_{\lfloor \psi N \rfloor \text{ times}}, 1, \ldots, 1) \qquad (12.11)$$

with probability $1 - N^{-\gamma}$ resp. $N^{-\gamma}$ for some fixed parameter $\psi \in (0,1]$

and $\gamma > 0$. The idea is of course that, from time to time, an exceptionally large family is produced, which recruits a (non-negligible) fraction ψ of the next generation.

This is appealing as being presumably the conceptually simplest model of this phenomenon. On the other hand, while one may be willing to accept the assumption that in a species with high reproductive potential and variability, such extreme reproductive events can occur, the stipulation that these generate *always the same* fraction ψ is certainly an over-simplification.

A more realistic model would allow *"random"* ψ, where the parameter ψ is chosen according to some (probability) measure F. So far, the question which F are "natural" for which biological applications is largely open.

A plausible class of Cannings models for scenarios with (asymptotically) heavy-tailed offspring distributions has been introduced and studied by Schweinsberg [48]: in each generation, individuals generate *potential* offspring as in a supercritical Galton–Watson process, where the tail of the offspring distribution varies regularly with index α, more precisely the probability to have more than k children decays like Const. $\times k^{-\alpha}$. Among these, N are sampled without replacement to survive and form the next generation. The parameter $\alpha \in (1, 2]$ governing the tail of individual litter sizes characterizes the limit process, and intuitively smaller α corresponds to more extreme variability among offspring numbers.

Mathematically, the situation is well understood (see [45], [41]): for the discussion of limit processes, we first specialize to the situation of two types only. Consider the Markov chain (12.2) on the time scale $1/c_N$, where c_N is defined in (12.3). If $c_N \to 0$, for some probability measure F on $[0, 1]$

$$\frac{N}{c_N} \mathbb{P}\{\nu_1 > Nx\} \longrightarrow \int_{(x,1]} \frac{1}{y^2} F(dy) \qquad (12.12)$$

for all $x \in (0, 1)$ with $F(\{x\}) = 0$ and

$$\frac{\mathbb{E}[\nu_1(\nu_1 - 1)\, \nu_2(\nu_2 - 1)]}{N^2} \cdot \frac{1}{c_N} \longrightarrow 0, \quad \text{as } N \to \infty, \qquad (12.13)$$

then the processes $\{X^N_{\lfloor t/c_N \rfloor}/N\}$ converge weakly to a Markov process

$\{Y_t\}$ in $[0,1]$ with generator

$$
Lf(y) = \frac{F(\{0\})}{2} y(1-y)\frac{d^2}{dy^2}f(y)
$$
$$
+ \int_{(0,1]} \Big(yf\big((1-r)y+r\big) + (1-y)f\big((1-r)y\big) - f(y) \Big)\frac{1}{r^2}F(dr)
$$

$$(12.14)$$

for $f \in C^2([0,1])$. The moment condition (12.13) has a natural interpretation in terms of the underlying genealogy; see the remark about simultaneous multiple collisions on page 349. Alternatively, $\{Y_t\}$ can be described as the solution of

$$
dY_t = \sqrt{F(\{0\})Y_{t-}(1-Y_{t-})}\, dB_t
$$
$$
+ \int_{(0,t]\times(0,1]\times[0,1]} \Big(1_{\{u\leq Y(t-)\}}r(1-Y_{t-}) - 1_{\{u>Y(t-)\}}rY_{t-} \Big)\, N(ds\,dr\,du),
$$

$$(12.15)$$

where $\{B_t\}$ is a standard Brownian motion and N is an independent Poisson process on $[0,\infty) \times (0,1] \times [0,1]$ with intensity measure $dt \otimes r^{-2}F_0(dr) \otimes du$ with $F_0 = F - F(\{0\})\delta_0$. Here, $r^{-2}F_0(dr)$ is the intensity with which exceptional reproductive events replacing a fraction r of the total population occur in the limiting process.

The class considered by Eldon and Wakeley [17] leads to $F = \delta_0$ for $\gamma > 2$,

$$
F = \frac{2}{2+\psi^2}\delta_0 + \frac{\psi^2}{2+\psi^2}\delta_\psi \quad \text{for } \gamma = 2 \tag{12.16}
$$

and δ_ψ for $1 < \gamma < 2$. The models considered by Schweinsberg in [48] yield Beta measures, namely

$$
F(dr) = \frac{\Gamma(2)}{\Gamma(2-\alpha)\Gamma(\alpha)} r^{1-\alpha}(1-r)^{\alpha-1}\, dr. \tag{12.17}
$$

In [5], these processes have been characterized as time-changes of α-stable continuous-mass branching processes renormalized to have total mass 1 at any time.

For the situation with infinitely many possible types, the corresponding limiting generalized Fleming–Viot process can be considered as a measure-valued diffusion with càdlàg paths whose generator, on test functions of the form (12.9) with ϕ two times continuously differentiable, is

$$F(\{0\})\mathcal{L}\Phi(\mu) + \int_E \int_{(0,1]} \Big(\Phi\big((1-r)\mu + r\delta_a\big) - \Phi(\mu) \Big) \frac{F_0(\mathrm{d}r)}{r^2} \, \mu(\mathrm{d}a),$$

(12.18)

where $\mathcal{L}$ is defined in (12.8).

12.2.3 Introducing mutation

We now introduce another major evolutionary 'player', which counteracts the levelling force of random genetic drift. Indeed, when on the right scale, see (12.22) below, mutation continuously introduces new types to a population, leading to reasonable levels of genetic variability.

Example: The two-alleles case. For our pre-limiting Cannings-models, imagine the following simple mechanism. At each reproduction event, particles retain the type of their parents with high probability. However, with a small probability, the type can change according to some mutation mechanism. In the situation of the two-allele model given by the types $\{a, A\}$, suppose that, independently for each child, a mutation from parental type a to A happens with probability $p_{a \to A}^{(N)}$, and denote by $p_{A \to a}^{(N)}$ the corresponding probability for a mutation from A to a.

Let c_N, as defined in (12.3), tend to zero. If we assume, in addition to (12.12), (12.13), that

$$\frac{p_{a \to A}^{(N)}}{c_N} \to \mu_{a \to A} \quad \text{and} \quad \frac{p_{A \to a}^{(N)}}{c_N} \to \mu_{A \to a}, \qquad (12.19)$$

then the process describing the fraction of the a-population converges to a limit which has generator, for a suitable test-function $f \in C^2$, given by

$$Lf(y) + \Big(-y\mu_{a \to A} + (1-y)\mu_{A \to a} \Big) \frac{\mathrm{d}}{\mathrm{d}y} f(y), \qquad (12.20)$$

where L is given by (12.14).

General mutation mechanisms. Here, we come back to consider measure-valued diffusions on some type space E. Let E be a compact metric space (we will later usually assume $E = [0, 1]^{\mathbb{N}}$ or $[0, 1]$). To describe a mutation mechanism, let $q(x, \mathrm{d}y)$ be a Feller transition function on $E \times \mathcal{B}(E)$, and define the bounded linear operator B on the set of bounded function on E by

$$Bf(x) = \int_E \left(f(y) - f(x)\right) q(x, \mathrm{d}y). \qquad (12.21)$$

Denote the individual mutation probability per individual in the Nth stage of the population approximation by r_N and assume that

$$\frac{r_N}{c_N} \to r \in [0, \infty), \qquad (12.22)$$

where c_N is defined in (12.3). Note that the scaling depends on the class of Cannings models considered. For example, for the models in the domain of attraction of a Beta-coalescent (see the considerations leading to (12.17)), the choice of α fixes the scaling of the individual mutation probability μ per generation: in a population of (large) size N, this translates to a rate

$$r = C_\alpha N^{\alpha - 1} \mu \qquad (12.23)$$

with which mutations appear. In the case $\alpha = 2$, this is the familiar formula $r \, (= \theta/2) = 2N\mu$.

Then, the empirical process $\{Z_t^N\}$, describing the distribution of types on E and defined in analogy to (12.7), converges to a limiting Markov process Z, whose evolution is described by the generator (using the notation from (12.9))

$$\mathcal{L}_{B,F}\Phi(\mu) = r \sum_{i=1}^{p} \int_{E^p} B_i(\phi(a_1, \ldots, a_p))\mu^{\otimes p}(\mathrm{d}a_1 \ldots \mathrm{d}a_p) + \mathcal{L}_F \Phi(\mu),$$

$$(12.24)$$

where $\mathcal{L}_F$ is defined by (12.18), and $B_i \phi$ is the operator B, defined in (12.21), acting on the ith coordinate of ϕ. This process is called the F-generalized Fleming–Viot process with individual mutation process B. Note that in the nomenclature of [4], this would be a ν-generalized FV process with $\nu(\mathrm{d}r) = F(\mathrm{d}r)/r^2$.

General Moran model with mutation. While the Cannings class uses discrete generations, the phenomena discussed above can also be expressed in terms of a continuous-time model, which is a natural generalization of the classical Moran model. For a given (fixed) total population size N let $\mathcal{B}_N$ be a Poisson process on $[0, \infty) \times \{1, 2, \ldots, N-1\}$ with intensity measure $\mathrm{d}t \otimes \mu_N$, where μ_N is some finite measure. If

(t, k) is an atom of $\mathcal{B}_N$, then at time t, a 'k-birth event' takes place: k uniformly chosen individuals die and are immediately replaced by the offspring of another individual, which is picked uniformly among the remaining $N - k$. "Extreme" reproductive events can thus be included by allowing μ_N to have suitable mass on ks comparable to N. The classical Moran model, in which only single birth events occur, corresponds to $\mu_N = N\delta_1$.

Additionally, assume that individuals have a type in E, and each particle mutates during its lifetime independently at rate $r_N \geq 0$ according to the jump process with generator B given by (12.21). Write $X_i^{(N)}(t)$ for the type of individual i at time t.

Let us denote the empirical process for the N-particle system by

$$Z_N(t) := \frac{1}{N} \sum_{i=1}^{N} \delta_{X_i^{(N)}(t)}. \qquad (12.25)$$

We will further on assume that $X_i^{(N)}(0) = X_i$, $i = 1, \ldots, N$, where the X_i are exchangeable and independent of $\mathcal{B}_N$, so in particular $\lim_{N \to \infty} Z_N(0)$ exists a.s. by de Finetti's Theorem.

For a reasonable large population limit, one obviously has to impose assumptions on μ_N and r_N. To connect to the formulation in [15], note that $\mathcal{B}_N$ can be equivalently described by the "accumulated births" process

$$A_N(t) := \sum_{(s,k) \in \mathrm{supp}(\mathcal{B}_N), \, s \leq t} k, \qquad t \geq 0, \qquad (12.26)$$

which is a compound Poisson process. We write

$$[A_N](t) = \sum_{s \leq t} \left(\Delta A_N(s)\right)^2$$

for the quadratic variation of A_N. Assume

$$N r_N \to r \qquad (12.27)$$

and

$$\frac{[A_N](Nt) + A_N(Nt)}{N^2} =: U_N(t) \Rightarrow U(t). \qquad (12.28)$$

Note that the limit process U must necessarily be a subordinator with generator

$$G_U f(x) = \int_{[0,1]} \left(f(x + u) - f(x)\right) \tilde{\nu}(\mathrm{d}u) + a f'(x). \qquad (12.29)$$

If (12.27) and (12.28) hold, the time-rescaled empirical processes converge:

$$\{Z_N(Nt), t \geq 0\} \Rightarrow Z, \quad \text{as } N \to \infty,$$

where $\{Z(t)\}$ is the solution of the well-posed martingale problem corresponding to (12.24); see [15], Theorems 3.2 and 1.1. The relation between G_U and F appearing in (12.24) is as follows:

$$a = 2F(\{0\}), \quad \tilde{\nu} \text{ is the image measure of } \frac{1}{r^2}F(dr) \text{ under } r \mapsto \sqrt{r}.$$
$$(12.30)$$

The latter is owed to the fact that "substantial" birth events, where k is of order N, appear with their squared relative size as jumps of U_N.

While the Assumption (12.28) is quite general, it is instructive (and will be useful later) to specialize to a particular class of approximating birth event rates $\mu_N(\{k\})$ which is closely related to the limiting operators (12.18): for a given $F \in \mathcal{M}_1([0,1])$, put

$$\mu_N(\{k\}) = NF(\{0\})\mathbf{1}_{\{k=1\}}$$
$$+ \frac{1}{N}\int_{(0,1]} \binom{N}{k+1} r^{k+1}(1-r)^{N-k-1}\frac{1}{r^2}F(dr),$$
$$k = 1, \ldots, N-1. \quad (12.31)$$

Then, (12.28) is fulfilled and the limiting U is described by (12.29) and (12.30). This is the (randomized) "Moran equivalent" of the "random ψ" discussed in Section 12.2.2, and will turn out to be the natural mechanism of the first N levels of the lookdown construction; see below. A way to think about the second term in (12.31) is that particles participate in an "r-extreme birth event" independently with probability r. Note that (12.31) implies, for any $x \in (0,1)$ with $F(\{x\}) = 0$,

$$N \sum_{k \geq \lfloor xN \rfloor}^{N-1} \mu_N(\{k\}) \xrightarrow{N \to \infty} \int_x^1 \frac{1}{r^2}F(dr), \quad (12.32)$$

so, in the limiting process, "x-reproductive events" occur at rate $dt \otimes x^{-2}F(dx)$. As in a k-birth event, the probability for a given particle to die is k/N; (12.31) implicitly defines the average lifetime of an individual in the Nth approximating model. The individual death rate of a "typical" particle in the N-particle model is

$$d_N = \sum_{k=1}^{N-1} \frac{k}{N}\mu_N(\{k\}) = F(\{0\}) + \int_{(0,1]} \frac{1-(1-r)^{N-1}}{r}F(dr). \quad (12.33)$$

If $1/r$ is not in $L_1(F)$, this will diverge as $N \to \infty$. In the last paragraph of the remark about "coming down from infinity" on page 350, we will see a relation to structural properties of the corresponding coalescents. Also note that (12.27) and (12.33) implicitly determine the mutation rate per "lifetime unit" in the Nth model, similarly as in (12.23).

Popular mutation models. Having the full generator (12.24) at hand, it is now easy to specialize to the following classical mutation models.

(1) *Finitely-many alleles.* In this model, we assume a general finite type space, say, $E = \{1, \ldots, d\}$. Then, the mutation mechanism can always be written as a stochastic transition matrix $P = (P_{ij})$ times the overall mutation rate $r \in (0, \infty)$. That is,

$$Bf(i) = r \sum_{j=1}^{d} P_{ij} \big(f(j) - f(i) \big).$$

(2) *Infinitely-many alleles.* Here, one assumes that each mutation leads to an entirely new type. Technically, one simply assumes that $E = [0, 1]$ and that each mutation, occuring at rate $r > 0$, independently picks a new type $x \in [0, 1]$, according to the uniform distribution on $[0, 1]$, i.e.

$$Bf(y) = r \int_{[0,1]} \big(f(x) - f(y) \big) dx.$$

Note that this the paradigm example of a parent-independent mutation model. After one mutation step, all information about the ancestral type is lost.

(3) *Infinitely-many sites model.* One thinks of a long part of a DNA sequence, so that each new mutation occurs at a different site. Hence, in principle, the information about the ancestral type is retained. Moreover, it is possible to speak about the "distance" between two types (e.g. by counting the pairwise differences).

As a rule of thumb, if the number of mutations observed is small compared to the square-root of the length of the sequence, this assumption is reasonable. For a mathematical formulation, one may set $E = [0, 1]^{\mathbb{N}}$ and define the mutation operator by

$$Bf(x_1, x_2, \ldots) = \int_{[0,1]} f(u, x_1, x_2, \ldots) - f(x_1, x_2, x_3, \ldots) \, du.$$

For a type vector $\bar{x} = (x_1, x_2, \ldots)$, one can interpret x_1 as the most recently mutated site, x_2 as the second most recently mutated site and so on. This additional information about the temporal order of mutations, which is usually not present in real sequence data, is "factored out" afterwards by considering appropriate equivalence classes.

For a sufficiently "old" population, which can be assumed to be in equilibrium, it is an interesting question whether for each pair of types $\bar{x}, \bar{y}$ visible in the population,

$$\text{there exist indices } i, j, \text{ such that } x_{i+k} = y_{j+k} \text{ for each } k \in \mathbb{N}. \quad (12.34)$$

The condition means that there is a most recent common ancestor for all the types. This question is a prototype of a question for which the evolution of a population should be studied *backwards* in time. We will come back to this in Section 12.3.

The infinitely-many sites model has an interesting combinatorial structure; see e.g. [27] or [6], Section 2. For example, in practice, one frequently does not know which of the bases visible at a segregating site is the mutant. This can be handled by considering appropriate equivalence classes.

12.2.4 Lookdown

The lookdown construction of Donnelly and Kurtz (see [15]) provides a unified approach to all the limiting population models which we have discussed so far, providing a clever nested coupling of approximating generalized Moran models in such a way that the measure-valued limit process is recovered as the empirical distribution process of an exchangeable system of countably many particles. However, its full power will become clearer when we consider genealogies of samples and follow history backwards in time in the next section. We present here a version suitable for populations of fixed total size. The construction is very flexible and works for many scenarios, including (continuous-mass) branching processes.

Note that Donnelly and Kurtz [15] call what follows the 'modified' lookdown construction, in order to distinguish it from the construction of the classical Fleming–Viot superprocess introduced by the same authors in [14]. Here we drop the prefix 'modified'.

Let $F \in \mathcal{M}_1([0, 1])$. The lookdown construction leading to an empirical process with generator (12.24) works as follows.

We consider a countably infinite system of particles, each of them being identified by a level $j \in \mathbb{N}$. We equip the levels with types ξ_t^j, $j \in \mathbb{N}$ in some type space E (and we think of E as being either $\{1, \dots, d\}$ or $[0, 1]$ or $[0, 1]^{\mathbb{N}}$, depending on our choice of mutation model). Initially, we require the types $\xi_0 = (\xi_0^j)_{j \in \mathbb{N}}$ to be an exchangeable random vector, so that

$$\lim_{N \to \infty} \frac{1}{N} \sum_{j=1}^{N} \delta_{\xi_0^j} = \mu,$$

for some finite measure μ on E. The point is that the construction will preserve exchangeability.

There are two sets of ingredients for the reproduction mechanism of these particles, one corresponding to the "finite variance" part $F(\{0\})$, and the other to the "extreme reproductive events" described by $F_0 = F - F(\{0\})\delta_0$. Restricted to the first N levels, the dynamics is that of a very particular permutation of the generalized Moran model described by (12.31), with the property that the particle with the highest level is always the next to die.

For the first part, let $\{L_{ij}(t), \ 1 \le i < j < \infty\}$ be independent Poisson processes with rate $F(\{0\})$. Intuitively, at jump times t of L_{ij}, the particle at level j "looks down" at level i and copies the type there, corresponding to a single birth event in a(n approximating) Moran model. Types on levels above j are shifted accordingly, in formulae

$$\xi_k(t) = \begin{cases} \xi_k(t-) & \text{if } k < j, \\ \xi_i(t-) & \text{if } k = j, \\ \xi_{k-1}(t-) & \text{if } k > j, \end{cases} \qquad (12.35)$$

if $\Delta L_{ij}(t) = 1$. This mechanism is well defined because, for each k, there are only finitely many processes L_{ij}, $i < j \le k$, at whose jump times ξ_k has to be modified.

For the second part, which corresponds to multiple birth events, let $\mathcal{B}$ be Poisson point process on $\mathbb{R}^+ \times (0, 1]$ with intensity measure $dt \otimes r^{-2}F_0(dr)$. Note that for almost all realisations $\{(t_i, y_i)\}$ of $\mathcal{B}$, we have

$$\sum_{i : t_i \le t} y_i^2 < \infty \quad \text{for all } t \ge 0. \qquad (12.36)$$

The jump times t_i in our point configuration $\mathcal{B}$ correspond to reproduction events. Let U_{ij}, $i, j \in \mathbb{N}$, be i.i.d. uniform($[0, 1]$). Define for $J \subset \{1, \dots, l\}$ with $|J| \ge 2$,

$$L_J^l(t) := \sum_{i : t_i \le t} \prod_{j \in J} 1_{U_{ij} \le y_i} \prod_{j \in \{1, \dots, l\} - J} 1_{U_{ij} > y_i}. \qquad (12.37)$$

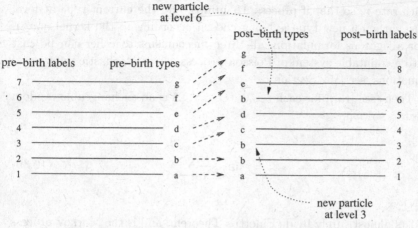

Fig. 12.1. Relabelling after a birth event involving levels 2, 3 and 6.

$L^l_J(t)$ counts how many times, among the levels in $\{1, \dots, l\}$, exactly those in J were involved in a birth event up to time t. Note that for any configuration $\mathcal{B}$ satisfying (12.36), since $|J| \geq 2$, we have

$$\mathbb{E}\big[L^l_J(t) \,\big|\, \mathcal{B}\big] = \sum_{i \,:\, t_i \leq t} y_i^{|J|}(1 - y_i)^{l-|J|} \leq \sum_{i \,:\, t_i \leq t} y_i^2 < \infty,$$

so that $L^l_J(t)$ is a.s. finite.

Intuitively, at a jump t_i, each level tosses a uniform coin, and all the levels j with $U_{ij} \leq y_i$ participate in this birth event. Each participating level adopts the type of the smallest level involved. All the other individuals are shifted upwards accordingly, keeping their original order with respect to their levels (see Fig. 12.1). More formally, if $t = t_i$ is a jump time and j is the smallest level involved, i.e. $U_{ij} \leq y_i$ and $U_{ik} > y_i$ for $k < j$, we put

$$\xi^k_t = \begin{cases} \xi^k_{t-} & \text{for } k \leq j, \\ \xi^j_{t-} & \text{for } k > j \text{ with } U_{ik} \leq y_i, \\ \xi^{k-J^k_t}_{t-} & \text{otherwise,} \end{cases} \tag{12.38}$$

where $J^k_{t_i} = \#\{m < k : U_{im} \leq y_i\} - 1$.

So far, we have treated the reproductive mechanism of the particle system. We now turn our attention to the third ingredient, the mutation steps.

For a given mutation rate r and mutation operator B, as defined in (12.21), define for each level $\in \mathbb{N}$ an independent Poisson process M_i

with rate r, so that if process M_i jumps, and the current type at level i is x, then a new type is being chosen according to the kernel $q(x, \cdot)$. For a rigorous formulation, all three mechanisms together can be cast into a countable system of Poisson process-driven stochastic differential equations, see [15], Section 6.

Then (see [15]), for each $t > 0$, $(\xi_t^1, \xi_t^2, \ldots)$ is an exchangeable random vector, so that

$$Z_t = \lim_{N \to \infty} \frac{1}{N} \sum_{j=1}^{N} \delta_{\xi_t^j} \tag{12.39}$$

exists almost surely by de Finetti's Theorem, and is the Markov process with generator (12.24) and initial condition $Z_0 = \mu$.

Remark 12.2. An alternative and very elegant way to encode the genealogy of a Fleming–Viot process with generator (12.18) is via a flow of bridges, as described in [4]. However, unlike the situation for the lookdown construction, it seems unclear how to incorporate mutation in this approach.   □

12.3 Neutral genealogies: beyond Kingman's coalescent

After having spent a considerable number of pages on models for the evolution of the type distribution of a population forwards in time, we now turn to the fruitful approach of looking backwards in time by analysing the genealogies of samples drawn at present. An important advantage of this approach is that in a neutral situation, this allows one to think of a stochastic two-step procedure, first simulating a genealogy, and then independently superimposing the mutation events on the given genealogical tree. This point of view has many computational and conceptual advantages. We will see below how the lookdown construction, introduced in Section 12.2.4, provides a unified framework by simultaneously describing the forwards evolution and all the genealogical trees of the approximating particle systems.

12.3.1 Genealogies and coalescent processes

A way to describe the genealogy of a sample of size n from a (haploid) population is to introduce a family of partitions of $\{1, \ldots, n\}$ as follows:

$$i \sim_t j \text{ iff } i \text{ and } j \text{ have the same ancestor time } t \text{ before present.}$$
$$(12.40)$$

Obviously, if $t \geq t'$, then $i \sim_{t'} j$ implies $i \sim_t j$, i.e. the ancestral partition becomes coarser as t increases.

For neutral population models of fixed population size in the domain of attraction of the classical Fleming–Viot process, such as the Wright–Fisher- and the Moran model, the (random) genealogy of a finite sample can be (approximately) described by the now classical Kingman-coalescent, which we introduce briefly, followed by the more recently discovered and much more general Λ-coalescents.

Kingman's coalescent. Let $\mathcal{P}_n$ be the set of partitions of $\{1, \ldots, n\}$ and let $\mathcal{P}$ denote the set of partitions of $\mathbb{N}$. For each $n \in \mathbb{N}$, Kingman [34] introduced the so-called n-*coalescent*, which is a $\mathcal{P}_n$-valued continuous-time Markov process $\{\Pi_t^{(n)}, t \geq 0\}$, such that $\Pi_0^{(n)}$ is the partition of $\{1, \ldots, n\}$ into singleton blocks, and then each pair of blocks merges at rate 1. Given that there are b blocks at present, this means that the overall rate to see a merger between blocks is $\binom{b}{2}$. Note that only *binary mergers* are allowed. Kingman [34] also showed that there exists a $\mathcal{P}$-valued Markov process $\{\Pi_t, t \geq 0\}$, which is now called the (standard) *Kingman-coalescent*, and whose restriction to the first n positive integers is the n-coalescent. To see this, note that the restriction of any n-coalescent to $\{1, \ldots, m\}$, where $1 \leq m \leq n$, is an m-coalescent. Hence the process can be constructed by an application of the standard extension theorem.

Λ-coalescents. Pitman [43] and Sagitov [45] introduced and discussed coalescents which allow *multiple collisions*, i.e. more than just two blocks may merge at a time. Again, such a coalescent with multiple collisions (called a Λ-*coalescent* in Pitman's terminology) is a $\mathcal{P}$-valued Markov-process $\{\Pi_t, t \geq 0\}$, such that, for each n, its restriction to the first n positive integers is a $\mathcal{P}_n$-valued Markov process (the "n-Λ-coalescent") with the following transition rates. Whenever there are b blocks in the partition at present, each k-tuple of blocks (where $2 \leq k \leq b \leq n$) is merging to form a single block at rate $\lambda_{b,k}$, and no other transitions are possible. The rates $\lambda_{b,k}$ do not depend on either n or the structure of the blocks. Pitman showed that in order to be consistent, which means that for all $2 \leq k \leq b$,

$$\lambda_{b,k} = \lambda_{b+1,k} + \lambda_{b+1,k+1},$$

such transition rates must necessarily satisfy

$$\lambda_{b,k} = \int_0^1 x^k (1-x)^{b-k} \frac{1}{x^2} \Lambda(dx), \tag{12.41}$$

for some finite measure Λ on the unit interval. We exclude the (trivial) case $\Lambda = 0$. By a trivial time transformation, one can always assume that Λ is a probability measure. In [45], the corresponding measure is termed $F (\doteq \Lambda/\Lambda([0,1])$, and this is the F appearing throughout Section 12.2.2.

Note that (12.41) sets up a one-to-one correspondence between coalescents with multiple collisions and finite measures Λ. Indeed, it is easy to see that the $\lambda_{b,k}$ determine Λ by an application of Hausdorff's moment problem, which has a unique solution in this case.

Owing to the restriction property, the Λ-coalescent on $\mathcal{P}$ (with rates obtained from the measure Λ as described above) can be constructed from the corresponding n-Λ-coalescents via extension.

Approximation of genealogies in finite population models. Consider a sample of size n from a (stationary) Cannings model of size $N \gg n$, without mutation, and define an ancestral relation process $\{R_k^{(N,n)} : k = 0, 1, \dots\}$ via (12.40). Recalling that c_N, as defined in -(12.3), is the probability for a randomly picked pair of individuals to have the same ancestor one generation ago, it seems reasonable to rescale time and define

$$\Pi_t^{(N,n)} := R_{\lfloor t/c_N \rfloor}^{(N,n)}, \quad t \geq 0, \tag{12.42}$$

as then (if $c_N \to 0$) for a sample of size 2, the time to the most recent common ancestor is approximately exponentially distributed with rate 1.

Indeed, [45] and [41] have shown that if $c_N \to 0$ and (12.4) holds true, then $\{\Pi_t^{(N,n)} : t \geq 0\}$ converges weakly to Kingman's n-coalescent, while (12.12) and (12.13) imply that the limit is a Λ-coalescent with transition rates given by (12.41), where $\Lambda = F$, with F from the right-hand side of (12.12).

Obviously, there is a close relation between multiple merger events in the genealogy of the sample and "extreme" reproductive events in the population, in which a non-negligible proportion, say $x \in (0, 1]$, of the population alive in the next generation goes back to a single ancestor in the current generation. In fact, the integrand in (12.41) can be interpreted as follows: when following b lineages backwards, in such an event, each of them flips a coin with success probability x and all the successful lineages subsequently merge.

On the other hand, although individuals *can* have more than two offspring, the moment condition (12.4) ensures that families are typically small compared to the total population size and thus implies that, in the limit, only binary mergers are visible in the genealogy.

Remark 12.3 (Simultaneous multiple collisions)

It should be pointed out that Möhle and Sagitov [41] provide a complete classification of possible limits of genealogies in Cannings models, in particular if the condition (12.13) is violated. In this case, the resulting genealogies contain *simultaneous* multiple collisions, which have been studied independently and termed "Ξ-coalescents" by Schweinsberg in [46], in which several groups of lineages can merge at exactly the same time. Note that the first factor in (12.13) is the probability of observing two simultaneous mergers in one generation in a sample of size four, whereas the second factor is the inverse of the pair coalescence probability.

Since a corresponding theory of forward population models in the spirit of Section 12.2.2 is not yet completely established and our space is limited, we restrict ourselves here to the "Λ-world". $\qquad\square$

Analytic duality. Consider an F-generalized Fleming–Viot process $\{Z_t\}$ with generator (12.18) starting from $Z_0 = \mu \in \mathcal{M}_1(E)$. The idea that the type distribution in an n-sample from the population at time t can be obtained by "colouring" t-ancestral partitions independently according to Z_0 has the following explicit analytical incarnation: For bounded measurable $f : E^n \to \mathbb{R}$,

$$\mathbb{E}\left[\int_E \cdots \int_E f_{\Pi_0}(a_1, \ldots, a_{|\Pi_0|}) \, Z_t(\mathrm{d}a_1) \cdots Z_t(\mathrm{d}a_p)\right]$$
$$= \mathbb{E}\left[\int_E \cdots \int_E f_{\Pi_t}(b_1, \ldots, b_{|\Pi_t|}) \, Z_0(\mathrm{d}b_1) \cdots Z_0(\mathrm{d}b_{|\Pi_t|})\right], \quad (12.43)$$

where Π is the n-F-coalescent starting at $\pi_0 = \{\{1\}, \ldots, \{n\}\}$, and, for any partition $\pi = \{C_1, ..., C_q\}$ of $\{1, ..., n\}$,

$$f_\pi(b_1, ..., b_q) := f(a_1, ..., a_p)$$

with $a_i := b_k$ if $i \in C_k$. This is classical for the Kingman case, and was first explicitly formulated in [4] for the Λ-case. Note that specializing (12.43) in the case $F = \delta_0$ to a two-point space yields the well-known moment duality between the Wright–Fisher diffusion (12.5) and the block-counting process of Kingman's coalescent, which is a pure death process with death rate $\binom{n}{2}$.

Remark 12.4 ("Coming down from infinity")

1. Not all Λ-coalescents seem to be reasonable models for the genealogies of biological populations, since some do not allow for a finite "time to the most recent common ancestor" of the entire population (T_{MRCA}) in the sense of "coming down from infinity in finite time". This means that any initial partition in $\mathcal{P}$ (and for all $\varepsilon > 0$, the partition Π_ε) a.s. consists of finitely many blocks only. Schweinsberg [46] established the following necessary and sufficient condition: if either Λ has an atom at 0 or Λ has no atom at zero and

$$\lambda^* := \sum_{b=2}^{\infty} \left(\sum_{k=2}^{b} (k-1) \binom{b}{k} \lambda_{b,k} \right)^{-1} < \infty, \qquad (12.44)$$

where $\lambda_{b,k}$ is given by (12.41), then the corresponding coalescent does come down from infinity (and if so, the time to come down to only one block has finite expectation). Otherwise, it stays infinite for all times. For the corresponding generalized $\frac{\Lambda}{\Lambda([0,1])}$-Fleming–Viot process $\{Z_t\}$ without mutation, (12.44) means that the size of the support of Z_t becomes one in finite time – the process fixes on the type of the population's "eve".

2. An important example for a coalescent, which (only just) does not come down from infinity is the Bolthausen–Sznitman coalescent, where $\Lambda(dx) = dx$ is the uniform distribution on $[0, 1]$. This is the Beta$(2 - \alpha, \alpha)$-coalescent with $\alpha = 1$, and it plays an important role in statistical mechanics models for disordered systems (see e.g. [8] for an introduction).

3. However, it should be observed that all n-Λ-coalescents (for finite n) do have an a.s. finite T_{MRCA}.

4. Note that by Kingman's theory of exchangeable partitions, for each $t > 0$, asymptotic frequencies of the classes exist. If a Λ-coalescent does *not* come down from infinity, it may or may not be the case that these

frequencies sum to 1 ("proper frequencies"). Pitman [43] showed that the latter holds iff $\int_{0+} r^{-1} \Lambda(dr) = \infty$. Note that if $\int_{[0,1]} r^{-1} \Lambda(dr) < \infty$, we see from (12.33) that $\lim_{N \to \infty} d_N < \infty$. Hence in the lookdown construction, at each time $t \geq 0$ there is a positive fraction of levels which have not yet participated in any lookdown event. These correspond to "dust". $\qquad\qquad\square$

Examples for coalescents which satisfy (12.44) are Kingman's coalescent, the process considered in [17], corresponding to (12.16) (but note that [17] also considers $F = \delta_\psi$ with $\psi \in (0,1)$, for which (12.44) fails), and the so-called Beta$(2-\alpha, \alpha)$-coalescents with $\alpha \in (1,2)$, with $\Lambda = F$ given by (12.17). Note that, even though (12.17) makes no sense for $\alpha = 2$, Kingman's coalescent can be included in this family as the weak limit Beta$(2-\alpha, \alpha) \to \delta_0$ as $\alpha \to 2$.

Coalescents and the modified lookdown construction. We now make use of the explicit description of the modified construction to determine the coalescent process embedded in it. Fix a (probability) measure F on $[0,1]$. Recall the Poisson processes L_{ij} and L_K^l from (12.37) in Section 12.2.4 above. For each $t \geq 0$ and $k = 1, 2, \ldots$, let $N_k^t(s), 0 \leq s \leq t$, be the level at time s of the ancestor of the individual at level k at time t. In terms of the L_K^l and L_{ij}, the process $N_k^t(\cdot)$ solves, for $0 \leq s \leq t$,

$$
\begin{aligned}
N_k^t(s) = k &- \sum_{1 \leq i < j \leq k} \int_{s-}^t \mathbf{1}_{\{N_k^t(u) > j\}} \, dL_{ij}(u) \\
&- \sum_{1 \leq i < j \leq k} \int_{s-}^t (j-i) \mathbf{1}_{\{N_k^t(u) = j\}} \, dL_{ij}(u) \\
&- \sum_{K \subset \{1, \ldots, k\}} \int_{s-}^t (N_k^t(u) - \min(K)) \mathbf{1}_{\{N_k^t(u) \in K\}} \, dL_K^k(u) \\
&- \sum_{K \subset \{1, \ldots, k\}} \int_{s-}^t (|K \cap \{1, \ldots, N_k^t(u)\}| - 1) \\
&\qquad\qquad \times \mathbf{1}_{\{N_k^t(u) > \min(K),\, N_k^t(u) \notin K\}} \, dL_K^k(u).
\end{aligned}
$$

$$(12.45)$$

Fix $0 \leq T$ and, for $t \leq T$, define a partition $\Pi^T(t)$ of $\mathbb{N}$ such that k and l are in the same block of $\Pi^T(t)$ if and only if $N_k^T(T-t) = N_l^T(T-t)$. Thus, k and l are in the same block if and only if the two levels k and l

at time T have the same ancestor at time $T - t$. Then ([15], Section 5),

the process $\{\Pi_t^T : 0 \leq t \leq T\}$ is an F-coalescent run for time T.

Note that by employing a natural generalization of the lookdown construction using driving Poisson processes on $\mathbb{R}$ and e.g. using $T = 0$ above, one can use the same construction to find an F-coalescent with time set $\mathbb{R}_+$. We would like to emphasize that the lookdown construction provides a realisation-wise coupling of the type distribution process $\{Z_t\}$ and the coalescent describing the genealogy of a sample, thus extending (12.43), which is merely a statement about one-dimensional distributions.

Superimposing mutations. Consider now an F-generalized Fleming–Viot process $\{Z_t\}$ with "individual" mutation operator rB, described by the generator $\mathcal{L}_{B,F}$ given by (12.24), starting from $Z_0 = \mu$. The lookdown construction easily allows us to prove that, for each t, the distribution of a sample of size n from Z_t can be equivalently described as follows: run an n-F-coalescent for time t, interpret this as a forest with labelled leaves. "Colour" each root independently according to μ, then run the Markov process with generator rB independently along the branches of each tree, and finally read off the types at the leaves.

Remark 12.5. If (12.44) is fulfilled and the individual mutation process with generator B has a unique equilibrium, one can let $t \to \infty$ in the above argument to see that $\{Z_t\}$ has a unique equilibrium, and the distribution of an n-sample from this equilibrium can be obtained by running an n-F-coalescent until it hits the trivial partition. Then colour this most recent common ancestor randomly according to the stationary distribution of B, and run the mutation process along the branches as above.

This approach is very fruitful in population genetics applications. For example, under condition (12.44), (12.34) will be satisfied for t large enough, irrespective of the initial condition.

12.4 Population genetic inference

Populations with extreme reproductive behaviour. Recently, biologists have studied the genetic variation of certain marine species with rather extreme reproductive behaviour; see e.g. Árnason [1] (Atlantic Cod) and Boom et al. [7] (Pacific Oyster). In this situation, one would like to decide which coalescent is suitable, based upon observed genetic types in a sample from the population.

Eldon and Wakeley [17] analysed the sample described in [7] and proposed a one-parameter family of Λ-coalescents, which comprises Kingman's coalescent as a boundary case, namely those described by (12.16), as models for their genealogy. Inference is then based on a simple *summary statistic*, the number of *segregating sites* and *singleton polymorphisms*. They conclude that ([17], p. 2622):

For many species, the coalescent with multiple mergers might be a better null model than Kingman's coalescent.

In this section, we obtain recursions for the type probabilities of an n-sample from a general Λ-coalescent under a general finite alleles model. We present two approaches, one based on the lookdown construction, the other on direct manipulations with the generator $\mathcal{L}_{B,F}$. We discuss how this recursion can then be used to derive a Monte-Carlo scheme to compute likelihoods of model parameters in Λ-coalescent scenarios given the observed types, in the spirit of [25]; see also [6] for the infinite-sites case. These can be used e.g. for maximum likelihood estimation.

Remark 12.6. Analogous recursions for the probability of configurations in the infinite-alleles model have been obtained in [39]. Exact asymptotic expressions for certain summary statistics for the infinite-alleles and infinite-sites models under Beta-coalescents [recall (12.17)] have been obtained in [3]. □

12.4.1 Finite-alleles recursion I: Using the lookdown construction

Recall that in the finite-alleles model, type changes, or mutations, occur at rate r, and $P = (P_{ij})$ is an irreducible stochastic transition matrix on the finite type space E. Note that silent mutations are allowed (i.e. $P_{jj} \geq 0$), denote the unique equilibrium of P by μ. We assume that the reproduction mechanism is described by some $F = \Lambda \in \mathcal{M}_1([0,1])$.

Suppose the system, described by the lookdown construction, is in equilibrium. Consider the first n levels at time 0 and let τ_{-1} be the last instant before 0 when at least one of the types at levels $1, \ldots, n$ changes.

Then, $-\tau_{-1}$ is exponentially distributed with rate

$$r_n = nr + \sum_{k=2}^{n} \binom{n}{k} \lambda_{n,k}. \tag{12.46}$$

Denote by p the distribution of the types of the first n levels in the stationary lookdown construction, say, at time 0. Later, due to exchangeability, we will merely be interested in the type frequency probability $p^0(\mathbf{n})$. Decomposing according to which event occurred at time τ_{-1}, we obtain

$$p\big((y_1,\ldots,y_n)\big)$$

$$= \frac{r}{r_n} \sum_{i=1}^{n} \sum_{z \in E} p\big((y_1,\ldots,y_{i-1},z,y_{i+1},\ldots,y_n)\big) P_{zy_i}$$

$$+ \frac{1}{r_n} \sum_{\substack{K \subset \{1,\ldots,n\} \\ |K| \geq 2}} \lambda_{n,|K|} \mathbf{1}_{\{\text{all } y_j \text{ equal for } j \ \in \ K\}} p\big(\gamma_K(y_1,\ldots,y_n)\big),$$

$$\tag{12.47}$$

where $\gamma_K(y_1,\ldots,y_n) \in E^{n-|K|+1}$ is that vector of types of length $n - |K| + 1$ which $\big(\xi_1(\tau_{-1}-),\ldots,\xi_{n-|K|+1}(\tau_{-1}-)\big)$ must be in order that a resampling event involving exactly the levels in K among levels $1,\ldots,n$ generates $\big(\xi_1(\tau_{-1}),\ldots,\xi_n(\tau_{-1})\big) = (y_1,\ldots,y_n)$. Formally,

$$\gamma_K(y_1,\ldots,y_n)_i = y_{i+\#((K\setminus\{\min K\})\cap\{1,\ldots,i\})}, \ 1 \leq i \leq n - |K| + 1.$$

As the type at level 1 is the stationary Markov process with generator rB, we have the boundary condition $p\big((y_1)\big) = \mu(y_1)$, $y_1 \in E$. Note that, by exchangeability,

$$p\big((y_1,\ldots,y_n)\big) = p\big((y_{\pi(1)},\ldots,y_{\pi(n)})\big)$$

for any permutation π of $\{1,\ldots,n\}$. So, the only relevant information is (of course) how many samples were of which type. For $\mathbf{n} = (n_1,\ldots,n_d) \in \mathbb{Z}_+^d$ we write $\#\mathbf{n} := n_1 + \cdots + n_d$ for the 'length', and

$$\kappa(\mathbf{n}) = \big(\underbrace{1,1,\ldots,1}_{n_1},\underbrace{2,\ldots,2}_{n_2},\ldots,\underbrace{d,\ldots,d}_{n_d}\big) \in E^{\#\mathbf{n}}$$

for a 'canonical representative' of the (absolute) type frequency vector $\mathbf{n}$. Let

$$p^0(\mathbf{n}) := \binom{\#\mathbf{n}}{n_1,n_2,\ldots,n_d} p\big(\kappa(\mathbf{n})\big) \tag{12.48}$$

be the probability that, in a sample of size $\#\mathbf{n}$, there are exactly n_j of type j, $j = 1, \ldots, d$. We abbreviate $n := \#\mathbf{n}$, and write e_k for the kth canonical unit vector of $\mathbb{Z}^d$. Noting that

$$n_j \binom{\#\mathbf{n}}{n_1, n_2, \ldots, n_d} p(\mathbf{n} - \mathbf{e}_j + \mathbf{e}_i) = (n_i + 1 - \delta_{ij}) p^0(\mathbf{n} - \mathbf{e}_j + \mathbf{e}_i)$$

and that (for $n_j \geq k$, otherwise the term is 0)

$$\binom{n_j}{k} \binom{\#\mathbf{n}}{n_1, n_2, \ldots, n_d} p(\mathbf{n} - (k-1)\mathbf{e}_j)$$

$$= \binom{n}{k} \frac{n_j - k + 1}{n - k + 1} p^0(\mathbf{n} - (k-1)\mathbf{e}_j),$$

(12.47) translates into the following recursion for p^0:

$$p^0(\mathbf{n}) = \frac{r}{r_n} \sum_{j=1}^{d} \sum_{i=1}^{d} (n_i + 1 - \delta_{ij}) P_{ij} p^0(\mathbf{n} - \mathbf{e}_j + \mathbf{e}_i)$$

$$+ \frac{1}{r_n} \sum_{\substack{j=1 \\ n_j \geq 2}}^{d} \sum_{k=2}^{n_j} \binom{n}{k} \lambda_{n,k} \frac{n_j - k + 1}{n - k + 1} p^0(\mathbf{n} - (k-1)\mathbf{e}_j) \quad (12.49)$$

with boundary conditions $p^0(\mathbf{e}_j) = \mu_j$.

Remark 12.7. In the Kingman case, we have $\lambda_{n,k} = \mathbf{1}(n \geq 2 = k)$, $r_n = n\theta/2 + n(n-1)/2 = n(n-1+\theta)/2$ (and we assume $r = \theta/2$ as "usual"), hence (12.49) becomes the well-known

$$p^0(\mathbf{n}) = \frac{\theta}{n-1+\theta} \sum_{j=1}^{d} \sum_{i=1}^{d} \frac{n_i + 1 - \delta_{ij}}{n} P_{ij} p^0(\mathbf{n} - \mathbf{e}_j + \mathbf{e}_i)$$

$$+ \frac{n-1}{n-1+\theta} \sum_{\substack{j=1 \\ n_j \geq 2}}^{d} \frac{n_j - 1}{n-1} p^0(\mathbf{n} - \mathbf{e}_j). \quad (12.50)$$

12.4.2 Finite-alleles recursion II: Generator approach

An alternative method to obtain the recursion for the type probabilities in the finite-alleles case is by using a generator approach; see [12]. Let $f \in C_2$ and $\Delta_d = \{(x_1, \ldots, x_d) : x_i \geq 0, x_1 + \cdots + x_d = 1\}$ and consider the mutation operator

$$\widetilde{B} f(x_1, \ldots, x_d) = r \sum_{i=1}^{d} \left(\sum_{j=1}^{d} x_j P_{ji} - x_i P_{ij} \right) \frac{\partial f}{\partial x_i}(x_1, \ldots, x_d).$$

For the resampling operator, we distinguish the Kingman and non-Kingman components. First, assume $\Lambda(\{0\}) = 0$ (non-Kingman). Consider

$$R_1 f(x_1, \ldots, x_d) = \sum_{i=1}^{d} \int x_i \Big(f\big(\bar{r}x_1, \ldots, \bar{r}x_{i-1}, \bar{r}x_i + r, \bar{r}x_{i+1}, \ldots, \bar{r}x_d\big)$$
$$- f(x_1, \ldots, x_d)\Big) r^{-2} \Lambda(dr),$$

(12.51)

where $\bar{r} = 1 - r$. For the Kingman part ($\Lambda = \delta_0$) of the resampling operator, we have

$$R_2 f(x_1, \ldots, x_d) = \frac{1}{2} \sum_{i,j=1}^{d} x_i (\delta_{ij} - x_j) \frac{\partial^2 f}{\partial x_i \partial x_j}(x_1, \ldots, x_d).$$

Finally, for general Λ and $a \geq 0$, write $R = R_1 + aR_2$, where R_1 uses $\Lambda_0 = \Lambda - \Lambda(\{0\})\delta_0$. Now, let $X(t) = (X_1(t), \ldots, X_d(t))$ be the stationary process with generator $L = \tilde{B} + R$ [note that $X_i(t) = Z_t(\{i\})$, where $\{Z_t\}$ is the stationary process with generator (12.24)]. Write $X = X(0)$. Let $\mathbf{n} = (n_1, \ldots, n_d)$, $n = n_1 + \cdots + n_d$. Then,

$$\mathbb{E}\left[\prod_{i=1}^{d} X_i^{n_i} \right]$$

is the probability of observing in a sample of size n from the equilibrium population type i precisely n_i times in a particular order (e.g. first n_1 samples of type 1, next n_2 samples of type 2, etc.). Put

$$f_{\mathbf{n}}(\mathbf{x}) := \mathbf{x}^{\mathbf{n}} := \prod_{i=1}^{d} x_i^{n_i}.$$

Then,

$$g(\mathbf{n}) := \binom{n}{n_1 \ldots n_d} \mathbb{E}[f_{\mathbf{n}}(X)]$$

is the probability of observing type i exactly n_i times, $i = 1, \ldots, d$, regardless of the order. Note that

$$\tilde{B} f_{\mathbf{n}}(x_1, \ldots, x_d) = r \sum_{i=1}^{d} \Big(\sum_{j=1}^{d} x_j P_{ji} - x_i P_{ij} \Big) n_i f_{\mathbf{n}-\mathbf{e}_i}(x_1, \ldots, x_d)$$

$$= r \sum_{i,j=1}^{d} n_i P_{ji} f_{\mathbf{n}-\mathbf{e}_i+\mathbf{e}_j}(\mathbf{x}) - rn f_{\mathbf{n}}(\mathbf{x})$$

and

$$f_{\mathbf{n}}((1-r)\mathbf{x} + r\mathbf{e_i}) = (1-r)^{n-n_i} \prod_{j\neq i}^{d} x_j^{n_j} \times ((1-r)x_i + r)^{n_i}$$

$$= (1-r)^{n-n_i} \prod_{j\neq i}^{d} x_j^{n_j} \times \sum_{k=0}^{n_i} \binom{n_i}{k} r^k (1-r)^{n_i-k} x_i^{n_i-k}$$

$$= \sum_{k=0}^{n_i} \binom{n_i}{k} r^k (1-r)^{n-k} \left(x_i^{n_i-k} \prod_{j\neq i}^{d} x_j^{n_j} \right),$$

so the term inside the integral in the expression (12.51) for R_1 can be written as

$$\sum_{i=1}^{d} \sum_{k=0}^{n_i} \binom{n_i}{k} r^k (1-r)^{n-k} x_i^{n_i-k+1} \prod_{j\neq i}^{d} x_j^{n_j} - \sum_{k=0}^{n} \binom{n}{k} r^k (1-r)^{n-k} \prod_{\ell=1}^{d} x_\ell^{n_\ell}$$

$$= \sum_{i:n_i\geq 2} \sum_{k=2}^{n_i} \binom{n_i}{k} r^k (1-r)^{n-k} x_i^{n_i-k+1} \prod_{j\neq i}^{d} x_j^{n_j}$$

$$- \sum_{k=2}^{n} \binom{n}{k} r^k (1-r)^{n-k} \prod_{\ell=1}^{d} x_\ell^{n_\ell},$$

observing that the terms with $k = 0$ and $k = 1$ cancel since $x_1 + \cdots + x_d = 1$ and $n_1 + \cdots + n_d = n$. Recalling the definition of $\lambda_{n,k}$ from (12.41), we obtain

$$R_1 f_{\mathbf{n}}(\mathbf{x}) = \sum_{i:n_i\geq 2} \sum_{k=2}^{n_i} \binom{n_i}{k} \lambda_{n,k} f_{\mathbf{n}-(k-1)\mathbf{e_i}}(\mathbf{x}) - \sum_{k=2}^{n} \binom{n}{k} \lambda_{n,k} f_{\mathbf{n}}(\mathbf{x}).$$

$$(12.52)$$

Furthermore

$$R_2 f_{\mathbf{n}}(\mathbf{x}) = \frac{1}{2} \sum_{i,j=1}^{d} x_i (\delta_{ij} - x_j) n_i (n_j - \delta_{ij}) f_{\mathbf{n}-\mathbf{e_i}-\mathbf{e_j}}(\mathbf{x})$$

$$= \sum_{i=1}^{d} \frac{n_i(n_i-1)}{2} f_{\mathbf{n}-\mathbf{e_i}}(\mathbf{x}) - \sum_{i,j=1}^{d} \frac{n_i(n_j - \delta_{ij})}{2} f_{\mathbf{n}}(\mathbf{x})$$

$$= \sum_{i=1}^{d} \frac{n_i(n_i-1)}{2} f_{\mathbf{n}-\mathbf{e_i}}(\mathbf{x}) - \frac{n(n-1)}{2} f_{\mathbf{n}}(\mathbf{x}). \qquad (12.53)$$

Combining the terms from R_1 and R_2 (using (12.52) and (12.53) above, and replacing Λ by Λ_0 in (12.51)), we have

$$Rf_{\mathbf{n}}(\mathbf{x}) = \sum_{i:n_i \geq 2} \sum_{k=2}^{n_i} \binom{n_i}{k} \lambda_{n,k} f_{\mathbf{n}-(k-1)\mathbf{e}_i}(\mathbf{x}) - \sum_{k=2}^{n} \binom{n}{k} \lambda_{n,k} f_{\mathbf{n}}(\mathbf{x}).$$

Thus we obtain from the stationarity condition $\mathbb{E}\,Lf_{\mathbf{n}}(X) = 0$ that

$$r_n \mathbb{E}f_{\mathbf{n}}(X) = r \sum_{i,j=1}^{d} n_i P_{ji} \mathbb{E}f_{\mathbf{n}-\mathbf{e}_i+\mathbf{e}_j}(X)$$

$$+ \sum_{i:n_i \geq 2} \sum_{k=2}^{n_i} \binom{n_i}{k} \lambda_{n,k} \mathbb{E}f_{\mathbf{n}-(k-1)\mathbf{e}_i}(X),$$

where r_n is defined in (12.46). Multiplying with $\binom{n}{n_1 \ldots n_d}/r_n$ and some algebra gives

$$g(\mathbf{n}) = \frac{r}{r_n} \sum_{i,j=1}^{d} (n_j + 1 - \delta_{ij}) P_{ji} g(\mathbf{n} - \mathbf{e}_i + \mathbf{e}_j)$$

$$+ \frac{1}{r_n} \sum_{i:n_i \geq 2} \sum_{k=2}^{n_i} \binom{n}{k} \lambda_{n,k} \frac{n_i - k + 1}{n - k + 1} g(\mathbf{n} - (k-1)\mathbf{e}_i),$$

which agrees with (12.49).

12.4.3 A Monte Carlo scheme for sampling probabilities

Recursion (12.49) can be used to estimate $p^0(\mathbf{n})$ for a given $\mathbf{n} \in \mathbb{Z}_+^d$ using a Markov chain, in the spirit of [25], as follows:

Let $\{X_k\}$ be a Markov chain on $\mathbb{Z}_+^d$ with transitions

$$\mathbf{n} \to \begin{cases} \mathbf{n} - \mathbf{e}_j + \mathbf{e}_i & \text{w. p.} \quad \frac{r}{r_n f(\mathbf{n})}(n_i + 1 - \delta_{ij}) P_{ij} & \text{if } n_j > 0, \\ \mathbf{n} - (k-1)\mathbf{e}_i & \text{w. p.} \quad \frac{1}{r_n f(\mathbf{t},\mathbf{n})} \binom{n}{k} \lambda_{n,k} \frac{n_i - k + 1}{n - k + 1} & \text{if } 2 \leq k \leq n_i, \end{cases}$$

where (with r_n as defined in (12.46))

$$f(\mathbf{n}) = \frac{1}{r_n} \left(\sum_{\substack{i,j=1 \\ n_j > 0}}^{d} r(n_i + 1 - \delta_{ij}) P_{ij} + \sum_{\substack{1 \leq i \leq d \\ n_i \geq 2}} \sum_{k=2}^{n_i} \binom{n}{k} \lambda_{n,k} \frac{n_i - k + 1}{n - k + 1} \right).$$

$$(12.54)$$

Then,

$$p^0(\mathbf{n}) = \mathbb{E}_{(\mathbf{n})} \prod_{l=0}^{\tau} f(\mathbf{t}(l), \mathbf{n}(l)). \tag{12.55}$$

Remark 12.8 (Inference for Kingman's coalescent)

Likelihood-based inference methods for Kingman's coalescent, some solving recursion (12.50) approximately via Monte–Carlo methods, others using MCMC, have been developed since the beginning of the 1990s; see [18], [24], [25], [26], [28], [29], [23], [12], [49]. In [49], Stephens and Donnelly provide proposal distributions for importance sampling, which are optimal in some sense, and compare them to various other methods. Their importance sampling scheme seems, at present, to be the most efficient tool for inference for relatively large datasets, but heavily uses the fact that Kingman's coalescent allows only binary mergers. It is at present unclear what an analogous strategy in the general Λ-case ought to be. □

12.4.4 Simulating samples

Let $E, (P_{ij}), \mu, r$ be the parameters of a finite-alleles model. Then, one may obtain the type configuration in an n-sample as follows:

Let $\{Y_t^{(n)}\}_{t\geq 0}$ be the *block counting process* corresponding to an n-Λ-coalescent, i.e. $Y_t^{(n)} = \#\{\text{blocks of } \Pi_t\}$ is a continuous-time Markov chain on $\mathbb{N}$ with jump rates

$$q_{ij} = \binom{i}{i-j+1} \lambda_{i,i-j+1}, \quad i > j \geq 1$$

starting from $Y_0^{(n)} = n$. Its Green function is

$$g(n,m) := \mathbb{E}\left[\int_0^\infty \mathbf{1}_{\{Y_s^{(n)}=m\}} \, ds \right] \quad \text{for } n \geq m \geq 2, \tag{12.56}$$

which can easily be computed recursively; see [6], Section 7.1. Denoting by $\tau := \inf\{t : Y_t^{(n)} = 1\}$ the time required to come down to only one class and by ∂ a "cemetery state", it follows from Nagasawa's Formula (see e.g. [44], (42.4)) that the time-reversed path

$$\tilde{Y}_t^{(n)} := \begin{cases} Y_{(\tau-t)-}^{(n)} & 0 \leq t < \tau, \\ \partial & \tau \leq t, \end{cases} \tag{12.57}$$

is a continuous-time Markov chain on $\{2, \ldots, n\} \cup \{\partial\}$ with jump rate matrix

$$\tilde{q}_{ji}^{(n)} = \frac{g(n,i)}{g(n,j)} q_{ij}, \quad j < i \leq n,$$

$$-\tilde{q}_{jj}^{(n)} = \sum_{i=j+1}^{n} \tilde{q}_{ji}^{(n)} = \sum_{\ell=1}^{j-1} q_{j\ell}, \qquad \tilde{q}_{n\partial}^{(n)} = -q_{nn}$$

and initial distribution $\mathbb{P}\{\tilde{Y}_0^{(n)} = k\} = g(n,k)q_{k1}$, $k = 2, 3, \ldots, n$. Note that, unless Λ is concentrated on $\{0\}$, the dynamics does depend on n. We write $\tilde{p}_{ji}^{(n)} := \tilde{q}_{ji}^{(n)}/(-\tilde{q}_{jj}^{(n)})$, $j < i \leq n$ for the transition matrix of the skeleton chain of $Y^{(n)}$.

In view of Remark 12.5 it is clear that the following algorithm generates an n-sample from the stationary distribution of the process with generator $\mathcal{L}_{B,F}$ given by (12.24):

Algorithm (generating samples)

(i) Generate K with $\mathbb{P}\{K = k\} = g(n,k)q_{k1}$, $k = 2, \ldots, n$ begin with $\eta = K\delta_X$, where $X \sim \mu$.

(ii) Draw $U \sim \text{Unif}([0,1])$.

If $U \leq \frac{kr}{kr + (-\tilde{q}_{kk}^{(n)})}$:
Replace one of the present types by a P-step from it, i.e. replace $\eta := \eta - \delta_x + \delta_y$ with probability $\frac{\eta_x}{\#\eta} P_{xy}$ (for $x \neq y$), where $\#\eta$ is the total mass of η.
Otherwise:
If $\#\eta = n$: Output η and stop.
Else, pick $J \in \{\#\eta, \ldots, n\}$ with $\mathbb{P}\{J = j\} = \tilde{p}_{\#\eta,j}^{(n)}$. Choose one of the present types (according to their present frequency), and add $J - \#\eta$ copies of this type, i.e. replace $\eta := \eta + (J - \#\eta)\delta_x$ with probability $\frac{\eta_x}{\#\eta}$.

(iii) Repeat (ii).

Remark 12.9

Ordered samples can be obtained from a realization of η by random re-ordering. In the case of parent-independent mutation, i.e. if $P_{ij} = P_j$ for all i, j, it is possible to simplify the procedure by simulating "backwards in time". "Active" ancestral lineages are lost either by (possibly multiple) coalescence or when hitting their "defining" mutation, in which case one simply assigns a random type drawn according to P_j.   □

Bibliography

[1] Árnason, E.: Mitochondrial cytochrome b DNA variation in the high-fecundity Atlantic Cod: trans-Atlantic clines and shallow gene genealogy, *Genetics* **166**, 1871–1885 (2004).

[2] Berestycki, N.; Berestycki, J.; Schweinsberg, J.: Small time behaviour of Beta-coalescents. *Ann. Inst. H. Poincaré Probab. Statist.* **44**, 214–138 (2008).

[3] Berestycki, N.; Berestycki, J.; Schweinsberg, J.: Beta-coalescents and continuous stable random trees. *Ann. Probab.* **35**, 1835–1887 (2007).

[4] Bertoin, J.; Le Gall, J.-F.: Bertoin, J.; Le Gall, J.-F.: Stochastic flows associated to coalescent processes. *Probab. Theor. Rel. Fields* **126**, no. 2, 261–288 (2003).

[5] Birkner, M.; Blath, J.; Capaldo, M.; Etheridge, A.; Möhle, M.; Schweinsberg, J.; Wakolbinger, A.: Alpha-stable branching and Beta-coalescents. *Electron. J. Probab.* **10**, 303–325 (2005).

[6] Birkner, M; Blath, J: Computing likelihoods for coalescents with multiple collisions in the infinitely-many-sites model. *J. Math. Biol.* **57**, 435–465 (2008).

[7] Boom, J.D.G.; Boulding, E. G.; Beckenbach, A. T.: Mitochondrial DNA variation in introduced populations of Pacific oyster, Crassostrea gigas, in British Columbia. *Can. J. Fish. Aquat. Sci.* 51:16081614 (1994).

[8] Bovier, A.: *Statistical Mechanics of Disordered System. A Mathematical Perspective.* Cambridge University Press (2006).

[9] Cannings, C.: The latent roots of certain Markov chains arising in genetics: a new approach, I. Haploid models. *Adv. Appl. Prob.* **6**, 260–290 (1974).

[10] Cannings, C.: The latent roots of certain Markov chains arising in genetics: a new approach, II. Further haploid models. *Adv. Appl. Prob.* **7**, 264–282 (1975).

[11] Dawson, D.: *Lecture Notes in Mathematics* **1541**, Ecole d'Eté de Probabilités de Saint-Flour XXI, Berlin: Springer (1993).

[12] De Iorio, M. and Griffiths, R. C.: Importance sampling on coalescent histories I. *Adv. Appl. Prob.* **36**, 417-433, (2004).

[13] De Iorio, M. and Griffiths, R. C.: Importance sampling on coalescent histories II: Subdivided population models. *Adv. Appl. Prob.* **36**, 434-454 (2004).

[14] Donnelly, P.; Kurtz, T.: A countable representation of the Fleming–Viot measure–valued diffusion. *Ann. Probab.* **24**, no. 2, 698–742 (1996).

[15] Donnelly, P.; Kurtz, T.: Particle representations for measure–valued population models. *Ann. Probab.* **27**, no. 1, 166–205 (1999).

[16] Durrett, R.; Schweinsberg, J.: A coalescent model for the effect of advantageous mutations on the genealogy of a population. *Stoch. Proc. Appl.* **115**, 1628–1657 (2005)

[17] Eldon, B.; Wakeley, J.: Coalescent processes when the distribution of off-spring number among individuals is highly skewed. *Genetics* **172**, 2621–2633 (2006).

[18] Ethier, S.; Griffiths, R.C.: The infinitely-many-sites model as a measure-valued diffusion. *Ann. Probab.* **15**, no. 2, 515–545 (1987).

[19] Ethier, S.; Kurtz, T.: *Markov Processes: Characterization and Convergence.* Wiley (1986).

[20] Ethier, S.; Kurtz, T.: Fleming-Viot processes in population genetics. *SIAM J. Control Optim.* **31**, no. 2, 345–386 (1993).

[21] Ewens, W.: *Mathematical Population Genetics. I. Theoretical Introduction.* Second edition. Springer (2004)

[22] Fisher, R.A.: On the dominance ratio. *Proc. R. Soc. Edin.*, **42**, 321–431 (1922).

[23] Felsenstein, J., Kuhner, M.K., Yamato, J. and Beerli, P.: Likelihoods on coalescents: a Monte Carlo sampling approach to inferring parameters from population samples of molecular data. *IMS Lecture Notes* Monograph Series **33**, 163–185 (1999).

[24] Griffiths, R.C. and Tavaré, S.: Simulating probability distributions in the coalescent. *Theor. Pop. Biol.* **46**, 131–159 (1994).

[25] Griffiths, R.C. and Tavaré, S.: Ancestral Inference in population genetics. *Stat. Sci.* **9**, 307–319 (1994).

[26] Griffiths, R.C. and Tavaré, S.: Sampling theory for neutral alleles in a varying environment. *Phil. Trans. R. Soc. Lon.*, Series B, **344**, 403–410 (1994).

[27] Griffiths, R.C. and Tavaré, S.: Unrooted genealogical tree probabilities in the infinitely-many-sites model. *Math. Biosci.* **127**, 77–98 (1995).

[28] Griffiths, R.C. and Tavaré, S.: Markov chain inference methods in population genetics. *Math. Comput. Modeling* **23**, no. 8/9, 141–158 (1996).

[29] Griffiths, R.C. and Tavaré, S.: Computational methods for the coalescent. *Progress in Population Genetics and Human Evolution*, 165–182, Berlin: Springer (1997).

[30] Griffiths, R.C.; Tavaré, S.: The age of a mutation in a general coalescent tree. *Comm. Statist. Stoch. Models* **14**, 273–29 (1998).

[31] Griffiths, R.C.; Tavaré, S.: The ages of mutations in gene trees. *Ann. Appl. Probab.* **9**, no. 3, 567–590 (1999).

[32] Hudson, R.R.: Gene genealogies and the coalescent process. *Oxford Surv. Evol. Bio.*, **7**, 1–44 (1990).

[33] Hein, J.; Schierup, M.H.; Wiuf, C.: *Gene Genealogies, Variation and Evolution – A Primer in Coalescent Theory.* Oxford University Press (2005).

[34] Kingman, J.F.C.: The coalescent. *Stoch. Proc. Appl.* **13**, 235–248 (1982).

[35] Kuhner, M.K., Yamato, J. and Felsenstein, J.: Estimating effective population size and mutation rate from sequence data using Metropolis-Hastings sampling. *Genetics* **140**, 1421–1430 (1995).

[36] Kuhner, M.K., Yamato, J. and Felsenstein, J.: Maximum likelihood estimation of population growth rates based on the coalescent. *Genetics* **149**, 429–434 (1998).

[37] Möhle, M.: Simulation algorithms for integrals of a class of sampling distributions arising in population genetics. *J. Stat. Comp. Simul.* **75**, 731–749 (2005).

[38] Möhle, M.: On the number of segregating sites for populations with large family sizes. *J. Appl. Prob.* **38**, 750–767 (2006).

[39] Möhle, M.: On sampling distributions for coalescent processes with simultaneous multiple collisions. *Bernoulli* **12**, 35–53 (2006).

[40] Möhle, M.: On a class of non-regenerative sampling distributions. *Combin. Probab. Comput.* **16**, 435–444 (2007)

[41] Möhle, M.; Sagitov, S.: A classification of coalescent processes for haploid exchangeable population models. *Ann. Probab.* **29**, 1547–1562 (2001).

[42] Nordborg, M.: Coalescent theory. In D.J. Balding, M. Bishop, and C. Cannings, eds. *Handbook of Statistical Genetics*, 179–208. Wiley (2001).

[43] Pitman, J.: Coalescents with multiple collisions. *Ann. Probab.* **27**, 1870–1902 (1999).

[44] Rogers, L.C.G.; Williams, D.: *Diffusions, Markov Processes and Martingales*, Vol. 1, 2nd edn. Chichester: Wiley (1994)

[45] Sagitov, S.: The general coalescent with asynchronous mergers of ancestral lines. *J. Appl. Probab.* **36** 1116–1125 (1999).

[46] Schweinsberg, J.: A necessary and sufficient condition for the Λ-coalescent to come down from infinity. *Electron. Comm. Probab.* **5**, 1–11 (2000).

[47] Schweinsberg, J.: Coalescents with simultaneous multiple collisions. *Electron. J. Probab.* **5**, paper no. 12, 50 pp. (2000).

[48] Schweinsberg, J.: Coalescent processes obtained from supercritical Galton–Watson processes. *Stoch. Proc. Appl.* **106**, 107–139 (2003).

[49] Stephens, M.; Donnelly, P.: Inference in molecular population genetics. *J. R. Stat. Soc. B.* **62**, 605–655 (2000).

[50] Tavaré, S.: *Ancestral Inference in Population Genetics*. Lecture Notes in Mathematics **1837**, Berlin: Springer (2001).

[51] Wakeley, J.: *Coalescent Theory: An Introduction*. Robert & Company Publishers, Greenwood Village, Colorado (2008).

[52] Wright, S.: Evolution in Mendelian populations. *Genetics* **16**, 97–159 (1931).

13

How often does the ratchet click? Facts, heuristics, asymptotics

Alison M. Etheridge

University of Oxford
Department of Statistics
1, South Parks Road
Oxford OX1 3TG, UK

Peter Pfaffelhuber

Department of Biology II
University of Munich
Großhaderner Straße 2
82152 Planegg-Martinsried, Germany

Anton Wakolbinger

J.W. Goethe-Universität
Fb. Informatik und Mathematik
60054 Frankfurt am Main, Germany

Abstract

The evolutionary force of recombination is lacking in asexually reproducing populations. As a consequence, the population can suffer an irreversible accumulation of deleterious mutations, a phenomenon known as *Muller's ratchet*. We formulate discrete and continuous time versions of Muller's ratchet. Inspired by Haigh's (1978) analysis of a dynamical system which arises in the limit of large populations, we identify the parameter $\gamma = N\lambda/(Ns \cdot \log(N\lambda))$ as most important for the speed of accumulation of deleterious mutations. Here N is population size, s is the selection coefficient and λ is the deleterious mutation rate. For large parts of the parameter range, measuring time in units of size N, deleterious mutations accumulate according to a power law in $N\lambda$ with exponent γ if $\gamma \geq 0.5$. For $\gamma < 0.5$ mutations cannot accumulate. We obtain diffusion approximations for three different parameter regimes, depending on the speed of the ratchet. Our approximations shed new light on analyses of Stephan et al. (1993) and Gordo and Charlesworth (2000). The heuristics leading to the approximations are supported by simulations.

13.1 Introduction

Muller's ratchet is a mechanism that has been suggested as an explanation for the evolution of sex [13]. The idea is simple; in an asexually reproducing population chromosomes are passed down as indivisible blocks and so the number of deleterious mutations accumulated along any ancestral line in the population can only increase. When everyone in the current 'best' class has accumulated at least one additional bad mutation then the minimum mutational load in the population increases: the ratchet clicks. In a sexually reproducing population this is no longer the case; because of recombination between parental genomes a parent carrying a high mutational load can have offspring with *fewer* deleterious mutations. The high cost of sexual reproduction is thus offset by the benefits of inhibiting the ratchet. Equally the ratchet provides a possible explanation for the degeneration of Y chromosomes in sexual organisms (see e.g. [2], [4], [5], [3], [15]).

However, in order to assess its real biological importance one should establish under what circumstances Muller's ratchet will have an evolutionary effect. In particular, *how many generations will it take for an asexually reproducing population to lose its current best class?* In other words, what is the rate of the ratchet?

In spite of the substantial literature devoted to the ratchet (see [11] for an extensive bibliography), even in the simplest mathematical models a closed form expression for the rate remains elusive. Instead various approximations have been proposed which fit well with simulations for particular parameter regimes. The analysis presented here unifies these approximations into a single framework and provides a more detailed mathematical understanding of their regions of validity.

The simplest mathematical model of the ratchet was formulated in the pioneering work of Haigh [8]. Consider an asexual population of constant size N. The population evolves according to classical Wright–Fisher dynamics. Thus each of the N individuals in the $(t + 1)$th generation independently chooses a parent from the individuals in the tth generation. The probability that an individual which has accumulated k mutations is selected is proportional to its *relative* fitness, $(1 - s)^k$. The number of mutations carried by the offspring is then $k + J$ where J is an (independent) Poisson random variable with mean λ.

Haigh identifies $n_0 = Ne^{-\lambda/s}$ as an approximation (at large times) for

Trends in Stochastic Analysis, ed. J. Blath, P. Mörters and M. Scheutzow.
Published by Cambridge University Press. ©Cambridge University Press 2008.

the total number of individuals carrying the least number of mutations and finds numerical evidence of a linear relationship between n_0 and the average time between clicks of the ratchet, at least for 'intermediate' values of n_0 (which he quantifies as $n_0 > 1$ and less than 25, say). On the other hand, for increasing values of n_0 Stephan et al. [16] note the increasing importance of s for the rate of the ratchet. The simulations of Gordo and Charlesworth [7] also suggest that for n_0 fixed at a large value the ratchet can run at very different rates. They focus on parameter ranges that may be the most relevant to the problem of the degeneration of large non-recombining portions of chromosomes.

In our approach we use diffusion approximations to identify another parameter as being an important factor in determining the rate of the ratchet. We define

$$\gamma := \frac{N\lambda}{Ns\log(N\lambda)}. \tag{13.1}$$

Notice that $n_0 = N(N\lambda)^{-\gamma}$. In these parameters one can reinterpret Haigh's empirical results as saying that if we measure time in units of size N then the rate of the ratchet follows a power law in $N\lambda$. In fact our main observation in this note is that for a substantial portion of parameter space (which we shall quantify a little more precisely later) we have the following

Rule of thumb. *The rate of the ratchet is of the order $N^{\gamma-1}\lambda^{\gamma}$ for $\gamma \in (1/2, 1)$, whereas it is exponentially slow in $(N\lambda)^{1-\gamma}$ for $\gamma < 1/2$.*

There are two novelties here: first, the abrupt change in behaviour at $\gamma = 1/2$ and second the power-law interpretation of the rate for $\gamma \in (1/2, 1)$. As an appetiser, Figure 13.1 illustrates that this behaviour really is reflected in simulated data; see also Section 13.6.

The rule of thumb breaks down for two scenarios: first, if $\gamma > 1$ then for large $N\lambda$ we have $n_0 < 1$ and so our arguments, which are based on diffusion approximations for the size of the best class, will break down. This parameter regime, which leads to very frequent clicks of the ratchet, was studied by Gessler [6]. Second, if $N\lambda$ is *too* large then we see the transition from exponentially rare clicks to frequent clicks takes place at larger values of γ.

The rest of this paper is laid out as follows. In Section 13.2 we review the work of Haigh [8]. Whereas Haigh's work focuses on discrete time dynamics, in Section 13.3 we write down instead a (continuous–time)

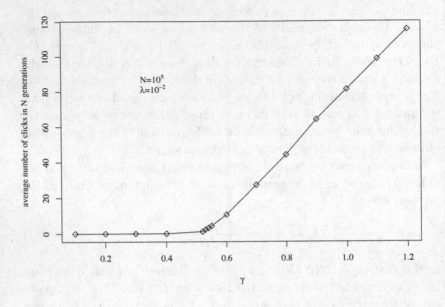

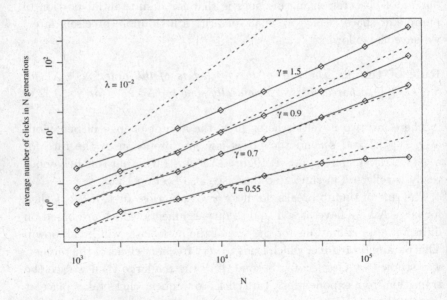

Fig. 13.1. (A) We plot the rate of the ratchet against γ, where time is measured in units of N generations. As predicted by our rule of thumb we see a sharp change of behaviour around $\gamma = 0.5$. (B) We see the power-law behaviour of the rate for various values of γ. The solid lines are given by simulation of a Wright–Fisher model. The dashed lines fit the prediction that the time between clicks is a constant times $N(N\lambda)^{-\gamma}$ for $\gamma > 0.5$. Note that this breaks down for $\gamma > 1$.

Fleming–Viot diffusion approximation to the model whose behaviour captures the dynamics of large populations when s and λ are small. We then pass in Section 13.4 to the infinite population (deterministic) limit. This system can be solved exactly and by, in the spirit of Haigh, using this deterministic system to estimate the behaviour of the 'bulk' of the population we obtain in Section 13.5 a one-dimensional diffusion which approximates the size of the best class in our Fleming–Viot system. The drift in this one-dimensional diffusion will take one of three forms depending upon whether the ratchet is clicking rarely, at a moderate rate or frequently per N generations (but always rarely per generation). Performing a scaling of this diffusion allows us to predict the relationship between the parameters $N\lambda$ and Ns of the biological model and the value of γ at which we can expect to see the phase transition from rare clicking to power-law behaviour of the rate of the ratchet. In Section 13.6 we compare our predictions to simulated data, and in Section 13.7 we discuss the connection between our findings and previous work of Stephan et al. [16], Stephan and Kim [17] and Gordo and Charlesworth [7].

13.2 The discrete ratchet – Haigh's approach

The population dynamics described in the introduction can be reformulated mathematically as follows. Let N be a fixed natural number (the population size), $\lambda > 0$ (the mutation parameter) and $s \in (0, 1)$ (the selection coefficient). The population is described by a stochastic process taking values in $\mathcal{P}(\mathbb{N}_0)$, the simplex of probability weights on $\mathbb{N}_0$. Suppose that $\mathbf{x}(t) = (x_k(t))_{k=0,1,\dots} \in \mathcal{P}(\mathbb{N}_0)$ is the vector of *type frequencies* (or *frequency profile*) in the tth generation (so for example $Nx_k(t)$ individuals in the population carry exactly k mutations). Let H be an $\mathbb{N}_0$-valued random variable with $\mathbf{P}[H = k]$ proportional to $(1 - s)^k x_k(t)$, let J be a Poisson(λ)-random variable independent of H and let $K_1, \dots, K_N$ be independent copies of $H + J$. Then the random type frequencies in the next generation are

$$X_k(t+1) = \frac{1}{N}\#\{i : K_i = k\}. \tag{13.2}$$

We shall refer to this as the *ratchet dynamics in discrete time*.

First consider what happens as $N \to \infty$. By the law of large numbers, (13.2) results in the deterministic dynamics

$$\mathbf{x}(t+1) := \mathbf{E}_{\mathbf{x}(t)}[\mathbf{X}(t+1)]. \tag{13.3}$$

An important property of the dynamics (13.3) is that vectors of Poisson weights are mapped to vectors of Poisson weights. To see this, note that when $N \to \infty$ the right-hand side of (13.3) is just the *law* of the random variable $H + J$. If $\mathbf{x}(t) = \text{Poisson}(\alpha)$, the Poisson distribution with mean α, then H is Poisson $(\alpha(1-s))$-distributed and consequently $\mathbf{x}(t+1)$ is the law of a Poisson$(\alpha(1-s) + \lambda)$ random variable. We shall see in Section 13.4 that the same is true of the continuous time analogue of (13.3) and indeed we show there that for every initial condition with $x_0 > 0$ the solution to the continuous time equation converges to the stationary point

$$\pi := \text{Poisson}(\theta)$$

as $t \to \infty$, where

$$\theta := \frac{\lambda}{s}.$$

Haigh's analysis of the finite population model focuses on the number of individuals in the best class. Let us write k^* for the number of mutations carried by individuals in this class. In a finite population, k^* will increase with time, but the profile of frequencies *relative to* k^*, $\{X_{k+k^*}\}$, forms a recurrent Markov chain. We set

$$\mathbf{Y} := (Y_k)_{k=0,1,\dots} := (X_{k^*+k})_{k=0,1,\dots}$$

and observe that, since fitness is always measured relative to the mean fitness in the population, between clicks of the ratchet the equation for the dynamics of $\mathbf{Y}$ is precisely the same as that for $\mathbf{X}$ when $X_0 > 0$. Suppose that after t generations there are $n_0(t) = N y_0(t)$ individuals in the best class. Then the probability of sampling a parent from this class *and* not acquiring any additional mutations is

$$p_0(t) := (y_0(t)/W(t)) e^{-\lambda},$$

where

$$W(t) = \sum_{i=0}^{\infty} y_i(t)(1-s)^i. \tag{13.4}$$

Thus, given $y_0(t)$, the size of the best class in the next generation has a binomial distribution with N trials and success probability $p_0(t)$, and so the evolution of the best class is determined by $W(t)$, the mean fitness of the population. We shall see this property of Y_0 reflected in the diffusion approximation of Section 13.3.

A principal assumption of Haigh's analysis is that, immediately before a click, the individuals of the current best class have all been distributed upon the other classes, in proportion to their Poisson weights. Thus immediately after a click he takes the type frequencies (relative to the new best class) to be

$$\tilde{\pi} := \frac{1}{1 - \pi_0}(\pi_1, \pi_2, \ldots) = \frac{1}{1 - e^{-\theta}}\left(\theta e^{-\theta}, \frac{\theta^2}{2}e^{-\theta}, \ldots\right). \tag{13.5}$$

The time until the next click is then subdivided into two phases. During the first phase the deterministic dynamical system decays exponentially fast towards its Poisson equilibrium, swamping the randomness arising from the finite population size. At the time that he proposes for the end of the first phase the size of the best class is approximately $1.6\pi_0$. The mean fitness of the population has decreased by an amount which can also be readily estimated and, combining this with a Poisson approximation to the binomial distribution, Haigh proposes that (at least initially) during the second (longer) phase the size of the best class should be approximated by a Galton–Watson branching process with a Poisson offspring distribution.

Haigh's original proposal was that since the mean fitness of the population (and consequently the mean number of offspring in the Galton Watson process) changes only slowly during the second phase it should be taken to be constant throughout that phase. Later refinements have modified Haigh's approach in two key ways. First, they have worked with a diffusion approximation so that the Galton–Watson process is replaced by a Feller diffusion and, second, instead of taking a constant drift, they look for a good approximation of the mean fitness *given the size of the best class*, resulting in a Feller diffusion with *logistic* growth. Our aim in the rest of this paper is to unify these approximations in a single mathematical framework and discuss them in the light of simulations. A crucial building block will be the following extension of (13.5), which we call the *Poisson profile approximation* (or PPA) of $\mathbf{Y}$ based on Y_0:

$$\Pi(Y_0) := \left(Y_0, \frac{1 - Y_0}{1 - \pi_0}(\pi_1, \pi_2, \ldots)\right). \tag{13.6}$$

As a first step, we now turn to a diffusion approximation for the full ratchet dynamics (13.2).

13.3 The Fleming–Viot diffusion

For large N and small λ and s, the following stochastic dynamics on $\mathcal{P}(\mathbb{N}_0)$ in continuous time captures the conditional expectation and variance of the discrete dynamics (13.2):

$$dX_k = \left(\sum_j s(j-k)X_j X_k + \lambda(X_{k-1} - X_k) \right) dt$$

$$+ \sum_{j \neq k} \sqrt{\frac{1}{N} X_j X_k} \, dW_{jk}, \quad k = 0, 1, 2, \ldots, \qquad (13.7)$$

where $X_{-1} := 0$, and $(W_{jk})_{j>k}$ is an array of independent standard Wiener processes with $W_{kj} := -W_{jk}$. This is, of course, just the infinite-dimensional version of the standard multi-dimensional Wright–Fisher diffusion. Existence of a process solving (13.7) can be established using a diffusion limit of the discrete dynamics of §13.2. The coefficient $s(j-k)$ is the fitness difference between type k and type j, $\lambda(X_{k-1} - X_k)$ is the flow into and out of class k due to mutation and the diffusion coefficients $\frac{1}{N} X_j X_k$ reflect the covariances due to multinomial sampling.

Remark 13.1 *Often when one passes to a Fleming–Viot diffusion approximation, one measures time in units of size N and correspondingly the parameters s and λ appear as Ns and $N\lambda$. Here we have not rescaled time, hence the factor of $\frac{1}{N}$ in the noise and the unscaled parameters s and λ in the equations.* $\qquad \square$

Writing

$$M_1(\mathbf{X}) := \sum_j j X_j$$

for the first moment of $\mathbf{X}$, (13.7) translates into

$$dX_k = \left(s(M_1(\mathbf{X}) - k) - \lambda)X_k + \lambda X_{k-1} \right) dt + \sum_{j \neq k} \sqrt{\frac{1}{N} X_j X_k} \, dW_{jk}$$

$$(13.8)$$

In exactly the same way as in our discrete stochastic system, writing k^* for the number of mutations carried by individuals in the fittest class, one would like to think of the population as a travelling wave with profile $Y_k = X_{k+k^*}$. Notice in particular that

$$dY_0 = (sM_1(\mathbf{Y}) - \lambda)Y_0 dt + \sqrt{\frac{1}{N} Y_0(1 - Y_0)} dW_0,$$

where W_0 is a standard Wiener process. Thus, just as in Haigh's setting, the frequency of the best class is determined by the mean fitness of the population.

Substituting into (13.7) one can obtain a stochastic equation for M_1,

$$dM_1 = (\lambda - sM_2)dt + dG$$

where $M_2 = \sum_j (j - M_1)^2 X_j$ and the martingale G has quadratic variation

$$d\langle G \rangle = \frac{1}{N} M_2 dt.$$

Thus the speed of the wave is determined by the variance of the profile. Similarly,

$$dM_2 = (-\tfrac{1}{N} M_2 + (\lambda - sM_3))dt + dH$$

where $M_3 = \sum_j (j - M_1)^3 X_j$ and the martingale H has quadratic variation $d\langle H \rangle = \frac{1}{N}(M_4 - M_2)dt$ with M_4 denoting the fourth centred moment and so on.

These equations for the centred moments are entirely analogous to those obtained in [9] except that in the Fleming–Viot setting they are exact. As pointed out there, Bürger [1] obtained similar equations to study the evolution of polygenic traits. The difficulty in using these equations to study the rate of the ratchet is of course that they are not closed: the equation for M_k involves M_{k+1} and so on. Moreover, there is no obvious approximating closed system. By contrast, the infinite population limit, in which the noise is absent, turns out to have a closed solution.

13.4 The infinite population limit

The continuous time analogue of (13.3) is the deterministic dynamical system

$$dx_k = \Big((s(M_1(\mathbf{x}) - k) - \lambda)x_k + \lambda x_{k-1} \Big)dt, \quad k = 0, 1, 2, \ldots \quad (13.9)$$

where $x_{-1} = 0$, obtained by letting $N \to \infty$ in our Fleming–Viot diffusion (13.8). Our goal in this section is to solve this system of equations. Note that Maia et al. [12] have obtained a complete solution of the corresponding discrete system following (13.3).

As we shall see in Proposition 13.2, the stationary points of the system are exactly the same as for (13.3), that is, $\mathbf{x} = \pi$ and all its right shifts $(\pi_{k-k^*})_{k=0,1,\ldots}$, $k^* = 0, 1, 2, \ldots$. Since the Poisson distribution can be characterized as the only distribution on $\mathbb{N}_0$ with all cumulants equal, it is natural to transform (13.9) into a system of equations for the cumulants, $\kappa_k, k = 1, 2, \ldots$, of the vector $\mathbf{x}$. The cumulants are defined by the relation

$$\log \sum_{k=0}^{\infty} x_k e^{-\xi k} = \sum_{k=1}^{\infty} \kappa_k \frac{(-\xi)^k}{k!}. \tag{13.10}$$

We assume $x_0 > 0$ and set

$$\kappa_0 := -\log x_0. \tag{13.11}$$

Proposition 13.2 For $\kappa_k, k = 0, 1, 2, \ldots$ as in (13.10) and (13.11) the system (13.9) is equivalent to

$$\dot{\kappa}_k = -s\kappa_{k+1} + \lambda, \qquad k = 0, 1, 2, \ldots$$

Setting $\underline{\kappa} := (\kappa_0, \kappa_1, \ldots)$ this system is solved by

$$\underline{\kappa} = B\underline{\kappa}(0)^\top + \frac{\lambda}{s}(1 - e^{-st})\underline{1}, \quad B = (b_{ij})_{i,j=0,1,\ldots},$$

$$b_{ij} = \begin{cases} \frac{(-st)^{j-i}}{(j-i)!} & j \geq i \\ 0 & \text{otherwise.} \end{cases} \tag{13.12}$$

In particular,

$$x_0(t) = e^{-\kappa_0(t)} = x_0(0) \frac{\exp\left(-\frac{\lambda}{s}(1 - e^{-st})\right)}{\left(\sum_{k=0}^{\infty} x_k(0)e^{-stk}\right)} \tag{13.13}$$

and

$$\kappa_1(t) = \sum_{k=0}^{\infty} k x_k(t) = -\frac{\partial}{\partial \xi} \log \sum_{k=0}^{\infty} x_k(0)e^{-\xi k}\bigg|_{\xi=st} + \frac{\lambda}{s}(1 - e^{-st}). \tag{13.14}$$

Remark 13.3 If $\mathbf{x}(0)$ is a Poisson(μ) distribution then substituting into (13.12) we see that $\mathbf{x}(t)$ is a Poisson distribution with parameter $\lambda/s + e^{-st}(\mu - \lambda/s)$. In other words, just as for the discrete dynamical system considered by Haigh, vectors of Poisson weights are mapped to vectors of Poisson weights. In particular $\pi := Poisson(\lambda/s)$ is once again a stationary point of the system. Moreover, this proposition shows that for any vector $\mathbf{x}(0)$ with $x_0(0) > 0$, the solution converges to this stationary point. The corresponding convergence result in the discrete *case is established in [12]. More generally, if k^* is the smallest value of k for which $x_k(0) > 0$ then the solution will converge to $(\pi_{k-k^*})_{k=0,1,2,\dots}$.*

Proof [Proof of Proposition 13.2]. Using (13.9) we have

$$\frac{d}{dt} \log\Big(\frac{1}{x_0} \sum_{k=0}^{\infty} x_k e^{-\xi k}\Big)$$

$$= \frac{x_0}{\sum_{k=0}^{\infty} x_k e^{-\xi k}} \Big(-\frac{\sum_{k=0}^{\infty} x_k e^{-\xi k}}{x_0^2} \dot{x}_0 + \frac{1}{x_0} \sum_{k=0}^{\infty} \dot{x}_k e^{-\xi k}\Big)$$

$$= -s \sum_{j=0}^{\infty} j x_j + \lambda - s \frac{\sum_{k=0}^{\infty} k x_k e^{-\xi k}}{\sum_{k=0}^{\infty} x_k e^{-\xi k}} + s \sum_{j=0}^{\infty} j x_j$$

$$\qquad + \frac{\lambda}{\sum_{k=0}^{\infty} x_k e^{-\xi k}} \Big(e^{-\xi} \sum_{k=1}^{\infty} x_{k-1} e^{-\xi(k-1)} - \sum_{k=0}^{\infty} x_k e^{-\xi k}\Big)$$

$$= s \frac{d}{d\xi} \log\Big(\sum_{k=0}^{\infty} x_k e^{-\xi k}\Big) + \lambda e^{-\xi}.$$

Thus, (13.10) gives

$$\frac{d}{dt} \sum_{k=0}^{\infty} \kappa_k \frac{(-\xi)^k}{k!} = -s \sum_{k=0}^{\infty} \kappa_{k+1} \frac{(-\xi)^k}{k!} + \lambda e^{-\xi}.$$

Comparing coefficients in the last equation we obtain

$$\dot{\kappa}_k = -s\kappa_{k+1} + \lambda, \qquad k = 0, 1, \dots.$$

This linear system can readily be solved. We write

$$D := (\delta_{i+1,j})_{i,j=0,1,2,\dots}, \qquad \underline{1} = (1, 1, \dots),$$

so that

$$\underline{\dot{\kappa}}^\top = -sD\underline{\kappa}^\top + \lambda \underline{1}^\top. \tag{13.15}$$

Since

$$(e^{-Dst})_{ij} = \begin{cases} \frac{(-st)^{j-i}}{(j-i)!} & j \geq i \\ 0 & \text{otherwise,} \end{cases}$$

the linear system (13.15) is solved by

$$\underline{\kappa}(t)^\top = e^{-Dst}\underline{\kappa}(0)^\top + \lambda \int_0^t e^{-Dsu} \underline{1} du = e^{-Dst}\underline{\kappa}(0)^\top + \frac{\lambda}{s}(1 - e^{-st})\underline{1}^\top.$$

$\square$

Remark 13.4 *With the initial condition* $\mathbf{x}(0) := \tilde{\pi}$ *given by* (13.5), *equations* (13.13) *and* (13.14) *become*

$$x_0(t) = e^{-\theta} \frac{\theta e^{-st}}{1 - e^{-\theta e^{-st}}} \tag{13.16}$$

and

$$\kappa_1(t) = \theta - 1 + \frac{\theta e^{-st}}{e^{\theta e^{-st}} - 1}. \tag{13.17}$$

At time

$$\tau := \frac{\log \theta}{s}, \tag{13.18}$$

we have $x_0(\tau) = e^{-\theta} \frac{1}{1-e^{-1}} \approx 1.6\pi_0$. *Comparing with Section 13.2 we see that in our continuous time setting* τ *is precisely the counterpart of the time proposed by Haigh as the end of 'phase one'.*

In Section 13.5 our prediction for $M_1(\mathbf{Y})$ given Y_0 will require the value of $M_1(\mathbf{y}(\tau))$ for $\mathbf{y}$ solving (13.9) when started from a Poisson profile approximation. This is the purpose of the next proposition.

Proposition 13.5 *For* $y_0 \in (0,1)$, *let* $\mathbf{y}(t)$ *be the solution of* (13.9) *with the initial state* $\mathbf{y}(0) := \Pi(y_0)$ *defined in* (13.6), *and let* τ *be Haigh's relaxation time defined in* (13.18). *Then for* $A \geq 0$ *with* $\eta := \theta^{1-A}$

$$M_1(\mathbf{y}(A\tau)) = \theta + \frac{\eta}{e^{\eta} - 1}\left(1 - \frac{y_0(A\tau)}{\pi_0}\right).$$

Proof. Since

$$\sum_{k=0}^{\infty} \pi_k e^{-\xi k} = \exp\left(-\theta(1 - e^{-\xi})\right) = \pi_0^{1-e^{-\xi}},$$

we have

$$\sum_{k=0}^{\infty} y_k e^{-\xi k}\bigg|_{\xi = sA\tau} = y_0 + \frac{1 - y_0}{1 - \pi_0}\pi_0\left(e^{\theta e^{-\xi}} - 1\right)\bigg|_{\xi = sA\tau}$$

$$= y_0 + \frac{1 - y_0}{1 - \pi_0}\pi_0(e^{\eta} - 1)$$

and

$$-\frac{\partial}{\partial \xi}\sum_{k=0}^{\infty} y_k e^{-\xi k}\bigg|_{\xi = sA\tau} = \frac{1 - y_0}{1 - \pi_0}\pi_0^{1-e^{-\xi}}\theta e^{-\xi}\bigg|_{\xi = sA\tau} = \frac{1 - y_0}{1 - \pi_0}\pi_0 e^{\eta}\eta.$$

Using the solution (13.13) and (13.14) and $y_0(0) = y_0$

$$y_0(A\tau) = y_0 \frac{\pi_0 e^{\eta}}{y_0 + \frac{1-y_0}{1-\pi_0}\pi_0(e^{\eta}-1)}$$

$$= y_0 \frac{\pi_0 e^{\eta}(1-\pi_0)}{y_0(1-\pi_0 e^{\eta}) + \pi_0(e^{\eta}-1)}, \qquad (13.19)$$

$$M_1(\mathbf{y}(A\tau)) = \frac{\frac{1-y_0}{1-\pi_0}\pi_0 e^{\eta}\eta}{y_0 + \frac{1-y_0}{1-\pi_0}\pi_0(e^{\eta}-1)} + \theta - \eta$$

$$= \theta + \eta \frac{\pi_0 - y_0}{y_0(1-\pi_0 e^{\eta}) + \pi_0(e^{\eta}-1)}. \qquad (13.20)$$

From (13.19),

$$y_0 = \frac{y_0(A\tau)\pi_0(e^{\eta}-1)}{\pi_0 e^{\eta}(1-\pi_0) - y_0(A\tau)(1-\pi_0 e^{\eta})}$$

and thus

$$\pi_0 - y_0 = \frac{\pi_0 e^{\eta}(\pi_0 - y_0(A\tau))(1-\pi_0)}{\pi_0 e^{\eta}(1-\pi_0) - y_0(A\tau)(1-\pi_0 e^{\eta})},$$

$$y_0(1-\pi_0 e^{\eta}) + \pi_0(e^{\eta}-1)$$

$$= \pi_0(e^{\eta}-1) \frac{\pi_0 e^{\eta}(1-\pi_0)}{\pi_0 e^{\eta}(1-\pi_0) - y_0(A\tau)(1-\pi_0 e^{\eta})}.$$

Plugging the last two equations into (13.20) we find

$$M_1(\mathbf{y}(A\tau)) = \theta + \frac{\eta}{e^{\eta}-1}\left(1 - \frac{y_0(A\tau)}{\pi_0}\right).$$

$\square$

13.5 One-dimensional diffusion approximations

Recall from Section 13.3 that in our Fleming–Viot model the frequency Y_0 of the best class follows

$$dY_0 = \left(sM_1(\mathbf{Y}) - \lambda\right)Y_0 dt + \sqrt{\frac{1}{N}Y_0(1-Y_0)}\, dW_0, \qquad (13.21)$$

where W_0 is a standard Wiener process. The system of equations (13.8) is too complex for us to be able to find an explicit expression for $M_1(\mathbf{Y})$, which depends on the whole vector $\mathbf{Y}$ of class sizes. Instead we seek a good approximation of M_1 given Y_0. Substituting this into equation (13.21) will then yield a one-dimensional diffusion which we use as an approximation for the size of the best class. Of course this assumption of a functional dependence between Y_0 and M_1 is a weakness of

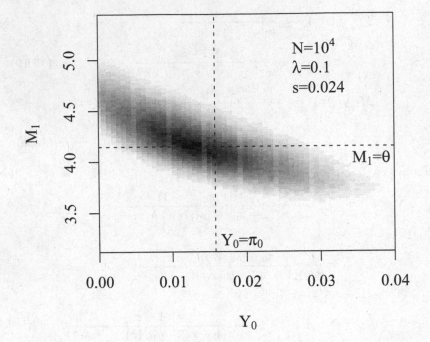

Fig. 13.2. Using simulations (see also Section 13.6) we plot (Y_0, M_1). There is a good fit to a linear relationship between Y_0 and M_1. Note that $\gamma = 0.6$ in the figure.

the one-dimensional diffusion approximation, but simulations show that there is a substantial correlation between Y_0 and M_1, see, for example, Figure 13.2.

To understand our approach to finding a map $Y_0 \mapsto M_1$, recall as a first step Haigh's approximation that immediately after a click of the ratchet the profile has the form (13.5). The reasoning is as follows. Deviations of $\mathbf{Y}$ from a Poisson profile can only be due to the randomness arising from resampling in a finite population. Since resampling has no tendency to increase or decrease the frequency of a given class, the average profile immediately after a click of the ratchet is approximated by the state where π_0 is distributed evenly over all other classes according to their equilibrium frequencies. During his short 'phase one', Haigh then allows this profile to 'relax' through the action of the discrete dynamical system (13.3) and it is the mean fitness in the population after this short relaxation time which determines the behaviour of the best class during 'phase two'.

A natural next step in extending this argument is to suppose that also *in between* click times the resampling distributes the mass $\pi_0 - Y_0$ evenly on all other classes. In other words, given Y_0, approximate the state of the system by $\mathbf{Y} = \Pi(Y_0)$ given by (13.6).

Of course in reality the dynamical system interacts with the resampling as it tries to restore the system to its Poisson equilibrium. If this restoring force is strong, just as in Haigh's approach one estimated mean fitness during phase two from the 'relaxed' profile, so here one should approximate the mean fitness M_1 not from the Poisson profile approximation (PPA) but from states which arise by evolving the PPA using the dynamical system for a certain amount of time. We call the resulting states *relaxed Poisson profile approximations* or RPPAs. There are three different parameter regimes with which we shall be concerned. Each corresponds to a different value of η in the functional relationship

$$M_1 = \theta + \frac{\eta}{e^\eta - 1}\left(1 - \frac{Y_0}{\pi_0}\right). \tag{13.22}$$

of Proposition 13.5. These can be distinguished as follows:

$$A \text{ small,} \qquad \eta \approx \theta, \qquad M_1 \approx \frac{\theta}{1 - \pi_0}(1 - Y_0), \tag{13.23a}$$

$$A = 1, \qquad \eta = 1, \qquad M_1 \approx \theta + 0.58\left(1 - \frac{Y_0}{\pi_0}\right), \tag{13.23b}$$

$$A \text{ large,} \qquad \eta \approx 0, \qquad M_1 \approx \theta + \left(1 - \frac{Y_0}{\pi_0}\right). \tag{13.23c}$$

The resulting maps $Y_0 \mapsto M_1$ are plotted in Figure 13.3. Observe that, for consistency, M_1 has to increase, on average, by 1 during one click of the ratchet.

Finally, before we can apply our one-dimensional diffusion approximation we must choose a starting value for Y_0 following equation (13.21). For A large, the system is already close to its new equilibrium at the time of a click and so we take $Y_0 = \pi_0$.

For $A = 1$, at the time of the click we observe a state which has relaxed for time τ from a state of the form $\tilde{\pi}$ from (13.5). We computed in Remark 13.4 that such a state comes with $Y_0 = 1.6\pi_0$.

For small values of A, observe that the profile of the population immediately after a click is approximately $\tilde{\pi}$ from (13.5). Since $\tilde{\pi}$ is not a

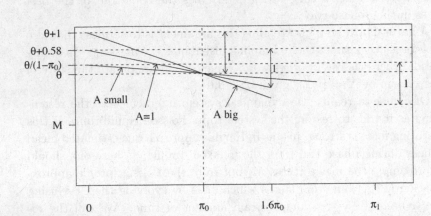

Fig. 13.3. Since simulations show a strong correlation between the first moment M_1 and Y_0, we use (13.23a)–(13.23c) to predict M_1 from Y_0 depending on the model parameters.

state of the form $\Pi(y_0)$ the arguments that led to Proposition 13.5 do not apply. Instead we follow Haigh in dividing the time between clicks into two phases. Consider first 'phase one'. Recall from the dynamical system that

$$\mathrm{d}Y_0 = (sM_1 - \lambda)Y_0 \, \mathrm{d}t.$$

We write

$$(sM_1 - \lambda)Y_0 = c \, (\pi_0 - Y_0),$$

where c (like Y_0 and M_1) depends on $r = \theta e^{-st}$. Starting in $\mathbf{Y}(0) = \tilde{\pi}$ we have from (13.16) and (13.17)

$$Y_0(t) = e^{-\theta} \frac{\theta e^{-st}}{1 - \exp(-\theta e^{-st})} = e^{-\theta} \frac{r}{1 - e^{-r}},$$

$$M_1(t) = \theta - 1 + \frac{\theta e^{-st}}{\exp(\theta e^{-st}) - 1} = \theta - 1 + \frac{r}{e^r - 1}.$$

We compute

$$\frac{c}{s} = \frac{(M_1 - \theta)Y_0}{\pi_0 - Y_0} = \frac{1 - \frac{r}{e^r - 1}}{1 - \frac{1 - e^{-r}}{r}} = \frac{r(1 - e^{-r}) - r^2 e^{-r}}{r(1 - e^{-r}) - (1 - e^{-r})^2}.$$

It can be checked that this expression lies between 1 and 1.25 for all $r > 0$ which suggests that the size of the best class in the initial phase after a click is reasonably described by the dynamics

$$dY_0 = s\,(\pi_0 - Y_0)\,dt + \sqrt{\frac{1}{N}Y_0}\,dW_0 \qquad (13.24)$$

started from $\frac{\pi_1}{(1-\pi_0)}$. We allow Y_0 to evolve according to equation (13.24) until it reaches $1.6\pi_0$, say, and then use our estimate of M_1 from equation (13.23) to estimate the evolution of Y_0 during the (longer) 'phase two'.

We assume that states of the ratchet are RPPAs, i.e. Poisson profile approximations (13.6) which are relaxed for time $A\tau$, where $\tau = \frac{1}{s}\log\theta$, which leads to the functional relationship (13.22). Consequently, we suggest that (13.21) is approximated by the 'mean reversion' dynamics

$$dY_0 = s\frac{\eta}{e^\eta - 1}\Big(1 - \frac{Y_0}{\pi_0}\Big)Y_0 dt + \sqrt{\frac{1}{N}Y_0}\,dW_0, \qquad (13.25)$$

with $\eta = \theta^{1-A}$, where we have used a Feller noise instead of the Wright–Fisher term in (13.21). In other words, Y_0 is a *Feller branching diffusion with logistic growth*.

Using the three regimes from (13.23), we have the approximations

$$A \text{ small}, \qquad dY_0 = \lambda(\pi_0 - Y_0)Y_0 dt + \sqrt{\frac{1}{N}Y_0}dW, \qquad (13.26a)$$

$$A = 1, \qquad dY_0 = 0.58s\Big(1 - \frac{Y_0}{\pi_0}\Big)Y_0 dt + \sqrt{\frac{1}{N}Y_0}\,dW_0, \qquad (13.26b)$$

$$A \text{ large}, \qquad dY_0 = s\Big(1 - \frac{Y_0}{\pi_0}\Big)Y_0 dt + \sqrt{\frac{1}{N}Y_0}\,dW_0, \qquad (13.26c)$$

(where in the first equation we have used that $\frac{1}{1-\pi_0} \approx 1 + \pi_0$ and that $Y_0\pi_0$ is negligible).

An equation similar to (13.26b) was found (by different means) in [16] and further discussed in [7]. Stephan and Kim [17] analyse whether a prefactor of 0.5 or 0.6 in (13.26b) fits better with simulated data. We discuss the relationship with these papers in detail in Section 13.7.

The expected time to extinction of a diffusion following (13.25) is readily obtained from a Green function calculation similar to that in [10]. We refrain from doing this here, but instead use a scaling argument to identify parameter ranges for which the ratchet clicks and to give evidence for the rule of thumb formulated in the introduction.

Consider the rescaling

$$Z(t) = \frac{1}{\pi_0} Y_0 \left(N\pi_0 t \right).$$

For A small equation (13.26a) becomes

$$
\begin{aligned}
dZ &= N\lambda\pi_0^2(1 - Z)Zdt + \sqrt{Z}dW \\
&= (N\lambda)^{1-2\gamma}(1 - Z)Zdt + \sqrt{Z}dW.
\end{aligned}
\tag{13.27}
$$

For $A = 1$ on the other hand we obtain from (13.26b)

$$
\begin{aligned}
dZ &= 0.58 N s\pi_0(1 - Z)Zdt + \sqrt{Z}dW \\
&= 0.58 \frac{1}{\gamma \log(N\lambda)}(N\lambda)^{1-\gamma}(1 - Z)Zdt + \sqrt{Z}dW.
\end{aligned}
\tag{13.28}
$$

For A large we obtain from (13.26c) the same equation without the factor of 0.58.

From this rescaling we see that the equation that applies for small A, i.e. (13.27), is strongly mean reverting for $\gamma < 1/2$. Recall that the choice of small A is appropriate when the ratchet is clicking frequently and so this indicates that frequent clicking simply will not happen for $\gamma < 1/2$. To indicate the boundary between rare and moderate clicking, equation (13.28) is much more relevant than equation (13.27). At first sight, equation (13.28) looks strongly mean reverting for all $\gamma < 1$, which would seem to suggest that the ratchet will click only exponentially slowly in $(N\lambda)^{1-\gamma}$. However, the closer γ is to 1, the larger the value of $N\lambda$ we must take for this asymptotic regime to provide a good approximation. For example, in the table below we describe parameter combinations for which the coefficient in front of the mean reversion term in equation (13.28) is at least 5. We see that for $\gamma < 1/2$ this coefficient is large for most of the reasonable values of $N\lambda$, whereas for $\gamma > 1/2$ it is rather small over a large range of $N\lambda$.

γ	0.3	0.4	0.5	0.55	0.6	0.7	0.8	0.9
$N\lambda \geq$	20	10^2	$9 \cdot 10^2$	$4 \cdot 10^3$	$2 \cdot 10^4$	$4 \cdot 10^6$	$2 \cdot 10^{11}$	$8 \cdot 10^{26}$

Thus, for example, if $\gamma = 0.7$ we require $N\lambda$ to be of the order of 10^6 for the strong mean reversion of equation (13.28) to be evident. This is not a value of $N\lambda$ which will be observed in practice. Indeed, as a 'rule of thumb', for biologically realistic parameter values, we should expect the transition from no clicks to a moderate rate of clicks to take place at around $\gamma = 0.5$.

13.6 Simulations

We have argued that the one-dimensional diffusions (13.26) approximate the frequency in the best class and from this deduced the *rule of thumb* from Section 13.1. In this section we use simulations to test the validity of our arguments.

For a population following the dynamics (13.2), the $(t+1)$st generation is formed by multinomial sampling of N individuals with weights

$$p_k(t) = \sum_{j=0}^{k} \frac{x_{k-j}(t)(1-s)^{k-j}}{W(t)} e^{-\lambda} \frac{\lambda^j}{j!}, \qquad (13.29)$$

where $W(t)$ is the average fitness in the tth generation from (13.4) and it is this Wright–Fisher model which was implemented in the simulations.

To supplement the numerical results of Figure 13.1 we provide simulation results for the average time between clicks (where time is measured in units of N generations) for fixed N and γ and varying λ; see Figure 13.4. Note that, for fixed γ in equation (13.1), s is increasing with λ . We carry out simulations using a population size of $N = 10^5$ and λ varying from 10^{-4} to 1. For $\gamma = 0.5$ we observe that the power-law behaviour breaks down already for $N\lambda = 10^3$ and the diffusion (13.26b) predicts the clicking of the ratchet sufficiently well. For increasing γ, the power law breaks down only for larger values of $N\lambda$. For $\gamma = 0.7$, in our simulations we only observe the power-law behaviour but conjecture that for larger values of $N\lambda$ the power law would break down; compare with the table above.

For a finer analysis of which of the equations (13.26) works best, we study the resulting Green functions numerically; see Figure 13.5. In particular, we record the relative time spent in some dY_0 in simulations and compare this quantity to the numerically integrated, normalized Green functions given through the diffusions (13.26a) and (13.26b). (We do not consider (13.26c) because it only gives an approximation if the ratchet clicks rarely.) We see in (A) that for $\gamma = 0.5$ (13.26b) produce better estimates not only for the average time between clicks (Figure 13.4)

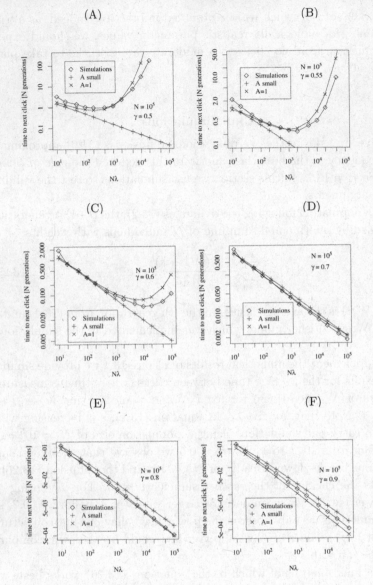

Fig. 13.4. The power-law behaviour of the rate of the ratchet with respect to γ is valid for a large portion of the parameter space. (A) For $\gamma = 0.5$ clicks become rare and the power law does not apply for $N\lambda > 10^3$. (B), (C) For $\gamma = 0.55$ and $\gamma = 0.6$, we have to explore a larger portion of the parameter space in order to see that the power law does not apply any more. (D), (E), (F) For $\gamma \geq 0.7$ we never observe a deviation from the power law. For every plot, we used $N = 10^5$ and simulations ran for 5×10^6 ($\gamma = 0.5$: 2×10^7) generations for each value of $N\lambda$.

(A) (B)

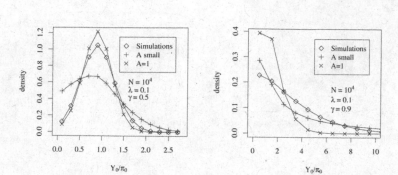

Fig. 13.5. We compare the plots for the occupation density of Y_0 from the simulations with theoretical curves corresponding to the Green functions for the cases of small A and $A = 1$ in (13.26). (A) If clicks are rare, $A = 1$ produces better results than small A. (B) If clicks are frequent, the simulated densities of Y_0 are better approximated by small A. Every plot is based on the simulation of $5 \cdot 10^5$ generations.

but also for the time spent around some point y_0. However, for $\gamma = 0.9$, clicks are more frequent and we expect (13.26a) to provide a better approximation. Indeed, although both (13.26a) and (13.26b) predict the power-law behaviour, as (B) shows, the first equation produces better estimates for the relative amount of time spent in some dY_0.

To support our claim that the states observed are relaxed Poisson profile approximations we use a phase-plane analysis; see Figure 13.6. At any point in time of a simulation, values for Y_0 and M_1 can be observed. The resulting plots indicate that we can distinguish the three parameter regimes introduced in Section 13.5. In the case of rapid clicking of the ratchet (so that the states we observe have not relaxed a lot and thus are approximately of the form $\Pi(y_0)$) we see in (A), (B) that the system is driven by the restoring force to $M_1 < \theta$. The reason is that M_1 is small at click times and these are frequent. However, the slope of the line relating Y_0 and M_1 is low, as predicted by (13.23a). (We used $(1 - Y_0)/(1 - \pi_0) \approx 1 - Y_0 + \pi_0$ in the plot here.) For the case $A = 1$ the system spends some time near $Y_0 = 0$ and thus the ratchet clicks, but not frequently. So, the dynamical system restores states partly to

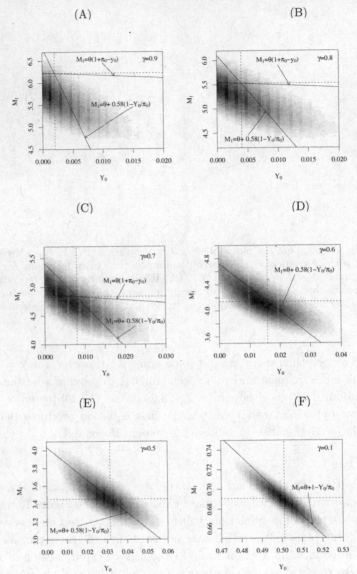

Fig. 13.6. There are three regimes for the relationship between Y_0 and M_1, as given in (13.23). If clicks are frequent, at least the slope of the relationship between Y_0 and M_1 fits roughly to equation (13.23a). If clicks occur reasonably often, (13.23b) gives a good approximation. If clicks are rare, (13.23c) gives a reasonable prediction. The plots show simulations for different values of γ. The dashed horizontal and vertical lines are $M_1 = \theta$ and $Y_0 = \pi_0$, respectively. For every plot we used $N = 10^4, \lambda = 0.1$ and simulations ran for 10^6 generations.

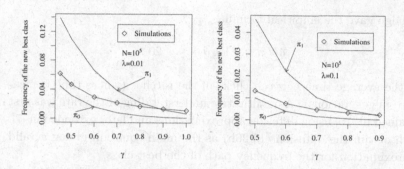

Fig. 13.7. Our heuristic that observed states come from a RPPA applies in particular at click times. (A) For small γ, i.e., rare clicking, the frequency of the best class is already close to π_0 while (B) it is close to π_1 for larger γ, i.e. frequent clicking. Every plot is based on the simulation of 5×10^6 generations.

equilibrium and we see that the slope given in (13.23b) gives the most reasonable prediction in (C), (D), (E). For rare clicking, i.e. A large, the dynamical system has even more time and (F) shows that the slope is as predicted by (13.23c).

Our prediction that we observe profiles which are well approximated by a relaxed PPA applies especially well at click times. We check this numerically by observing the frequency of the (new) best class at click times; see Figure 13.7. For small γ the ratchet clicks rarely and the system has some time to relax to its new equilibrium even before the click of the ratchet. As a consequence, we see that the frequency of the (new) best class at the time of the click is already close to π_0. However, if γ is large and the ratchet clicks frequently, the dynamical system has no time before the click to relax the system to the new equilibrium. Therefore, we observe that the frequency of the new best class is close to π_1.

13.7 Discussion

Haigh [8] was the first to attempt a rigorous mathematical analysis of the ideas of Muller [14]. However, in spite of the apparent simplicity of Haigh's mathematical formulation of the model, the exact rate of

Muller's ratchet remains elusive. In this paper, we have developed arguments in the spirit of [8], [16] and [7] to give approximations for this rate.

Haigh gave the empirical formula

$$4N\pi_0 + 7\log\theta + \frac{2}{s} - 20$$

for the average time between clicks of the ratchet (where time is measured in generations). A quantitative understanding of the rate was first obtained in [16] using diffusion approximations and later extended in [7]. Both obtain the diffusion (13.26b) as the main equation giving a valid approximation for the frequency path of the best class.

The reasoning leading to (13.26b) in these papers is twofold. [16] and [17] argue that although fitness decreases by $se^{-\lambda}$ during one 'cycle' of the discrete ratchet model from Section 13.2 (in which the system advances from one Poisson equilibrium to the next), at the actual click time only a fraction of the fitness has been lost. They suggest $kse^{-\lambda}$ for $k = 0.5$ or $k = 0.6$ as the loss of fitness at click times. In other words they predict the functional relationship $M_1(Y_0)$ discussed in Section 13.5 by linear interpolation between $M_1(\pi_0) = \theta$ and $M_1(0) = \theta + ke^{-\lambda} \approx \theta + k$; compare with Figure 13.3. On the other hand, Gordo and Charlesworth [7] use a calculation of Haigh which tells us that if the dynamical system (13.2) is started in $\tilde{\pi}$ from (13.5), then at the end of phase one (corresponding in the continuous setting, as we observed in Remark 13.4, to time $\frac{1}{s}\log\theta$) we have $sM_1 \approx 1 - e^{-\lambda}(1 + 0.42s)$. This leads to the approximation $M_1(1.6\pi_0) = \theta - 0.42$ and again interpolating linearly using $M_1(\pi_0) = \theta$ gives (13.26b).

Simulations show that (13.26b) provides a good approximation to the rate of the ratchet for a wide range of parameters; see e.g. [17]. The novelty in our work is that we derive (13.26b) explicitly from the dynamical system. In particular, we do not use a linear approximation, but instead derive a functional linear relationship in Proposition 13.5. In addition, we clarify the rôle of the two different phases suggested by Haigh. As simulations show, since phase one is fast, it is already complete at the time when phase two starts. Therefore, in practice, we observe states that are relaxed PPAs.

The drawback of our analysis is that we cannot give good arguments for the choice of $A = 1$ in (13.23b) and (13.26b). However, note that the choice of $A = 1$ is essential to obtain the prefactor of 0.58 in (13.23b). For example, if $\theta = 10$, the choice of $A = 0.5$ leads to a prefactor of 0.13

while $A = 2$ leads to the prefactor of 0.95, neither of which fits with simulated data; see Figure 13.6.

We obtain two more diffusion approximations, which are valid in the cases of frequent and rare clicking, respectively. In practice, both play little rôle in the prediction of the rate of the ratchet. For fast clicking, (13.26b) shows the same power-law behaviour as (13.26a) and rare clicks are never observed in simulations.

Of course from a biological perspective our mathematical model is very naive. In particular, it is unnatural to suppose that each new mutation confers the same selective disadvantage and, indeed, not all mutations will be deleterious. Moreover, if one is to argue that Muller's ratchet explains the evolution of sex, then one has to quantify the effect of recombination. Such questions provide a rich, but challenging, mathematical playground.

Acknowledgements We have discussed Muller's ratchet with many different people. We are especially indebted to Ellen Baake, Nick Barton, Matthias Birkner, Charles Cuthbertson, Don Dawson, Wolfgang Stephan, Jay Taylor and Feng Yu.

Bibliography

[1] Bürger, R. (1991). Moments, cumulants and polygenic dynamics. *J. Math. Biol.*, 30:199–213.

[2] Charlesworth, B. (1978). Model for evolution of Y chromosomes and dosage compensation. *Proc. Natl. Acad. Sci. USA*, 75(11):5618–5622.

[3] Charlesworth, B. (1996). The evolution of chromosomal sex determination and dosage compensation. *Curr. Biol.*, 6:149–162.

[4] Charlesworth, B. and Charlesworth, D. (1997). Rapid fixation of deleterious alleles can be caused by Muller's ratchet. *Genet. Res.*, 70:63–73.

[5] Charlesworth, B. and Charlesworth, D. (1998). Some evolutionary consequences of deleterious mutations. *Genetica*, 102/103:2–19.

[6] Gessler, D. D. (1995). The constraints of finite size in asexual populations and the rate of the ratchet. *Genet. Res.*, 66(3):241–253.

[7] Gordo, I. and Charlesworth, B. (2000). The degeneration of asexual haploid populations and the speed of Muller's ratchet. *Genetics*, 154(3):1379–1387.

[8] Haigh, J. (1978). The accumulation of deleterious genes in a population—Muller's Ratchet. *Theor. Popul. Biol.*, 14(2):251–267.

[9] Higgs, P. G. and Woodcock, G. (1995). The accumulation of mutations in asexual populations and the structure of genealogical trees in the presence of selection. *J. Math. Biol.*, 33:677–702.

[10] Lambert, A. (2005). The branching process with logistic growth. *Ann. Appl. Probab.*, 15(2):1506–1535.

[11] Loewe, L. (2006). Quantifying the genomic decay paradox due to Muller's ratchet in human mitochondrial DNA. *Genet. Res.*, 87:133–159.

[12] Maia, L. P., Botelho, D. F., and Fontatari, J. F. (2003). Analytical solution of the evolution dynamics on a multiplicative fitness landscape. *J. Math. Biol.*, 47:453–456.

[13] Maynard Smith, J. (1978). *The Evolution of Sex*. Cambridge University Press.

[14] Muller, H. (1964). The relation of recombination and mutational advance. *Mutat. Res.*, 106:2–9.

[15] Rice, W. (1994). Degeneration of a non-recombining chromosome. *Science*, 263:230–232.

[16] Stephan, W., Chao, L., and Smale, J. (1993). The advance of Muller's ratchet in a haploid asexual population: approximate solutions based on diffusion theory. *Genet. Res.*, 61(3):225–231.

[17] Stephan, W. and Kim, Y. (2002). Recent applications of diffusion theory to population genetics. In S. Montgomery and M. Venille (eds), *Modern Developments in Theoretical Population Genetics*, Oxford University Press, pp. 72–93.

Printed in the United States
by Baker & Taylor Publisher Services

Printed in the United States
by Baker & Taylor Publisher Services